Nonlinear Optics in Semiconductors II

SEMICONDUCTORS
AND SEMIMETALS
Volume 59

Semiconductors and Semimetals

A Treatise

Edited by R. K. Willardson
CONSULTING PHYSICIST
SPOKANE, WASHINGTON

Eicke R. Weber
DEPARTMENT OF MATERIALS SCIENCE
AND MINERAL ENGINEERING
UNIVERSITY OF CALIFORNIA AT
BERKELEY

Nonlinear Optics in Semiconductors II

SEMICONDUCTORS AND SEMIMETALS
Volume 59

Volume Editors

ELSA GARMIRE

THAYER SCHOOL OF ENGINEERING
DARTMOUTH COLLEGE
HANOVER, NEW HAMPSHIRE

ALAN KOST

HUGHES RESEARCH LABORATORIES
MALIBU, CALIFORNIA

ACADEMIC PRESS
San Diego   London   Boston
New York   Sydney   Tokyo   Toronto

This book is printed on acid-free paper.

COPYRIGHT © 1999 BY ACADEMIC PRESS

ALL RIGHTS RESERVED.
NO PART OF THIS PUBLICATION MAY BE REPRODUCED OR TRANSMITTED IN ANY FORM OR BY ANY MEANS, ELECTRONIC OR MECHANICAL, INCLUDING PHOTOCOPY, RECORDING, OR ANY INFORMATION STORAGE AND RETRIEVAL SYSTEM, WITHOUT PERMISSION IN WRITING FROM THE PUBLISHER.

The appearance of the code at the bottom of the first page of a chapter in this book indicates the Publisher's consent that copies of the chapter may be made for personal or internal use of specific clients. This consent is given on the condition, however, that the copier pay the stated per-copy fee through the Copyright Clearance Center, Inc. (222 Rosewood Drive, Danvers, Massachusetts 01923), for copying beyond that permitted by Sections 107 or 108 of the U.S. Copyright Law. This consent does not extend to other kinds of copying, such as copying for general distribution, for advertising or promotional purposes, for creating new collective works, or for resale. Copy fees for pre-1999 chapters are as shown on the title pages; if no fee code appears on the title page, the copy fee is the same as for current chapters. 0080-8784/99 $30.00

ACADEMIC PRESS
525 B Street, Suite 1900, San Diego, CA 92101-4495, USA
1300 Boylston Street, Chestnut Hill, Massachusetts 02167, USA
http://www.apnet.com

ACADEMIC PRESS LIMITED
24–28 Oval Road, London NW1 7DX, UK
http://www.hbuk.co.uk/ap/

International Standard Book Number: 0-12-752168-2
International Standard Serial Number: 0080-8784

PRINTED IN THE UNITED STATES OF AMERICA
98 99 00 01 02 QW 9 8 7 6 5 4 3 2 1

Contents

PREFACE ix
LIST OF CONTRIBUTORS xiii

Chapter 1 Second Order Nonlinearities and Optical Rectification
Jacob B. Khurgin

I. INTRODUCTION 2
II. SECOND-ORDER NONLINEAR EFFECTS IN BULK SEMICONDUCTORS 4
   1. *General Considerations* 4
   2. *Classical Definition of Higher-Order Susceptibility* 6
   3. *Semiclassical Theory of Second-Order Susceptibility* 12
   4. *Off-Resonant Second-Order Susceptibilities of Zinc-Blende Semiconductors* . 19
   5. *Calculation of Second-Order Susceptibility Using a Hybrid Bond Model* . . . 25
III. $\chi^{(2)}$ EXPERIMENTS IN BULK SEMICONDUCTORS 30
   1. *Measurements of $\chi^{(2)}$ Values* 30
   2. *Experimental Demonstration of Second-Order Nonlinear Processes in Semiconductors* 34
IV. SECOND-ORDER SUSCEPTIBILITY IN SEMICONDUCTOR QUANTUM WELLS AND SUPERLATTICES 42
   1. *Asymmetric Quantum Wells* 42
   2. *Intersubband Second-Order Nonlinearities: Theory* 49
   3. *Intersubband Second-Order Nonlinearities: Experiment* 55
   4. *Interband Second-Order Nonlinearities: Theory and Experiment* . . 58
V. OPTICAL RECTIFICATION AND TERAHERTZ EMISSION IN SEMICONDUCTORS . . . 63
   1. *Brief Review of the Optical Rectification Term in $\chi^{(2)}$* . . 63
   2. *Optical Rectification and Coherent Terahertz Emission in Bulk Semiconductors* 67
   3. *Optical Rectification and Terahertz Emission in Semiconductor Quantum Wells* 73
VI. CONCLUSIONS 76
LIST OF ABBREVIATIONS AND ACRONYMS 78
REFERENCES 79

Chapter 2 Nonlinearities in Active Media
Katherine L. Hall, Erik R. Thoen and Erich P. Ippen

I. INTRODUCTION	84
II. ACTIVE SEMICONDUCTOR MEDIA	85
1. *Linear Gain in Active Materials*	86
2. *Carrier Confinement and Waveguiding*	91
3. *Nonlinear Gain in Active Media*	94
III. MEASUREMENT TECHNIQUES	100
1. *Orthogonally Polarized Collinear Pump-Probe Technique*	101
2. *The Heterodyne Pump-Probe Technique*	106
IV. NONLINEAR INDEX OF REFRACTION IN ACTIVE MEDIA	112
V. DATA ANALYSIS AND INTERPRETATION	119
VI. SHAPING AND SATURATION OF SHORT PULSES IN ACTIVE WAVEGUIDES	134
VII. FOUR-WAVE MIXING	141
VIII. APPLICATIONS	145
1. *Wavelength Converters*	146
2. *All-Optical Switching*	148
IX. SUMMARY	153
REFERENCES	155

Chapter 3 Optical Responses of Quantum Wires/Dots and Microcavities
Eiichi Hanamura

I. THEORETICAL ASPECTS	161
1. *Superradiance of Excitons from Quantum Dots and Wires*	164
2. *Excitonic Optical Nonlinearity*	169
3. *Excitons in Quantum Wells*	178
4. *Coherent Light Emission from Quantum Wires*	182
II. EXPERIMENTAL RESULTS	189
1. *Superradiance of Excitons*	189
2. *Figure of Merit for Nonlinear Optical Responses*	191
3. *Biexcitons as a Nonlinear Medium*	195
4. *Weak Localization and Nonlinear Optical Responses*	201
LIST OF ACRONYMS	208
REFERENCES	208

Chapter 4 Semiconductor Nonlinearities for Solid-State Laser Modelocking and Q-Switching
Ursula Keller

I. INTRODUCTION	211
1. *Motivation*	211
2. *Semiconductor Saturable Absorbers: Historical Overview*	214
II. SEMICONDUCTOR SATURABLE ABSORBER MIRRORS (SESAMs)	221
1. *Macroscopic Properties of SESAMs*	221
2. *Requirements for Passive Modelocking*	223
3. *Requirements for Passive Q-Switching*	243
4. *Microscopic Properties*	245

III. SESAM Designs	249
1. *Antiresonant Fabry-Perot Saturable Absorber (A-FPSA)*	249
2. *Dispersive SESAMs*	259
3. *Antiresonant Fabry-Perot p-i-n Modulator*	259
4. *General SESAM Design*	260
IV. Passively Modelocked Solid State Lasers Using SESAMs	260
1. *Laser Design and Diode Pumping*	260
2. *Approaching Limits: Ti:Sapphire Lasers*	267
V. Passively Q-Switched Solid State Lasers Using SESAMs	272
1. *Laser Design d Diode Pumping*	260
2. *Approaching Limits: Microchip Lasers*	272
VI. Conclusions and Outlook	274
List of Abbreviations and Acronyms	275
References	276

Chapter 5 Transient Grating Studies of Carrier Diffusion and Mobility in Semiconductors
Alan Miller

I. Optical Nonlinearities and Carrier Transport	287
II. Transient Gratings	290
III. Bulk Semiconductors	292
IV. Exciton Saturation in MQWs	293
V. In-Well Transport in MQWs	297
VI. Cross-Well Transport in QWs	303
VII. Hetero-n-i-p-i Structures	306
VIII. Conclusions	309
List of Acronyms	310
References	310

Index	313
Contents of Volumes in This Series	317

Preface

This two-volume set is designed to bring together two streams of thought — semiconductors and nonlinear optics — and to bridge the gap between optics and electronics. Practical nonlinear optical devices in semiconductors are on the verge of becoming a reality, as switches, modulators, converters, and sensors.

A high level of direct band semiconductor technology has come about because of the many practical applications of semiconductor lasers, modulators, and high-speed detectors. In general, their performance has been enhanced by using quantum confinement, particularly quantum wells. This technological base has provided high quality materials and devices from which nonlinear optical studies have proceeded. In particular, epitaxial films grown by technologies such as molecular beam epitaxy and metalo-organic chemical vapor deposition have extended the range of options available for optical nonlinearities.

Nonlinear optics grew rapidly with the development of lasers, originating in the study of the interactions of laser light with dielectric media. Some early work was performed on semiconductors, but the high intensities initially needed for nonlinear optics required high power lasers and often caused permanent material damage. In the last twenty years, however, semiconductors have been shown to exhibit sensitive nonlinearities, which can be made particularly large by using quantum confinement or narrow gap materials. The steady increase in semiconductor materials technology has enabled a number of interesting applications for nonlinear optics and furthered the understanding of the basic physical principles taking place in both bulk and quantum confined systems. Optical nonlinearities in semiconductors can now be accessed by milliwatt lasers, e.g., laser diodes.

Why publish a review now? We believe that at the present time most of the basic concepts in semiconductor nonlinear optics are well understood, and that a review of the basic science and technology in one place would be very useful. We have put together this two-volume set to stand together as a respresentation of the major thrusts of nonlinear optics in semiconductors.

Most of the chapters contain enough basic background that they can be read without extensive additional study. Fundamental physical principles as well as engineering approaches and applications are presented. Most of the chapters have a balanced account of both experimental and theoretical advances. As often happens with books of this sort, some important areas were missed due to time and space constraints. We were not able to include nonlinear optics in semiconductor waveguides, microcavities, self-electro-optic effect devices, or to include fabrication technologies. Finally, we have chosen to concentrate on basic concepts more than applications, as these latter tend to become dated more rapidly. Volume 58 begins with a review of optical nonlinearities that arise from absorption-induced photocarriers and the effect of these electrons and holes on the absorption and refractive index near the band edge in semiconductors. Chapter one surveys local optical nonlinearities and models useful for device and applications designers. The resonant nonlinearities arise from screening and filling of available states in bulk and quantum wells, and include the role of excitons and free carriers. Methods for analyzing experimental measurements, figures of merit, and tables of published values are also presented.

The second chapter investigates how the transport of photo-carriers affects resonant absorption and related refraction, through lengthening the lifetime of local nonlinearities and through non-local nonlinearities arising from photo-carrier screening fields. Examples of carrier transport nonlinearities, a study of electro-absorption in quantum wells, the properties of n-i-p-i structures, and characteristics of typical device configurations are all presented. In the third chapter Daniel Chemla reviews the status of the present understanding of the temporal evolution of optical nonlinearities in semiconductors. The availability of femtosecond lasers, along with phase measurements, have made possible studies of purely coherent processes involving only virtual transitions such as the optical Stark effect, as well as transitions involving real electron-hole pairs. This chapter introduces the many-body concepts necessary to interpret experimental data, separating out excitonic effects, as well as two-particle and four-particle correlation effects. Non-resonant optical nonlinearities occuring in the transparent region of semiconductors are considered in the last two chapters of the first volume. Sheik-Bahae and Van Stryland review two-photon absorption and its related nonlinear refraction, presenting a simple two-band model that fits a wide range of direct band semiconductor data. They review experimental methods for obtaining data and some applications in optical switching and limiting.

Millerd, Ziari and Partovi review photorefractivity in semiconductors, a wave-mixing process that relates optically generated carriers and electro-optical nonlinearities. Resonant nonlinearities provide enhanced performance in some cases, especially in photorefractive quantum wells. Advanced applications using photorefractivity are also outlined.

Volume 59 continues the review of nonlinear optics in semiconductors with some of the newest research that promises an array of potential applications. Khurgin describes how the very large second-order nonlinearities in semiconductors can lead to harmonic generation, once advanced growth and fabrication techniques that enable phase-matching have been developed. Quantum wells and superlattices have large second order nonlinearities, particularly using intersubband transitions. Optical rectification and its resulting terahertz emission in semiconductors are also described.

Hall, Thoen and Ippen explore optical nonlinearities in active semiconductor gain media, separating out diffusion, carrier scattering and carrier heating effects using femtosecond pump-probe techniques. Cross-phase and cross-gain modulation, spectral hole burning, and two photon absorption are all identified and applied to understanding semiconductor lasers and optical amplifiers. Shaping and saturation of pulses in active waveguides, four-wave mixing for wavelength conversion, and all-optical switching, all have application to broadband optical communications and switching systems.

Hanamura provides a theoretical basis for the enhancement of the optical response in semiconductors due to quantum confinement. His analysis includes excitons and biexcitons in quantum wells, wires, and dots. The chapter provides experimental results and analysis on superradiance and coherent light emission as well as figures of merit for nonlinear optical response.

Keller discusses passive switching of solid state lasers by using semiconductor saturable absorption integrated directly into a mirror structure. Various cavity designs are presented, including the anti-resonant Fabry-Perot and dispersive devices. Designs and results for mode-locked and Q-switched solid state lasers using these devices are included.

Miller presents nonlinear optics using picosecond pulse measurements as a tool to better understand properties of semiconductor carrier transport. Determinations of carrier diffusion and mobility in bulk and quantum wells are made, separately considering lateral carrier motion in the wells and movement across the wells.

The purpose of this two-volume set is to review research into the nonlinear optics of semiconductors, which has led both to a much better understanding of semiconductors and to the demonstration of various applications, from phase conjugation to optical switching to ultrashort pulse generation to optical limiting. While semiconductor nonlinear optical devices have not yet reached the level of importance that semiconductor lasers and modulators have, a number of the semiconductor nonlinearities are important to the laser community and some have the potential to achieve wide deployment in applications. It is our hope that this book will serve those in the semiconductor community interested in nonlinear optics, those in nonlinear optics interested in semiconductors, and also as a general resource.

ELSA GARMIRE
ALAN KOST

List of Contributors

Numbers in parenthesis indicate the pages on which the authors' contribution begins.

KATHERINE L. HALL, (83) *Advanced Networks Group, Massachusetts Institute of Technology, Lincoln Laboratory, Lexington, Massachusetts*

EIICHI HANAMURA, (161) *Optical Sciences Center, The University of Arizona, Tucson, Arizona*

ERICH P. IPPEN, (83) *Department of Electrical Engineering and Computer Science, Massachusetts Institute of Technology, Cambridge, Massachusetts*

URSULA KELLER, (211) *Institute of Quantum Electronics, Physics Department, Swiss Federal Institute of Technology (ETH), Zurich, Switzerland*

JACOB B. KHURGIN, (1) *Department of Electrical and Computer Engineering, Johns Hopkins University, Baltimore, Maryland*

ALAN MILLER, (287) *School of Physics and Astronomy, University of St. Andrews, Fife, Scotland*

ERIK R. THOEN, (83) *Department of Electrical Engineering and Computer Science, Massachusetts Institute of Technology, Cambridge, Massachusetts*

CHAPTER 1

Second-Order Nonlinearities and Optical Rectification

Jacob B. Khurgin

DEPARTMENT OF ELECTRICAL AND COMPUTER ENGINEERING
JOHNS HOPKINS UNIVERSITY
BALTIMORE, MARYLAND

    I. INTRODUCTION 2
   II. SECOND-ORDER NONLINEAR EFFECTS IN BULK SEMICONDUCTORS 4
       1. *General Considerations* 4
       2. *Classical Definition of Higher-Order Susceptibility* 6
       3. *Semiclassical Theory of Second-Order Susceptibility* 12
       4. *Off-Resonant Second-Order Susceptibilities of Zinc-Blende Semiconductors* . . 19
       5. *Calculation of Second-Order Susceptibility Using a Hybrid Bond Model* . . . 25
 III. $\chi^{(2)}$ EXPERIMENTS IN BULK SEMICONDUCTORS 30
       1. *Measurements of $\chi^{(2)}$ Values* 30
       2. *Experimental Demonstration of Second-Order Nonlinear Processes in Semiconductors* 34
  IV. SECOND-ORDER SUSCEPTIBILITY IN SEMICONDUCTOR QUANTUM WELLS AND SUPERLATTICES 42
       1. *Asymmetric Quantum Wells* 42
       2. *Intersubband Second-Order Nonlinearities: Theory* 49
       3. *Intersubband Second-Order Nonlinearities: Experiment* 55
       4. *Interband Second-Order Nonlinearities: Theory and Experiment* 58
   V. OPTICAL RECTIFICATION AND TERAHERTZ EMISSION IN SEMICONDUCTORS 63
       1. *Brief Review of the Optical Rectification Term in $\chi^{(2)}$* 63
       2. *Optical Rectification and Coherent Terahertz Emission in Bulk Semiconductors* 67
       3. *Optical Rectification and Terahertz Emission in Semiconductor Quantum Wells* 73
  VI. CONCLUSIONS 76
     LIST OF ABBREVIATIONS AND ACRONYMS 78
     REFERENCES 79

List of Abbreviations and Acronyms can be located preceding the references to this chapter.

I. Introduction

As stressed throughout these two volumes, using semiconductor materials as nonlinear optical elements allows one fully to bridge the gap between optics and electronics. It opens the possibility for truly integrated optoelectronic circuits in which all major elements—sources and detectors of radiation, electronic signal processing circuitry, and nonlinear optical devices such as switches, modulators, and optical converters—are built from the same family of semiconductor materials, which are compatible with each other. In this chapter, we are interested in the second-order (quadratic) nonlinear phenomena. Being the lowest-order nonlinear optical effects, the second-order nonlinear phenomena are thus the ones observable at the lowest power densities, and their use is, in principle, preferable to the use of third- and higher-order effects. The most ubiquitous "bread-and-butter" optoelectronic materials—the III–V binary semiconductors, their alloys, and quantum-size layered structures (quantum wells and superlattices)—exhibit, as a result of the heteropolar nature of their bonding, a very high second-order susceptibility that is typically at least an order of magnitude higher than those of the competing materials. At the same time, many of the III–V materials (and, similar to them in many ways, the II–VI materials) have direct bandgaps. The values of the bandgaps cover the important range—from as low as 0.5 eV for antimonides to as high as 3.5 eV for nitrides. Furthermore, after enormous effort devoted to the refinement of their growth and fabrication methods, the high-optical-quality and high-damage-threshold III–V semiconductor materials can be routinely grown and formed in the waveguides, making them compatible with laser diodes and with optical fiber technology.

Yet, even with such an impressive array of properties, the semiconductors have not found use in many second-order nonlinear devices owing to their only, but very important, shortcoming: neither the zinc-blende nor hexagonal lattice structures in which all III–V and II–VI semiconductors crystallize have sufficient birefringence to compensate for the large dispersion inherent in these materials. This, until very recently, has made the phasematching necessary for efficient frequency conversion unattainable in semiconductors. Since phasematching is of such paramount importance, we will distinguish in this chapter between two types of second-order nonlinear processes—those that require phasematching (i.e., harmonic generation and sum/difference generation) and those that do not (i.e., linear electro-optic effect and optical rectification, including terahertz oscillation generation).

In the nearly four decades since the inception of nonlinear optics (Armstrong et al., 1962; Ward, 1965; Butcher and McLean, 1963/1964), several theories adequately describing the origin of the second-order nonlinearities in solids and the nature of interaction between linear and nonlinear waves have been developed, and several excellent textbooks and monographs on the subject have been published. It is, therefore, beyond the scope of this chapter to consider in great detail all the aspects of second-order nonlinearities, for which the reader is referred to full-scale treatments. Yet, when it comes to the study of the origin of nonlinearity and descriptions of the nonlinear characteristics of the materials, most of the popular texts concentrate on a few well-known examples of ionic crystals, such as $KTiPO_4$ (KTP), $LiNbO_3$, $KDHPO_4$ (KDP), and others. The nonlinearity of these insulating crystals can be treated using a semiclassical model of anharmonic oscillators, and yet, when one considers the nonlinear properties of semiconductors, one has to take into account the dispersion of the bands — a topic not dealt with in most texts. Besides, historically, when the second-order nonlinear phenomena have been mentioned, harmonic generation has been considered first and treated in the most detail. Yet, in recent years, as the new sources of radiation in the blue and ultraviolet (UV) regions of the spectrum have continued to improve, second harmonic generation has taken a back seat to optical parametric generation and difference frequency generation of infrared (IR) and far-IR light. At these frequencies, the nonlinear properties are quite different from those encountered in second harmonic generation, and in our opinion there is a lack of comprehensive literature dealing adequately with this subject. Furthermore, in the last one and a half decades, development of fundamentally new quantum-size semiconductor heterostructures — quantum wells and superlattices — has taken place. The large size (in comparison to the atomic size) of the wavefunctions in these structures, as well as the ability to manipulate the asymmetry of the structure, open the prospects of engineering artificially large nonlinearities. So much progress has taken place in the last few years in the area of quantum-size semiconductor nonlinear structures that there is a need for a book treating this subject.

The purpose of this chapter is to complement the existing literature on second-order nonlinear optics in the areas specific to the semiconductors and their heterostructures. It is structured as follows.

After this brief introduction, we begin in Section II with the specifics of calculating nonlinear optical susceptibilities in the system with nonlocalized states. In order to do this, both classical and quantum theories of $\chi^{(2)}$ are reviewed, and a few competing approaches are described. Special attention is given to the consequences of using $\mathbf{A} \cdot \mathbf{P}$ and $\mathbf{E} \cdot \mathbf{r}$ interaction Hamiltonians. We then present in some detail the simple bond-polarizability theory of nonresonant $\chi^{(2)}$ at frequencies below the bandgap and compare it with the classical estimate and with more detailed full bandstructure calculations.

In Section III, the results of the measurements of $\chi^{(2)}$ in bulk semiconductors are described, and comparisons with different theoretical predictions are made. Then the problems associated with phasematching are considered, and a short review of the experimental results in harmonic, difference frequency, and parametric generation in bulk semiconductors is given.

Section IV is devoted to the new and exciting area that has developed in the last decade—nonlinear optics of the semiconductor quantum wells and superlattices, which have large dipole moments and thus very high nonlinearities. The distinction is drawn between two types of processes—band-to-band and intersubband—and the latest experimental results are reviewed for both types of processes.

In Section V, we focus on the resonant or near-resonant second-order properties of semiconductors—optical rectification and THz generation—which, we shall show, require special treatment. We present a model for their calculation in bulk materials and quantum wells and then describe some experimental results.

The last section mentions a number of important topics that were not touched on during this brief review, and presents some conclusions.

II. Second-Order Nonlinear Effects in Bulk Semiconductors

1. General Considerations

We are interested in the responses (linear and nonlinear) of the crystalline materials to the electromagnetic field in the optical range, covering wavelengths ranging from near-IR to near-UV ($0.2\,\mu\text{m} \leq \lambda \leq 2\,\mu\text{m}$). This response is determined by the motion of bound electrons situated in the outer shells of the atoms constituting the crystal. These electrons are called valence electrons. There are two alternative ways of describing the motion of the valence electrons in the solid state. Owing to the presence of a periodic lattice, valence electrons are delocalized and are well described by the extended or Bloch states

$$\Psi_{\mathbf{k}}(\mathbf{r}) = u_{\mathbf{k},n}(\mathbf{r})e^{i\mathbf{k}\mathbf{r}} \qquad (1)$$

where n is an index of a band, $\mathbf{k}$ is a wavevector, and $u_{\mathbf{k},n}(\mathbf{r})$ is a so-called "periodic" atomic wavefunction. This description is the best for evaluating transport properties of the semiconductor, as well as its absorption. The reason behind this is very simple: the transport properties are determined by the relatively few states near the extrema of the bands that have free carriers—electrons near the bottom of the first conduction band and holes near the top of the valence band. Therefore, precise descriptions of these individual states, such as mobility and density of states, are necessary. Similarly, when absorption of the optical radiation is the subject of interest, for the photon of given energy $h\nu$, only a pair of states in the conduction and valence bands with energy difference $E_{k,c} - E_{k,v} = h\nu$ will participate in the absorption process, and a precise description of these states that includes the transition dipole strength and joint density of states must be provided.

When, however, the processes of interest are nonresonant ones—such as, for example, refraction of light and most of the nonlinear processes mentioned in this chapter—all the states of different energies within each band react collectively to the optical excitation. Therefore, dispersion of the energies in the momentum space can be disregarded and the collective response of all the N extended states in the band can be represented as the collective response of the N localized states, or, in other words, molecular orbitals. The points that calculations based on molecular orbital description are quite adequate in the below-the-bandgap frequency range where no real transitions occur and that the joint density of states is irrelevant for calculation have been stressed by Leman and Friedel (1962) for the elemental group IV semiconductors. Furthermore, it was first shown by Robinson (1967) that nonresonant nonlinear susceptibility can be estimated by considering only the ground state molecular orbitals. A number of earlier studies used this approach (Flytzanis and Ducuing, 1969; Jha and Bloembergen, 1968; Levine, 1973) and achieved good agreement with the experiment.

The early attempts to use the full-bandstructure calculations to determine $\chi^{(2)}$ provided an insight into how the inversion asymmetry alters the momentum matrix elements and results in second-order polarizability (Bell, 1972; Aspnes, 1972; Fong and Shen, 1975), but in the absence of good computer models the results were further away from the experimental data than the ones obtained using a simple bond approach. Only in the late 1980s did sophisticated computer techniques allow a full-band calculation of nonlinear susceptibility (Moss, Sipe, and Van Driel, 1987; Ghahramani, Moss, and Sipe, 1991; Sipe and Ghahramani, 1993).

In this section we will outline the most general considerations associated with the $\chi^{(2)}$ calculations, beginning with the primitive classical model and then working our way up through the more sophisticated ones. It is impossible to provide a full treatment of this subject in this short chapter and it is also entirely unnecessary, since many excellent books on the subject already exist (Boyd, 1992; Shen, 1984; Butcher and Cotter, 1990), but a short tour through the basics of $\chi^{(2)}$ calculations can be quite useful. By the end of this section the reader should have developed a good intuitive understanding of how the nonlinear susceptibility is related to such basic properties of semiconductor bonding as bond length, bond polarity (Phillips, 1968; Van Vechten, 1969), and the binding energy, and should be able to predict the magnitudes of nonlinear effects in various semiconductors.

2. Classical Definition of Higher-Order Susceptibility

Although nonlinear optical effects belong to the realm of quantum electronics, the nature of the nonlinear polarization can be best envisioned in what is essentially a classical framework, using the anharmonic oscillator model. Consider the anharmonic oscillator model of the tetragonal semiconductor (Fig. 1a). Each valence electron is situated in one of the four bonds between the cation (C) and anion (A). In the III–V semiconductors, C is a group III ion and A is a group V ion. Owing to different electronegativities of the C and A ions, the electron is more strongly attracted to the anion, and, in the absence of external fields, it assumes its equilibrium position closer to the anion. The potential $\tilde{U}(\mathbf{r})$ holding the electron in each bond, shown in Fig. 1(b), is asymmetrical, and it favors location of the wavefunc-

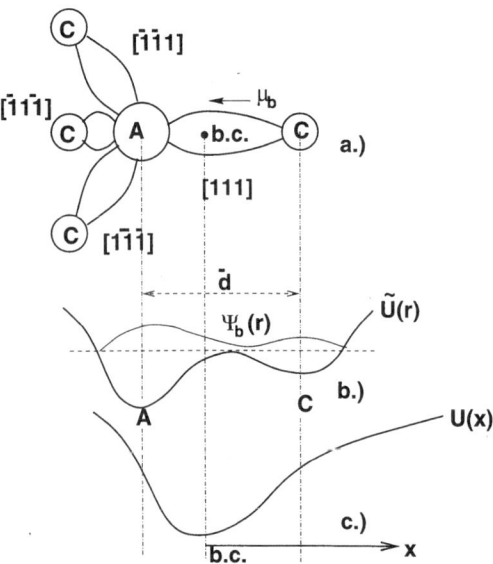

FIG. 1. Classical description of optical properties in zinc-blende semiconductors: (a) four tetragonal bonds, (b) binding potential $\bar{U}(\mathbf{r})$ and bonding orbital, and (c) classical anharmonic approximation.

1 Second-Order Nonlinearities and Optical Rectification

tion Ψ_b of the electron in a bond closer to the anion with the center of bond charge (b.c.) shown in the figure. In the classical description, we associate the coordinate x with the center of bond charge.

The electron is considered to be a point charge, which in the absence of external fields is kept at $x = 0$ by potential $U(x)$ (Fig. 1c). This one-dimensional potential $U(x)$ is, of course, not identical to the real potential $\tilde{U}(\mathbf{r})$, but is simply a "fit" that would make the classical electron behave in a manner similar to the real, wavelike electron. This is why it has its single minimum at the bond charge center, rather than two minima at two ions. In the Taylor expansion of the potential $U(x)$, the asymmetry is manifested by the presence of odd-power terms:

$$U = -\int F_r dx = \frac{1}{2}kx^2 + \frac{1}{3}ax^3; \qquad U(x) \neq U(-x) \tag{2}$$

where parameter a characterizes anharmonicity and thus (as we shall soon see) nonlinearity of the medium. By differentiating Eq. (2), we obtain the expression for the restoring force holding the electron near the equilibrium:

$$F_r = -kx - ax^2 \tag{3}$$

Since $U(x)$ is a binding potential keeping the electron bound to the ion, one can make a rough estimate of the value of the parameter k:

$$k \sim \frac{2U_b}{\bar{d}^2} \tag{4}$$

where $\bar{d}$ is the degree of electron localization, which is roughly on the order of a bond length but, as we shall see below, not necessarily equal to it. Let us introduce the eigenfrequency of oscillations:

$$\omega_0 = \sqrt{\frac{k}{m_0}} \tag{5}$$

where m_0 is a free electron mass so that

$$\bar{d}^2 \sim \frac{2U_b}{m_0 \omega_0^2} \tag{6}$$

For the semiconductors, binding energy is on the order of work function (i.e., $U_b \sim 5\text{--}10\,\text{eV}$), while the resonant energies are on the order of bandgap (i.e., $\hbar\omega_0 \sim 1\text{--}3$ eV). Therefore, average displacement of the bound electron from the ion is about $\bar{d} \sim 3\text{--}10$ Å. This is an important result, especially for the narrow-gap semiconductors. It shows that the electron is not confined to one specific bond (whose length is only about 2 Å) but is extended over a few bonds. Let us now consider the anharmonic term and its role. For the small amplitudes of displacement x we can write $ax \ll k = m_0\omega^2$. The upper bound for a can be found from the assumption that when the amplitude of displacement reaches $\bar{d}/2$, the anharmonic term approaches the order of magnitude of a harmonic one. Indeed, from Fig. 1 and Eq. (2) one can see that when $ax^2 = -kx$ the potential energy curve becomes flat—i.e., ionization takes place. Thus

$$|a| \leq k/\bar{d} \sim 2U_b/\bar{d}^3 \sim 100 \text{ meV/Å}^3 \tag{7}$$

Alternatively, using Eq. (5), we obtain

$$\frac{|a|}{m_0} \leq \frac{\omega_0^2}{\bar{d}} \tag{8}$$

Let us now write the equation of motion:

$$\frac{d^2x}{dt^2} + \gamma\frac{dx}{dt} + \omega_0^2 x + \frac{a}{m_0}x^2 = -\frac{e}{m_0}\mathscr{E}(t) \tag{9}$$

where γ is a damping constant and $\mathscr{E}$ is an external optical field. Assume that we apply an optical field

$$\mathscr{E}(t) = \frac{1}{2}(E_1 e^{-i\omega_1 t} + E_2 e^{-i\omega_2 t} + c.c.) \tag{10}$$

to our anharmonic oscillator. If the field is smaller thn the atomic field:

$$\mathscr{E} \ll E_a \sim \frac{U_b}{e\bar{d}} \sim \frac{\omega_0^2 m_0 \bar{d}}{e} \sim 10^8\text{--}10^9 \text{ V/cm} \tag{11}$$

one can look for the response of the anharmonic oscillator as a sum of harmonic terms [with superscript (1)] and second-order anharmonic terms [with superscript (2)]:

$$x(t) = \frac{1}{2}(x_1^{(1)}e^{-i\omega_1 t} + x_2^{(1)}e^{-i\omega_2 t} + c.c.)$$

$$+ \frac{1}{2}(x^{(2)}_{\omega_1+\omega_2}e^{-i(\omega_1+\omega_2)t} + x^{(2)}_{\omega_1-\omega_2}e^{-i(\omega_1-\omega_2)t} + x^{(2)}_{2\omega_1}e^{-i2\omega_1 t} \quad (12)$$

$$+ x^{(2)}_{2\omega_2}e^{-i2\omega_2 t} + c.c.) + x^{(2)}_0$$

with all the second-order terms being much smaller than the first-order terms. Then, in order to find the first-order terms, we can neglect all the anharmonic terms and find

$$x^{(1)}_i = -\frac{e/m_0}{\omega_0^2 - \omega_i^2 - i\omega_i\gamma} E_i \quad (13)$$

From Eqs. (13) and (11) we can see that as long as we are far from resonance, the response is indeed small:

$$x^{(1)}_i \sim \bar{d}\frac{E_i}{E_a} \ll \bar{d} \quad (14)$$

Substituting Eq. (13) into Eq. (9), we obtain the expressions for the second-order terms:

$$x^{(2)}_{\omega_1 \pm \omega_2} = -\frac{a/m_0}{\omega_0^2 - (\omega_1 \pm \omega_2)^2 - i\omega_i\gamma} x^{(1)}_1 x^{(1)}_2$$

$$= \frac{(e^2 a/m_0^3) E_1 E_2}{[\omega_0^2 - (\omega_1 \pm \omega_2)^2 - i(\omega_1 \pm \omega_2)\gamma](\omega_0^2 - \omega_1^2 - i\omega_1\gamma)(\omega_0^2 - \omega_2^2 - i\omega_2\gamma)}, \quad (15)$$

$$x^{(2)}_{2\omega_i} = -\frac{a/2m_0}{\omega_0^2 - 4\omega_i^2 - 2i\omega_i\gamma} [x^{(1)}_i]^2$$

$$= \frac{(e^2 a/2m_0^3) E_i^2}{(\omega_0^2 - 4\omega_i^2 - 2i\omega_i\gamma)(\omega_0^2 - \omega_1^2 - i\omega_1\gamma)(\omega_0^2 - \omega_2^2 - i\omega_2\gamma)}, \quad (16)$$

and

$$x^{(2)}_0 = -\frac{a/2m_0}{\omega_0^2}[(x^{(1)}_1)^2 + (x^{(2)}_2)^2]$$

$$= -\frac{e^2 a/4m_0^3}{\omega_0^2}\left[\frac{E_1^2}{(\omega_0^2 - \omega_1^2 - i\omega_1\gamma)^2} + \frac{E_2^2}{(\omega_0^2 - \omega_2^2 - i\omega_2\gamma^2)}\right] \quad (17)$$

Now, since $x_i^{(1)} \ll \bar{d}$ (Eq. 14) and assuming maximum anharmonicity (Eq. 8) $|a| \sim \omega_0^2 m_0/\bar{d}$, then, once again in the absence of resonance, we obtain

$$x_{\omega_i+\omega_j}^{(2)} \sim \frac{x_i^{(1)} x_j^{(1)}}{\bar{d}} \sim \bar{d}\frac{E_i E_j}{E_a^2} \ll x_i^{(1)} \qquad (18)$$

justifying both our perturbation approximation and the possibility for the truncation of the third- and higher-order terms.

As a next step, we can evaluate the linear and nonlinear components of the polarization vector:

$$\mathbf{P}(t) = \frac{1}{2}[\mathbf{P}^{(1)}(\omega_1)e^{-i\omega_1 t} + \mathbf{P}^{(1)}(\omega_2)e^{-i\omega_2 t} + c.c.]$$

$$+ \frac{1}{2}[\mathbf{P}^{(2)}(\omega_1+\omega_2)e^{-i(\omega_1+\omega_2)t} + \mathbf{P}^{(2)}(\omega_1-\omega_2)e^{-i(\omega_1-\omega_2)t} \qquad (19)$$

$$+ \mathbf{P}^{(2)}(2\omega_1)e^{-i2\omega_1 t} + \mathbf{P}^{(2)}(2\omega_2)^{(2)}e^{-i2\omega_2 t} + c.c.] + \mathbf{P}_{DC}^{(2)}$$

where

$$\mathbf{P}^{(1)}(\omega_i) = -eNx_i^{(1)}, \qquad (20)$$

$$\mathbf{P}^{(2)}(\omega_i \pm \omega_j) = -eNx_{\omega_i \pm \omega_j}^{(2)}, \qquad (21)$$

and N is the number of atoms. We now introduce linear and nonlinear susceptibilities as

$$\mathbf{P}^{(1)}(\omega_i) = \varepsilon_0 \chi^{(1)}(\omega_i)\mathbf{E}_i \qquad (22)$$

and

$$\mathbf{P}^{(2)}(\omega_i \pm \omega_j) = \varepsilon_0 \chi^{(2)}(\omega_i \pm \omega_j)\mathbf{E}_i \mathbf{E}_j \qquad (23)$$

where ε_0 — the dielectric permittivity of the vacuum — is introduced to have susceptibilities dimensionless, and we take into account the vector character of the fields. It follows that, in general, linear susceptibility $\chi^{(1)}$ is a second-rank tensor and second-order susceptibility $\chi^{(2)}$ is a third-rank tensor. In the simple model used here, the susceptibility is a scalar, so we can evaluate the susceptibilities as

$$\chi^{(1)}(\omega_i) = \frac{Ne^2}{m_0 \varepsilon_0} \frac{1}{\omega_0^2 - \omega_i^2 - i\omega_i \gamma}, \qquad (24)$$

$$\chi^{(2)}(\omega_1 \pm \omega_2)$$
$$= \frac{aNe^3}{m_0^3\varepsilon_0} \frac{1}{[\omega_0^2 - (\omega_1 \pm \omega_2)^2 - i(\omega_1 \pm \omega_2)\gamma](\omega_0^2 - \omega_1^2 - i\omega_1\gamma)(\omega_0^2 - \omega_2^2 - i\omega_2\gamma)}, \quad (25)$$

$$\chi^{(2)}(2\omega_i) = \frac{aNe^3}{2m_0^3\varepsilon_0} \frac{1}{(\omega_0^2 - 4\omega_i^2 - 2i\omega_i\gamma)(\omega_0^2 - \omega_i^2 - i\omega_i\gamma)(\omega_0^2 - \omega_i^2 - i\omega_i\gamma)}, \quad (26)$$

and

$$\chi^{(2)}(0) = \frac{aNe^3}{2m_0^3\varepsilon_0} \frac{1}{(\omega_0^2 - \omega_1^2 - i\omega_1\gamma)^2} \quad (27)$$

Using these expressions, one can evaluate the expected order of magnitude for nonresonant classical linear and nonlinear susceptibilities of most semiconductor materials. In the long-wavelength limit, we obtain for the linear susceptibility

$$\chi^{(1)}(0) = \frac{Ne^2}{m_0\varepsilon_0\omega_0^2} \sim \frac{Ne^2\bar{d}^2}{2\varepsilon_0 U_b} \quad (28)$$

Assuming that $N \sim 10^{23}$ cm^{-3}, $U_b \sim 5\text{--}10$ eV, and $\bar{d} = 1\text{--}2$ Å, we obtain $\chi^{(1)} \sim 5\text{--}15$, with higher-gap semiconductors having lower susceptibilities. As for second-order susceptibility, assuming the maximum value of asymmetry parameter a to be described in Eq. (8), we can obtain the long-wavelength limit:

$$\chi^{(2)}(0) = \frac{|a|Ne^3}{2m_0^3\varepsilon_0\omega_0^6} \leq \frac{Ne^3}{2m_0^2\varepsilon_0\omega_0^4\bar{d}} = \frac{Ne^3\bar{d}^3}{8\varepsilon_0 U_b^3} = \chi^{(1)}(\omega_0) \frac{e\bar{d}}{4U_b} \quad (29)$$

This equation makes perfect sense, because it shows that the second-order susceptibility increases in the extended systems with large $\bar{d}$ and also in the systems with low binding energy U_b, since it is easy to reach anharmonicity in weakly confining potential. One can also see that, in the off-resonant limit,

$$\chi^{(2)}(\omega_1 \pm \omega_2) \leq \chi^{(1)}(\omega_1)\chi^{(1)}(\omega_2)\chi^{(1)}(\omega_1 + \omega_2) \frac{1}{\chi^{(1)}(0)} \frac{\varepsilon_0}{2eN\bar{d}}$$
$$= \frac{\chi^{(1)}(\omega_1)\chi^{(1)}(\omega_2)\chi^{(1)}(\omega_1 + \omega_2)}{[\chi^{(1)}(0)]^2} \frac{e\bar{d}}{4U_b} \quad (30)$$

Equations (29) and (30) both indicate that, in general,

$$\chi^{(2)} \sim \frac{\chi^{(1)}}{E_a} \tag{31}$$

Thus, one can expect the semiconductors with bandgaps in the visible or IR range to have $\chi^{(2)}$ on the order of a few hundred picometers per volt. Usually, $\chi^{(2)}$ is somewhat smaller, generally because the nonlinear polarizations of different bonds tend partially to cancel each other, but using Eq. (31), gives a reasonble rough estimate.

Thus, we have derived the expressions for the sum and difference frequency coefficient, the second harmonic coefficient, and the optical rectification coefficient. There is another coefficient—the linear electro-optic effect—that is a special case of sum generation with one of the fields being of low frequency ω_2 relative to optical frequency $\omega_1 = \omega$. Then we obtain

$$\mathbf{P}^{(2)}(\omega_1 + 0) = \varepsilon_0 \chi^{(2)}(\omega_1 + 0)\mathbf{E}_1 \mathbf{E}_{DC} \tag{32}$$

where

$$\chi^{(2)}(\omega_1 + 0) = \frac{aNe^3}{m_0^3 \varepsilon_0} \frac{1}{\omega_0^2(\omega_0^2 - \omega_1^2 - i\omega_1\gamma)^2} \tag{33}$$

This is the linear electro-optic or Pockels effect. As one can see, at least away from the resonance, the expressions for the linear electro-optic effect and optical rectification are identical. As one gets closer to the resonance, the dispersion of the electro-optic effect and rectification is very strong, as a result of the double-resonant term in the denominator.

The classical theory gives us an opportunity to get a physical insight into the origin of the second-order nonlinearities and can provide us with the order-of-magnitude answer for the *maximum* possible values of $\chi^{(2)}$, but to understand it completely one needs to know the shapes of bond orbitals, and thus quantum mechanics must be invoked.

3. SEMICLASSICAL THEORY OF SECOND-ORDER SUSCEPTIBILITY

In this section we sketch the major steps leading to the semiclassical expression for the second-order susceptibility. The term "semiclassical" here is understood as a combination of quantum-mechanical treatment of the material system with classical treatment of electromagnetic fields—i.e., without invoking the secondary quantization mechanism. We consider a quantum-mechanical picture of the matter wherein each atom, or each bond, is represented by a set of eigenfunctions $\psi_1, \psi_2, \ldots, \psi_i, \ldots$, of the Hamiltonian

$$\mathscr{H}_0 = \begin{bmatrix} E_1 & 0 & \cdots & 0 & \cdots \\ 0 & E_2 & \cdots & 0 & \cdots \\ \cdots & \cdots & \cdots & \cdots & \cdots \\ 0 & 0 & \cdots & E_i & \cdots \\ \cdots & \cdots & \cdots & \cdots & \cdots \end{bmatrix} \tag{34}$$

The system is subject to the electromagnetic field

$$\mathscr{E}(t) = \frac{1}{2}[\mathbf{E}_1 e^{-i\omega_1 t} + \mathbf{E}_2 e^{-i\omega_2 t} + c.c.] \tag{35}$$

and the interaction Hamiltonian in the dipole approximation is

$$\mathscr{H}_\mathscr{E}(t) = -\boldsymbol{\mu} \cdot \mathscr{E}(t) = -\begin{bmatrix} 0 & \boldsymbol{\mu}_{12} \cdot \mathscr{E} & \cdots & \boldsymbol{\mu}_{1n} \cdot \mathscr{E} & \cdots \\ \boldsymbol{\mu}_{21} \cdot \mathscr{E} & \boldsymbol{\mu}_{22} \cdot \mathscr{E} & \cdots & \boldsymbol{\mu}_{2n} \cdot \mathscr{E} & \cdots \\ \cdots & \cdots & \cdots & \cdots & \cdots \\ \boldsymbol{\mu}_{n1} \cdot \mathscr{E} & \boldsymbol{\mu}_{n2} \cdot \mathscr{E} & \cdots & \boldsymbol{\mu}_{nn} \cdot \mathscr{E} & \cdots \\ \cdots & \cdots & \cdots & \cdots & \cdots \end{bmatrix} \tag{36}$$

where all the

$$\boldsymbol{\mu}_{mn} = -e\langle\psi_m|\mathbf{r}|\psi_n\rangle \tag{37}$$

are the matrix elements of the dipole moment. By virtue of choosing the origin of the coordinate at the "center of mass" of the ground state wavefunction, $\mu_{11} = 0$, whereas, owing to a lack of central symmetry, other diagonal moments are not equal to 0. We can describe the state of the system by its density matrix:

$$\rho = \begin{bmatrix} \rho_{11} & \rho_{12} & \cdots & \rho_{1n} & \cdots \\ \rho_{21} & \rho_{22} & \cdots & \rho_{2n} & \cdots \\ \cdots & \cdots & \cdots & \cdots & \cdots \\ \rho_{n1} & \rho_{n2} & \cdots & \rho_{nn} & \cdots \\ \cdots & \cdots & \cdots & \cdots & \cdots \end{bmatrix} \quad (38)$$

The equation of motion for the density matrix is

$$\frac{d}{dt}\rho_{mn} = -\frac{i}{\hbar}[\mathcal{H}\rho] - \gamma(\rho_{mn} - \bar{\rho}_{mn}) \quad (39)$$

where the last term represents phenomenologically the relaxation processes occurring in the system. For the off-diagonal terms, the equilibrium value $\bar{\rho}_{mn}$ is 0 and γ_{mn} is a dephasing rate. On the other hand, the equilibrium value of the diagonal matrix elements is not 0. The expected value of the dipole moment can be found as

$$\langle \boldsymbol{\mu} \rangle = \mathit{Tr}(\boldsymbol{\mu}\rho) \quad (40)$$

Initially the system is in one of the ground states l:

$$\rho_{mn}(0) = \delta_{ml}\delta_{nl}; \quad l = 1, 2, \ldots, l_{max} \quad (41)$$

The equation of motion (Eq. 39) for the diagonal matrix elements is

$$\frac{d}{dt}\rho_{nn} = -\frac{i}{\hbar}\sum_{m \neq n}(\boldsymbol{\mu}_{mn}\rho_{nm} - \boldsymbol{\mu}_{nm}\rho_{mn}) \cdot \mathcal{E} - \frac{\rho_{nn} - \rho_{nn}(0)}{T_n} \quad (42)$$

where T_n is a relaxation rate, and the equation for the off-diagonal matrix elements is

$$\frac{d}{dt}\rho_{mn} = -i\left(\omega_{mn} - \Delta\boldsymbol{\mu}_{mn} \cdot \frac{\mathcal{E}}{\hbar}\right)\rho_{mn} - \frac{i}{\hbar}(\rho_{mm} - \rho_{nn})\boldsymbol{\mu}_{mn} \cdot \mathcal{E}$$
$$- \frac{i}{\hbar}\sum_{k \neq m,n}(\rho_{mk}\boldsymbol{\mu}_{kn} - \rho_{kn}\boldsymbol{\mu}_{mk}) \cdot \mathcal{E} - \gamma\rho_{mn} \quad (43)$$

where $\Delta\mu_{mn} = \mu_{mm} - \mu_{nm}$, and we assume that all the nondiagonal matrix elements decay at the same rate γ. We can now proceed to solve these equations perturbatively, assuming that the density matrix elements up to the second order of perturbation are

$$\rho_{nn}(t) = \sigma_{nn}(0) + \sigma_{nn}(\omega_1 + \omega_2)e^{i(\omega_1 + \omega_2)t} + \sigma_{nn}(\omega_1 - \omega_2) \\ + \sigma_{nn}(2\omega_1) + \sigma_{nn}(2\omega_2) + c.c., \quad (44)$$

$$\rho_{mn}(t) = \sigma_{mn}(0) + \sigma_{mn}(\omega_1)e^{i\omega_1 t} + \sigma_{mn}(\omega_2)e^{i\omega_2 t} + \sigma_{mn}(\omega_1 + \omega_2)e^{i(\omega_1 + \omega_2)t} \\ + \sigma_{mn}(\omega_1 - \omega_2) + \sigma_{mn}(2\omega_1) + \sigma_{mn}(2\omega_2) + c.c., \quad (45)$$

and

$$\rho_{nm}(t) = \rho_{mn}^*(t) \quad (46)$$

Equations (42) and (43) can now be solved perturbatively. In the first order, the only nonzero terms are

$$\sigma_{ml}^{(1)}(\omega_s) = \frac{1}{2\hbar} \frac{\boldsymbol{\mu}_{m1} \cdot \mathbf{E}_s}{\omega_{ml} - \omega_s - i\gamma} \Delta\rho_{nl}^0 \quad (47)$$

and

$$\sigma_{ml}^{(1)}(-\omega_s) = \frac{1}{2\hbar} \frac{\boldsymbol{\mu}_{m1} \cdot \mathbf{E}_s}{\omega_{ml} - \omega_s - i\gamma} \Delta\rho_{nl}^0 \quad (48)$$

where $s = 1, 2$, $\Delta\rho_{nl}^0 = \rho_{nn}(0) - \rho_{ll}(0)$. Using Eq. (40) immediately yields the expression for the linear polarization:

$$\frac{1}{2}\mathbf{P}(\omega_s) = N \sum_{ml} \boldsymbol{\mu}_{lm}[\sigma_{ml}^{(1)}(\omega_s) + \boldsymbol{\mu}_{ml}^i \sigma_{lm}^{(1)}(\omega_s)] \\ = \frac{N}{\hbar} \sum_{ml} \left[\frac{\boldsymbol{\mu}_{lm}\boldsymbol{\mu}_{ml}}{\omega_{ml} - \omega_s - i\gamma} + \frac{\boldsymbol{\mu}_{ml}\boldsymbol{\mu}_{lm}}{\omega_{ml} + \omega_s + i\gamma} \right] \cdot \mathbf{E}_s \Delta\rho_{nl}^0 = \frac{1}{2}\varepsilon_0 \chi^{(1)}(\omega_s)\mathbf{E}_s(\omega_s) \quad (49)$$

From this equation, the expression for the tensor of linear susceptibility follows:

$$\chi_{ij}^{(1)}(\omega_s) = \frac{N}{\varepsilon_0 \hbar} \sum_{ml} \left[\frac{\mu_{lm}^i \mu_{ml}^j}{\omega_{ml} - \omega_s - i\gamma} + \frac{\mu_{ml}^i \mu_{lm}^j}{\omega_{ml} + \omega_s + i\gamma} \right] \Delta\rho_{nl}^0 \quad (50)$$

In the second order, we obtain a number of terms:

$$\sigma_{nn}^{(2)}(0) = -\sum_{s=1,2} \frac{1}{2} \frac{i}{\hbar} [\sigma_{nl}^{(1)}(\omega_s)\boldsymbol{\mu}_{ln} \cdot \mathbf{E}_s^* - c.c.] T_n$$

$$= \sum_{s=1,2} \frac{1}{2\hbar^2} \frac{\boldsymbol{\mu}_{nl} \cdot \mathbf{E}_s \boldsymbol{\mu}_{ln} \cdot \mathbf{E}_s^* \gamma T_n}{(\omega_{nl} - \omega_s)^2 + \gamma^2} \Delta\rho_{nl}^0 \tag{51}$$

These terms correspond to a linear absorption from the ground state $|l\rangle$ into the stake $|n\rangle$:

$$\sigma_{nn}^{(2)}(\omega_1 + \omega_2) = -\frac{1}{2}\frac{i}{\hbar}[\sigma_{nl}^{(1)}(\omega_1)\boldsymbol{\mu}_{ln} \cdot \mathbf{E}_2 + \sigma_{nl}^{(1)}(\omega_2)\boldsymbol{\mu}_{ln} \cdot \mathbf{E}_1] \frac{1}{T_n^{-1} - i(\omega_1 + \omega_2)}$$

$$= \frac{1}{4\hbar^2} \frac{\omega_1 + \omega_2 + 2i\gamma}{\omega_1 + \omega_2 + iT_n^{-1}} \frac{\boldsymbol{\mu}_{nl} \cdot \mathbf{E}_1 \boldsymbol{\mu}_{ln} \cdot \mathbf{E}_2}{(\omega_{nl} - \omega_1 + i\gamma)(\omega_{nl} - \omega_2 + i\gamma)} \Delta\rho_{nl}^0 \tag{52}$$

The first denominator in Eq. (52) is typically very large for optical frequencies, and thus we can disregard the fast changes in diagonal matrix elements: this is called "rotating wave approximation." We thus neglect $\sigma_{nn}^{(2)}(\omega_1 + \omega_2)$ as well as degenerate terms $\sigma_{nn}^{(2)}$ and $\sigma_{nn}^{(2)}(2\omega_2)$. We cannot however, disregard the oscillation of the population density at beat frequencies:

$$\sigma_{nn}^{(2)}(\omega_1 + \omega_2) = -\frac{1}{2}\frac{i}{\hbar}[\sigma_{nl}^{(1)}(\omega_1)\boldsymbol{\mu}_{1n} \cdot \mathbf{E}_2^* - \sigma_{ln}^{(1)}(-\omega_2)\boldsymbol{\mu}_{n1} \cdot \mathbf{E}_1] \frac{1}{T_n^{-1} - i(\omega_1 - \omega_2)}$$

$$= \frac{1}{4\hbar^2} \frac{\omega_1 - \omega_2 + 2i\gamma}{\omega_1 - \omega_2 + iT_n^{-1}} \frac{\boldsymbol{\mu}_{n1} \cdot \mathbf{E}_1 \boldsymbol{\mu}_{1n} \cdot \mathbf{E}_2^*}{(\omega_{nl} - \omega_1 - i\gamma)(\omega_{nl} - \omega_2 - i\gamma)} \Delta\rho_{nl}^0 \tag{53}$$

And for the nondiagonal matrix elements, we obtain

$$\sigma_{mn}^{(2)}(\omega_2 + \omega_1) = \frac{1}{2\hbar}\Bigg(\Delta\boldsymbol{\mu}_{mn} \cdot [\sigma_{mn}^{(1)}(\omega_1)\mathbf{E}_2 + \sigma_{mn}^{(1)}(\omega_2)\mathbf{E}_1]$$

$$+ \sum_{k \neq m,n} \{[\sigma_{kn}^{(1)}(\omega_1)\boldsymbol{\mu}_{mk} \cdot \mathbf{E}_2 + \sigma_{kn}^{(1)}(\omega_2)\boldsymbol{\mu}_{mk} \cdot \mathbf{E}_1]$$

$$- [\sigma_{mk}^{(1)}(\omega_1)\boldsymbol{\mu}_{kn} \cdot \mathbf{E}_2 + \sigma_{mk}^{(1)}(\omega_2)\boldsymbol{\mu}_{kn} \cdot \mathbf{E}_2]\}\Bigg)$$

$$\times \frac{1}{\omega_{mn} - \omega_1 - \omega_2 - i\gamma}$$

1 SECOND-ORDER NONLINEARITIES AND OPTICAL RECTIFICATION

$$= \frac{1}{4\hbar^2} \left[\left(\frac{\boldsymbol{\mu}_{mn} \cdot \mathbf{E}_1 \Delta\boldsymbol{\mu}_{mn} \cdot \mathbf{E}_2}{\omega_{mn} - \omega_1 - i\gamma} + \frac{\boldsymbol{\mu}_{mn} \cdot \mathbf{E}_2 \Delta\boldsymbol{\mu}_{mn} \cdot \mathbf{E}_1}{\omega_{mn} - \omega_2 - i\gamma} \right) \Delta\rho_{mn}^{(0)} \right.$$

$$+ \sum_{k \neq m,n} \left(\frac{\boldsymbol{\mu}_{kn} \cdot \mathbf{E}_1 \boldsymbol{\mu}_{mk} \cdot \mathbf{E}_2}{\omega_{kn} - \omega_1 - i\gamma} + \frac{\boldsymbol{\mu}_{kn} \cdot \mathbf{E}_2 \boldsymbol{\mu}_{mk} \cdot \mathbf{E}_1}{\omega_{kn} - \omega_2 - i\gamma} \right) \Delta\rho_{kn}^{(0)}$$

$$\left. - \left(\frac{\boldsymbol{\mu}_{mk} \cdot \mathbf{E}_1 \boldsymbol{\mu}_{kn} \mathbf{E}_2}{\omega_{mk} - \omega_1 - i\gamma} + \frac{\boldsymbol{\mu}_{mk} \cdot \mathbf{E}_2 \boldsymbol{\mu}_{kn} \mathbf{E}_1}{\omega_{mk} - \omega_2 - i\gamma} \right) \Delta\rho_{mk}^{(0)} \right]$$

$$\times \frac{1}{\omega_{mn} - \omega_1 - \omega_2 - i\gamma} \tag{54}$$

with similar expressions for other terms. From the expressions in Eq. (54), one can proceed to find the nonlinear polarization by invoking Eq. (40):

$$\frac{1}{2} \mathbf{P}^{(2)}(\omega_2 + \omega_1; \omega_1, \omega_2) = N \sum_{mn} [\boldsymbol{\mu}_{nm} \sigma_{mn}^{(2)}(\omega_2 + \omega_1) + \boldsymbol{\mu}_{mn} \sigma_{nm}^{(2)}(\omega_2 + \omega_1)]$$

$$= \frac{1}{2} \varepsilon_0 \chi^{(2)}(\omega_2 + \omega_1; \omega_1, \omega_2) \mathbf{E}_2(\omega_2) \mathbf{E}_1(\omega_1) \tag{55}$$

and performing summation over all levels. An excellent treatment of this process is given by Boyd (1992). The resulting polarization expression is rather involved and, in general, impossible to evaluate because various matrix elements are not known. We can, however, significantly simplify the expressions for $\chi^{(2)}$ by considering the resonant and nonresonant cases separately. If we consider only the nonresonant case, all the damping terms can be eliminated and changing all the dummy indices in the summation, following Boyd (1992), we obtain

$$\chi_{ijk}^{(2)}(\omega_1 + \omega_2; \omega_1, \omega_2) = \frac{N}{2\varepsilon_0 \hbar^2} \sum_{l,m,n} \frac{\mu_{ln}^i \mu_{nm}^j \mu_{ml}^k}{(\omega_{nl} - \omega_1 - \omega_2)(\omega_{ml} - \omega_1)}$$

$$+ \frac{\mu_{ln}^i \mu_{nm}^k \mu_{ml}^j}{(\omega_{nl} - \omega_1 - \omega_2)(\omega_{ml} - \omega_2)}$$

$$+ \frac{\mu_{ln}^j \mu_{nm}^i \mu_{ml}^k}{(\omega_{nl} + \omega_2)(\omega_{ml} - \omega_1)} + \frac{\mu_{ln}^k \mu_{nm}^i \mu_{ml}^j}{(\omega_{nl} + \omega_2)(\omega_{ml} - \omega_1)}$$

$$+ \frac{\mu_{ln}^k \mu_{nm}^j \mu_{ml}^i}{(\omega_{nl} + \omega_1 + \omega_2)(\omega_{ml} + \omega_1)}$$

$$+ \frac{\mu_{ln}^j \mu_{nm}^k \mu_{ml}^i}{(\omega_{nl} + \omega_1 + \omega_2)(\omega_{ml} + \omega_2)} \tag{56}$$

where the suction is over all the initial (l), intermediate (m), and final (n) states and N is the density of atoms or bonds.

If one is interested in the resonant susceptibilities, one can formally introduce the dephasing rate of the off-diagonal matrix elements $i\gamma$ into Eq. (56). It is easy to see that as one approaches the vicinity of the resonance, which can be either a one-photon resonance such as

$$\chi^{(2)} \sim (\delta - i\gamma)^{-1} = (\omega_{ml} - \omega_1 - i\gamma)^{-1} \tag{57}$$

or a two-photon resonance such as

$$\chi^{(2)} \sim (\delta - i\gamma)^{-1} = (\omega_{nl} - \omega_1 - \omega_2 - i\gamma)^{-1} \tag{58}$$

where δ is detuning, the magnitude of second-order susceptibility $|\chi^{(2)}|$ increases as δ^{-1}, while the absorption proportional to the imaginary part of $\chi^{(1)}$:

$$\Im[\chi^{(1)}] \sim \frac{\gamma}{\delta^2 + \gamma^2} \tag{59}$$

increases as δ^{-2}. As a result, absorption of one of the waves involved in the nonlinear process becomes significant, thus rendering the nonlinear process inefficient. One can introduce the figure of merit for the nonlinear process, the ratio $\chi^2/\Im(\chi^{(1)})$, which gets worse at the resonance. Furthermore, in the bulk solid state materials, energy levels are broadened into the bands, and it is impossible to be in resonance with all the levels at the same time. For this reason, just like most solid state materials, the bulk semiconductors are used for the nonlinear frequency conversion far away from any resonance. In the quantum wells there are narrow levels within the band, and thus there exists a possibility of optimizing the detuning — i.e., of finding a trade-off between the magnitude of $\chi^{(2)}$ and absorption.

There is, however, one important exclusion from this rule — the difference frequency generation of far-IR light, $\chi^{(2)}(\omega_1 - \omega_2, \omega_1, \omega_2)$, when the difference frequency is comparable to the broadening. Then the denominator in Eq. (53) becomes nearly double-resonant, and thus the figure of merit stays constant and independent of detuning, and, for the sake of compactness, it becomes advantageous to use resonant susceptibility, which becomes

$$\chi^{(2)}_{i,j,k}(\omega_1 - \omega_2; \omega_1, \omega_2) = \frac{N}{2\varepsilon_0 \hbar^2} \frac{\omega_1 - \omega_2 + 2i\gamma}{\omega_1 - \omega_2 + iT_n^{-1}} \frac{\Delta\mu^i_{n1}\mu^j_{n1}\mu^k_{1n}}{(\omega_{nl} - \omega_1 - i\gamma)(\omega_{nl} - \omega_2 + i\gamma)} \tag{60}$$

which for the degenerate case $\omega_1 - \omega_2$ becomes simply optical rectification:

$$\chi^{(2)}_{i,j,k}(0;\omega_1,\omega_2) = \frac{N}{2\varepsilon_0 \hbar^2} \frac{\Delta\mu^i_{n1}\mu^j_{n1}\mu^k_{1n}T_n\gamma}{(\omega_{nl} - \omega_1 - i\gamma)(\omega_{nl} - \omega_1 + i\gamma)} \quad (61)$$

These special cases will be considered further in Section V, but for now we will concentrate on the off-resonance case. In order to accomplish this, we will look into the specific bandstructure of zinc-blende or hexagonal semiconductors.

4. Off-Resonant Second-Order Susceptibilities of Zinc-Blende Semiconductors

In this model, we consider specifically only the states corresponding to the four bonds holding the semiconductor together. These states arise from the basis consisting of two S-type and six P-type orbitals that constitute the valence electronic shells of the cation (C) and anion (A) atoms:

$$|S_c\rangle, |X_c\rangle, |Y_c\rangle, |Z_c\rangle, \quad |S_a\rangle, |X_a\rangle, |Y_a\rangle, |Z_a\rangle \quad (62)$$

These eight states combine with each other to create eight bands — four filled bonding-type states that constitute the four valence bands, and four empty antibonding-type states that constitute our conduction bands. The complete bandstructure can be calculated using either the tight-binding (LCAO) methods or the pseudo-potential method (Cohen and Chelikowsky, 1988), while near the special high-symmetry point the bandstructure is best described by the $\mathbf{k}\cdot\mathbf{p}$ method (Kane, 1957). There are the following states, shown in Fig. 2:

- The lowest valence band, with symmetry Γ^v_6. This state is a bonding combination of the $|S_a\rangle$ and $|S_c\rangle$ and is so far away from the rest of the states that it contributes very little to $\chi^{(2)}$.
- Three higher valence bands with symmetry Γ^v_8 or Γ^v_7 that in the absence of spin-orbit splitting are degenerate at Γ point. The states in these bands near the Γ point are the bonding combinations of the P-type states of cation and anion. The spin-orbit interaction further splits the bands into the light and heavy holes band, Γ^v_8 that are degenerate at Γ point and the split-off band Γ^v_7 separated from them by the spin-orbit interaction energy Δ. This energy is typically less than 1 eV and the

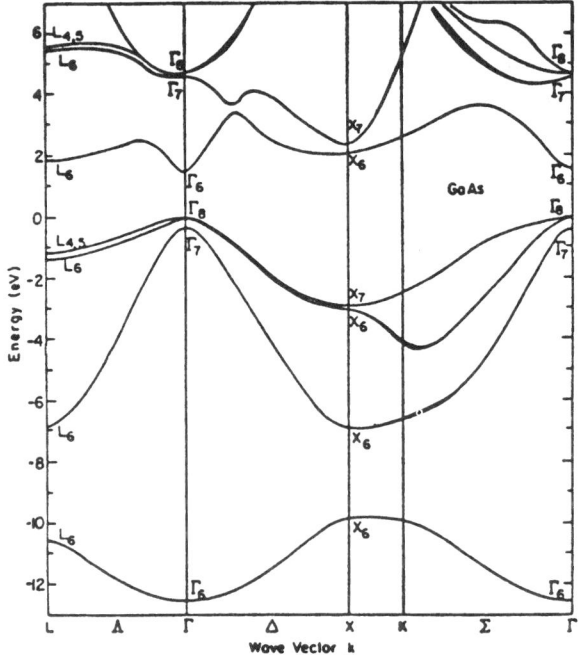

FIG. 2. Bandstructure of GaAs. The energy of the valence band maximum is taken to be zero. (Data from Cohen and Chelikowsky, 1988.)

spin-orbit interactions turns out to play a very small role in $\chi^{(2)}$; hence, we shall designate all three bands as $\Gamma^v_{7,8}$.
- The lowest conduction band Γ^c_6—the antibonding combination of $|S_a\rangle$ and $|S_c\rangle$.
- The higher conduction bands, arising from the three degenerate antibonding combinations of the P-type states. Spin-orbit interaction splits these bands into Γ^c_7 and double-degenerate Γ^c_8, but, as we have already mentioned, we are not taking this splitting into consideration and designate all three bands as $\Gamma^c_{7,8}$.

In order to calculate the nonlinear susceptibilities, one has to take into account the fact that the coordinates, and hence the matrix elements of the dipole moment, are not good numbers anymore, and instead of the $\mathbf{E} \cdot \mathbf{r}$ gauge one should use the $\mathbf{A} \cdot \mathbf{P}$ gauge, where $\mathbf{A}$ is a vector potential and $\mathbf{P}$ is a momentum operator. Then the expression for the second-order susceptibility becomes

1 Second-Order Nonlinearities and Optical Rectification

$\chi^{(2)}_{ijk}(\omega_1 + \omega_2; \omega_1, \omega_2)$

$$= \frac{i}{2\varepsilon_0 \hbar^2} \frac{e^3}{m_0^3 (\omega_1 + \omega_2)(\omega_1)(\omega_2)} \sum_{l,m,n} \sum_k \frac{P^i_{ln} P^j_{nm} P^k_{ml}}{(\omega_{nl} - \omega_1 - \omega_2)(\omega_{ml} - \omega_1)}$$

$$+ \frac{P^i_{ln} P^k_{nm} P^j_{ml}}{(\omega_{nl} - \omega_1 - \omega_2)(\omega_{ml} - \omega_2)} \frac{P^j_{ln} P^i_{nm} P^k_{ml}}{(\omega_{nl} + \omega_2)(\omega_{ml} - \omega_1)} \quad (63)$$

$$+ \frac{P^k_{ln} P^i_{nm} P^j_{ml}}{(\omega_{nl} + \omega_2)(\omega_{ml} - \omega_1)} + \frac{P^k_{ln} P^j_{nm} P^i_{ml}}{(\omega_{nl} + \omega_1 + \omega_2)(\omega_{ml} + \omega_1)}$$

$$+ \frac{P^j_{ln} P^k_{nm} P^i_{ml}}{(\omega_{nl} + \omega_1 + \omega_2)(\omega_{ml} + \omega_2)}$$

where $\mathbf{P}_{ml} = \langle m|i\hbar\nabla|n\rangle$ is the matrix element of the momentum operator. In Eq. (63), the ground state l must be occupied, while the intermediate state m can be either occupied or empty. If the second state is empty, we are dealing with the transitions by virtual electrons, which can be written as $v \to c \to c'$. If the second state is occupied, one can show by combining various terms in Eq. (63) that the process can be represented as a virtual hole transition and can be written as $c \to v \to v'$. This process involves two valence bands, and since the lowest valence band Γ^v_6 is separated from the rest of the bands by a large amount of energy, this band can be disregarded, and thus virtual hole transitions do not have to be considered. Thus, the initial states have to be that of $\Gamma^v_{7,8}$ symmetry, and this leaves us with only three independent momentum matrix elements:

1. $$p_{vc} = \langle \Gamma^v_{7,8} | p | \Gamma^c_6 \rangle = \frac{1}{\sqrt{1+V^2}} iP \quad (64)$$

   This is the Kane matrix element (Kane, 1957, 1966), which determines the strength of the absorption in the vicinity of the fundamental bandgap and is nonzero in both nonpolar and polar materials. (The meaning of the coefficient of P is explained below.)

2. $$p_{vc'} = \langle \Gamma^v_{7,8} | p | \Gamma^c_{7,8} \rangle = iQ \quad (65)$$

   This matrix element determines the strength of the absorption from the top of the valence band to the higher-lying conduction bands, and is also nonzero in both nonpolar and polar materials.

3. $$p_{cc'} = \langle \Gamma^c_6 | p | \Gamma^c_{7,8} \rangle \quad (66)$$

This matrix element determines the probability of the absorption from the lowest conduction band into the higher conduction bands. In the nonpolar materials, this matrix element is zero. In the polar materials, one can show (Aspnes, 1972) that the noncentrosymmetric potential mixes the symmetric $\Gamma^v_{7,8}$ and antisymmetric $\Gamma^c_{7,8}$ states, and as a result, as Aspnes (1972) has shown,

$$p_{cc'} \approx i \frac{V}{\sqrt{1+V^2}} P \tag{67}$$

where V is related to the polarity of the bond α_p as

$$\alpha_p = \frac{2V}{1+V^2} \tag{68}$$

with polarity defined here as the difference between the probabilities of finding the valence electron on the anion and the cation.

From the zinc-blende lattice symmetry, it is not difficult to understand that the only nonzero combination of the projections of three different interband momentum operators must include all three different components, and thus the only six nonzero elements of the $\chi^{(2)}$ tensor are

$$\chi^{(2)}_{xyz} = \chi^{(2)}_{xzy} = \chi^{(2)}_{yxz} = \chi^{(2)}_{yzx} = \chi^{(2)}_{zxy} = \chi^{(2)}_{zyx} = \chi^{(2)}_{14} \tag{69}$$

In fact, all the matrix elements can be extracted from the experimental measurements and then fitted into full-bandstructure calculations. Moss, Sipe, and Van Driel (1987) and Ghahramani, Moss, and Sipe (1991) have performed a complete set of such calculations and have obtained a number of results, shown in Fig. 3, that manifest as several features reflecting the complexity of the bandstructure above the two-photon edge—i.e., when $\omega_2 + \omega_1$ is larger than the bandgap. But below the resonance, the dispersion curve shows no features and in fact does not change much at all except for the smooth increase near the bandgap.

There is a simple explanation for this apparent lack of features: the majority of states in $\mathbf{k}$ space lie far away from the edge of the Brillouin zone (Fig. 2) and have transition energies that are much larger than the fundamental bandgap energy E_g. Therefore, as long as all the frequencies in Eq. (63) are smaller than $E_g/\hbar$, the denominators do not change much. It is therefore possible to assume that all the denominators in Eq. (63) are equal to some average energy $\bar{E}$. Furthermore, there are eight atoms in each cube

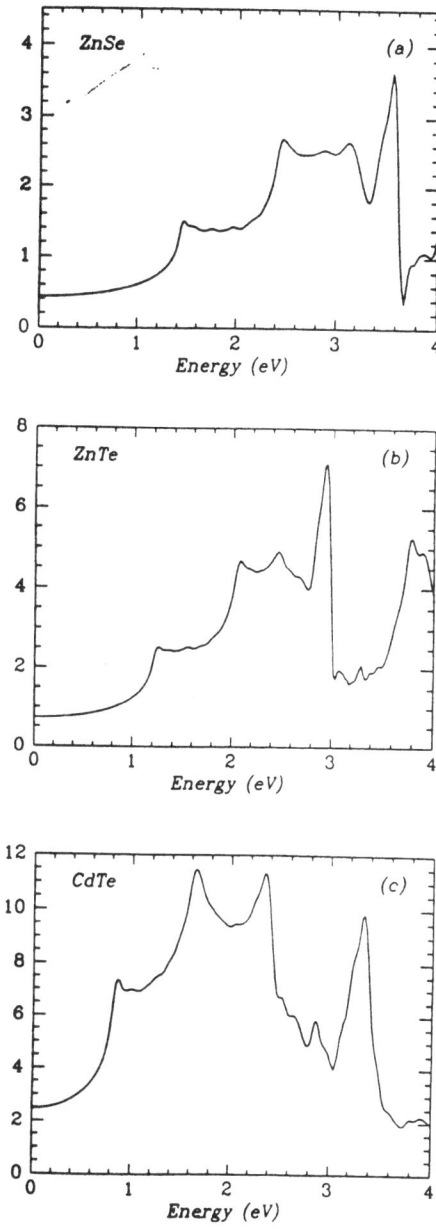

FIG. 3. Results of full-bandstructure calculations of $|\chi^{(2)}(2\omega;\omega,\omega)|$ for II–VI semiconductors: (a) ZnSe, (b) ZnTe, and (c) CdTe. The results are given in electrostatic units (1 esu = 4.19×10^{-4} m/V). (Data from Ghahramani, Moss, and Sipe, 1991.)

with volume a_0^{-3}, where a_0 is a lattice constant, and each atom contributes three electrons (disregarding the lowest valence band). The major problem with Eq. (63) is its divergence at $\omega_1 \to \infty$ or $\omega_2 \to \infty$. This is an unfortunate consequence of using the vector potential gauge, but this problem was successfully dealt with by Aspende (1972) and later by Sipe and Ghahramani (1993). Their approach featured inclusion of the free-carrier intraband transitions, expanding the denominators of Eq. (63) and then using the time inversion symmetry of the bandstructure $[E(\mathbf{k}) = E(-\mathbf{k})]$ to obtain this approximate expression for the second-order susceptibility in the long-wavelength limit:

$$\chi_{14}^{(2)}(0) \approx \frac{36e^3\hbar^3}{\varepsilon_0 m_0^3 a_0^3 \bar{E}^5} \alpha_p P^2 Q \tag{70}$$

The values of matrix elements P and Q can be obtained from the experimental measurements of optical transitions, and the average bandgap is roughly related to the Penn gap (Phillips, 1968), which can be obtained from measurement of the dielectric constant and is, according to Van Vechten (1969), on the order of 5 eV for most zinc-blende semiconductors. If one is interested in just an order of magnitude of $\chi^{(2)}$, one can simply assume that both P and Q are on the order of $\hbar/l_b$, where the bond length $l_b = a_0\sqrt{3/4}$. Then, assuming that $a_0 \sim 5.6$ Å, we obtain the expression

$$\chi_{14}^{(2)}(0) \sim \frac{450 e^3 \hbar^6}{\varepsilon_0 m_0^3 a_0^6 \bar{E}^5} \alpha_p \sim 35 \alpha_p \text{ pm/V} \tag{71}$$

Since α_p is less than 1 (about 0.5 for GaAs), one can see that the result is at least an order of magnitude smaller than the classical approximation (Eq. 30), and in fact Eq. (71) underestimates $\chi^{(2)}$ by a factor of a few. The reason for it is that the influence of the states near the center of the Brillouin zone is easy to underestimate, because the expression depends on the fifth order of the transition energy $\bar{E}$. A more precise expression can be obtained by using the same average gap assumption to develop the expression for the linear susceptibility:

$$\chi^{(1)}(0) = \varepsilon(\omega) - 1 \approx \frac{48 e^2 \hbar^2}{\varepsilon_0 m_0^2 a_0^3 \bar{E}^3} P^2 \tag{72}$$

Substituting Eq. (72) into Eq. (70), we obtain

$$\chi^{(2)}_{14}(0) \sim \frac{3}{4}\chi^{(1)}(\omega)\frac{eh Q}{m_0 \bar{E}^2}\alpha_p \tag{73}$$

which for GaAs yields $\chi^{(2)} \sim 80$ pm/V, which still substantially underestimates $\chi^{(2)}$. This equation actually is quite similar to the classical relation (Eq. 29), and if we assume that $Q \sim \hbar/l_b$ and $\bar{E} \sim \hbar^2/2m_0 l_b^2$, Eq. (73) is transformed into

$$\chi^{(2)}_{14}(0) \sim \frac{3}{2}\chi^{(1)}(0)\frac{el_b}{\bar{E}}\alpha_p \tag{74}$$

Thus we can see that the average gap plays the role of classical binding energy here.

It should be noted that the linear dependence on the bond polarity holds only for relatively small values of polarity, because at higher polarities the terms that are more than linear in polarity play a role. As mentioned before, much better agreement with experiment was achieved in full tight-binding bandstructure calculations by Moss, Sipe, and Van Driel (1987) and in a local-density approximation approach by Levine and Allan (1991), but it is important to note that even these calculations relied heavily on a number of parameters obtained by fitting the experimental results. It is therefore fair to say that Eq. (73) gives a very good intuitive idea about the dependence of $\chi^{(2)}$ on the polarity of the bond, as well as a decent order-of-magnitude estimate.

5. Calculation of Second-Order Susceptibility Using a Hybrid Bond Model

As we have just shown, in order to obtain the order-of-magnitude estimate of the nonresonant $\chi^{(2)}$, ones does not have to take into account the complicated bandstructure. The situation is further simplified if we part with the $\mathbf{A} \cdot \mathbf{P}$ gauge and all the complications arising because of the divergences and return to the $\mathbf{E} \cdot \mathbf{r}$ description—i.e., if we consider the bond orbitals. There are four hybrid orbitals directed in four different tetragonal directions for each cation or anion atom, each of them being a combination of the anion and cation states (Eq. 62):

$$|h^{(1)}_{c(a)}\rangle = \frac{1}{2}(|S_{c(a)}\rangle + |X_{c(a)}\rangle + |Y_{c(a)}\rangle + |Z_{c(a)}\rangle) \tag{75}$$

directed along $\langle 111 \rangle$,

$$|h_{c(a)}^{(2)}\rangle = \frac{1}{2}(|S_{c(a)}\rangle + |X_{c(a)}\rangle - |Y_{c(a)}\rangle - |Z_{c(a)}\rangle) \tag{76}$$

directed along $\langle 1\bar{1}\bar{1} \rangle$,

$$|h_{c(a)}^{(3)}\rangle = \frac{1}{2}(|S_{c(a)}\rangle - |X_{c(a)}\rangle + |Y_{c(a)}\rangle - |Z_{c(a)}\rangle) \tag{77}$$

directed along $\langle \bar{1}1\bar{1} \rangle$, and

$$|h_{c(a)}^{(4)}\rangle = \frac{1}{2}(|S_{c(a)}\rangle - |X_{c(a)}\rangle - |Y_{c(a)}\rangle + |Z_{c(a)}\rangle) \tag{78}$$

directed along $\langle \bar{1}\bar{1}1 \rangle$. From these orbitals, the bonding orbitals are constructed as

$$|b^{(i)}\rangle = \frac{1}{1+\beta^2}\frac{1}{2}(|h_a^{(i)}\rangle + \beta|h_c^{(i)}\rangle) \tag{79}$$

where the parameter β is related to the polarity as

$$\alpha_p = \frac{1-\beta^2}{1+\beta^2} \tag{80}$$

One of the bonding states is shown in Fig. 4. In the ground state, all the carriers are in the bonding state, and it is the state in which numerous reliable theoretical studies, confirmed by experiments, have been made (Cohen and Chelikowsky, 1988). For the off-resonant nonlinearities, as was first shown by Flytzanis and Ducuing (1969), it is possible to obtain the expression for the nonlinear susceptibility using only the data for the ground state. We consider the expressions for the linear and nonlinear susceptibilities for the ith bond. The ground state is $|b^{(i)}\rangle$, and for the simplicity of calculation, we assume that this state is long and narrow, so the dipole matrix elements are directed along the bond. Then, for all the bonds $|b^{(i)}\rangle$, we can write

$$\chi_{ii}^{(1)}(\omega) = \frac{N_i}{\varepsilon_0 \hbar}\sum_m \frac{\mu_{bm}^i \mu_{mb}^i 2\omega_{ml}}{\omega_{mb}^2 - \omega^2} \tag{81}$$

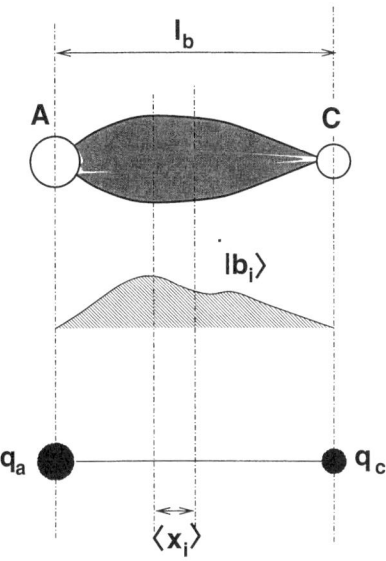

FIG. 4. Bonding orbital $|b_i\rangle$ and its primitive model.

If we assume that all the transition energies are equal, then, in the long-wavelength limit, we obtain

$$\chi_{ii}^{(1)}(0) = \frac{2N_i}{\varepsilon_0 \bar{E}} \sum_m \mu_{bm}^i \mu_{mb}^i = \frac{2e^2 N_i}{\varepsilon_0 \bar{E}} \sum_m \langle b_i | x_i - \langle x_i \rangle | m \rangle \langle m | x_i - \langle x_i \rangle | b_i \rangle \quad (82)$$

where $x_i - i$ is a coordinate taken along the direction of the ith bond, taken from anion to cation, and

$$\langle x_i \rangle = \langle b_i | x_i | b_i \rangle \quad (83)$$

is the coordinate of the center of charge of the state $|b_i\rangle$. Now, using the completeness of the set of states, $\Sigma_m |m\rangle\langle m| = 1$, we obtain

$$\chi_{ii}^{(1)}(0) = \frac{2N_i}{\varepsilon_0 \bar{E}} [\langle x_i^2 \rangle - \langle x_i \rangle^2] \quad (84)$$

Performing the averaging over four orbitals yields for the scalar of linear susceptibility:

$$\chi^{(1)}(0) \sim \frac{2Ne^2}{3\varepsilon_0 \bar{E}}[\langle x_i^2 \rangle - \langle x_i \rangle^2] \qquad (85)$$

Now, this result is quite similar to the classical result (Eq. 28), with the extent of the wavefunction playing the role of $\bar{d}$ and the average gap playing the role of the binding energy.

Similarly, for the second-order susceptibility, using Eq. (56) in the long-wavelength limit, we obtain

$$\chi^{(2)}_{iii}(0) = \frac{3N}{\varepsilon_0 \bar{E}^2} \sum_{m,n} \mu^i_{bn}\mu^i_{nm}\mu^i_{mb} = \frac{3Ne^3}{\varepsilon_0 \bar{E}^2} \langle b_i|(x_i - \langle x_i \rangle)^3|b_i \rangle \qquad (86)$$

If we now perform the averaging over the four bonds, we obtain

$$\chi^{(2)}_{14}(0) - \frac{Ne^3}{\sqrt{3}\varepsilon_0 \bar{E}^2} \langle b_i|(x_i - \langle x_i \rangle)^3|b_i \rangle \sim \frac{\sqrt{3}}{2} \chi^{(1)}(0) \frac{e\langle b_i|(x_i - \langle x_i \rangle)^3|b_i \rangle}{\bar{E}\langle (x_i - \langle x_i \rangle)^2 \rangle} \qquad (87)$$

This equation is a counterpart to the classical expressions (Eqs. 29 and 30), but now the origin of nonlinearity is traced to the nonzero third-order momentum of the bond, which can be calculated from a knowledge of the wavefunction of the ground state only. In order to see the connection between the polarity of the bond and nonlinear susceptibility, we can consider the most primitive model, also shown in Fig. 4, in which the charge on the anion is $q_a = -e[(1 + \alpha_p)/2]$ and the charge on the cation, separated from the anion by the distance $l_b = a_0\sqrt{3}/4$, is $q_c = -e[(1 - \alpha_p)/2]$. Then we can estimate the momenta relative to the center of the bond:

$$\langle x_i \rangle = -\frac{1}{2}\alpha_p l_b \qquad (88)$$

$$\langle (x_i - \langle x_i \rangle)^2 \rangle = \frac{l_b^2}{4}(1 - \alpha_p^2) \qquad (89)$$

$$\langle (x_i - \langle x_i \rangle)^3 \rangle = -\alpha_p \frac{l_b^3}{4}(1 - \alpha_p^2) \qquad (90)$$

The expression for the first-order susceptibility then becomes

$$\chi^{(1)}(0) \sim \frac{Ne^2 l_b^2}{6\varepsilon_0 \bar{E}} (1 - \alpha_p^2) \tag{91}$$

which at this point differs from the classical expression (Eq. 28) only by a factor of $\frac{1}{3}$, which of course comes simply from the averaging over the direction of the bonds. For the second-order susceptibility, we obtain

$$\chi^{(2)}_{14}(0) \sim -\frac{Ne^3 l_b^3}{4\sqrt{3}\,\varepsilon_0 \bar{E}^2} \alpha_p (1 - \alpha_p^2) \tag{92}$$

Now, since the density of valence electrons is $N = 32/a_0^3$,

$$\chi^{(2)}_{14}(0) \sim -\frac{3e^3}{8\varepsilon_0 \bar{E}^2} \alpha_p (1 - \alpha_p^2) \sim 10^4 (\bar{E})^{-2} \text{ pm V} \times \alpha_p (1 - \alpha_p^2) \tag{93}$$

where $\bar{E}$ is in electron-volts. This expression gives a correct order of magnitude for most of the tetragonal II–VI and III–V semiconductors, and, most important, it provides an explanation of the influence of the polarity of the bond. Obviously, in the absence of the bond polarity, the bond is symmetric and the second-order nonlinearity vanishes. On the other hand, when the bond polarity becomes too large, the cation and anion shells become essentially uncoupled, and the second-order susceptibility starts decreasing. One must also take into account the fact that the average bandgap $\bar{E}$ itself depends on the polarity. Indeed, it is not difficult to invoke the oscillator sum rule:

$$\sum_n \frac{2m}{\hbar^2} (E_n - E_b) |\langle b | x_i - \langle x_i \rangle | n \rangle|^2 = 1 \tag{94}$$

and, assuming that $E_n - E_b = \bar{E}$, obtain

$$\langle (x_i - \langle x_i \rangle)^2 \rangle = \frac{\hbar^2}{2m\bar{E}} = \frac{l_b^2}{4}(1 - \alpha_p^2) \tag{95}$$

or

$$\bar{E} = \frac{2\hbar^2}{m_0 l_b^2}(1 - \alpha_p^2)^{-1} \tag{96}$$

Then one can obtain the simplest expression for linear susceptibility:

$$\chi^{(1)}(0) = \varepsilon(0) - 1 \sim \frac{Nm_0 e^2 l_b^4}{12\hbar^2 \varepsilon_0}(1 - \alpha_p^2)^2 \sim \frac{3}{8}\pi \frac{a_0}{a_B}(1 - \alpha_p^2)^2 \quad (97)$$

where a_B is a Bohr radius. As for the second-order susceptibility, we derive

$$\chi_{14}^{(2)}(0) \sim \frac{27\pi^2}{512} \frac{a_0^2 \varepsilon_0}{e} \frac{a_0^2}{a_B^2} \alpha_p (1 - \alpha_p^2)^3$$

$$\sim \frac{a_0 a_B \varepsilon_0}{\pi e}[\varepsilon(0) - 1]^3 \frac{\alpha_p}{(1-\alpha_p^2)^3} \approx 0.1 a_0 [\varepsilon(0) - 1]^3 \frac{\alpha_p}{(1-\alpha_p^2)^3} \text{ pm/V} \quad (98)$$

where a_0 is given in angstroms. [Note that Eq. (98) does not diverge at $\alpha_p \to 1$, since $\varepsilon(\alpha_p \to 0) \to 1$]. This last expression (Eq. 98) includes the dielectric constant $\varepsilon(\omega)$, which can be measured experimentally and thus can be the basis for comparison with the experimentally available results.

III. $\chi^{(2)}$ Experiments in Bulk Semiconductors

1. Measurements of $\chi^{(2)}$ Values

There have been a number of measurements of second-order susceptibility in semiconductors, but since, as we shall soon see, it is impossible to achieve phasematching in these crystals, the conversion efficiency has been typically very low, and therefore the measurements have shown large variations. Since it is difficult to have an experimental arrangement with mutually coherent sources of frequencies ω_1 and ω_2, the measurements are always done only for the degenerate case of second harmonic generation, $\chi^{(2)}(2\omega; \omega, \omega)$, and the measured second harmonic coefficient is related to $\chi^{(2)}$ as

$$d_{36}(\omega) = \frac{1}{2}\chi_{14}(2\omega; \omega, \omega) \quad (99)$$

Typically, the measurement is performed by comparison with a well-known crystal, such as KDP or LiNbO$_3$. The results display a large degree of variation: consider, for instance, the case of the most prominent III–V material, GaAs. Soref and Moos (1964) used the transmission second harmonic measurements at 1.06 μm and obtained $\chi^{(2)} = 310$ pm/V. Soon afterward, Chang, Ducuing, and Bloembergen (1965) used the reflected SHG technique at the same wavelength and obtained quite different results: $\chi^{(2)} = 460$ pm/V. The measurements by Johnston and Kaminow (1969), which involved Raman scattering measurement, confirmed the results of Soref and Moos (1964) and obtained $\chi^{(2)} = 280$ pm/V at the same 1.06-μm wavelength. The measurements of Wynne and Bloembergen (1969) yielded a value of $\chi^{(2)} = 377$ pm/V in the far-IR range (10.6 μm), while later on Levine and Bethea (1972) measured only $\chi^{(2)} = 190$ pm/V at the same wavelength. Until recently, the most careful measurements, which used the wedge technique, were those performed by Choy and Byer (1976) at a wavelength of 2.12 μm. They obtained $\chi^{(2)} = 208$ pm/V. Similar discrepancies have been observed for other semiconductors — for example, GaP ($\chi^{(2)} = 50 \div 300$ pm/V) (Soref and Moos, 1964; Miller, 1964; Choy and Byer, 1976) or ZnSe ($\chi^{(2)} = 40 \div 100$ pm/V) (Soref and Moos, 1964; Hase et al., 1992).

Besides experimental difficulties associated with these measurements, in order to calculate the nonlinear susceptibility one has to know the values of the linear optical parameters of the given material, and in the 1960s and 1970s, when the majority of the measurements were performed, these values were not well known.

Recently, Shoji et al. (1997) performed systematic measurements of the second-order nonlinear optical coefficients of a large number of crystals, using several different techniques, such as second harmonic generation, difference frequency generation, and parametric fluorescence. The measurements were made at a variety of wavelengths in the near-IR range, using the experimental setup shown in Fig. 5. By using a sample in the form of a wedge, Shoji et al. (1997) managed to obtain the oscillatory dependence of second harmonic power on the sample thickness (Fig. 6) — evidence of phase mismatch, which is explained in the next subsection. Using these data, they were able to evaluate $\chi^{(2)}$ and coherence length. Their results are probably the most reliable up to this date, and they are summarized in Table I.

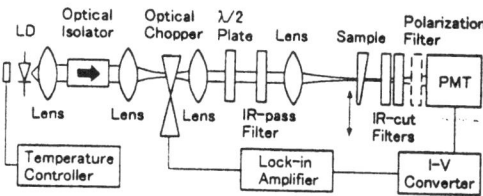

FIG. 5. Schematic of the experimental setup for absolute measurement by the wedge technique. LD, laser diode; PMT, photomultiplier tube. (Data from Shoji et al., 1997.)

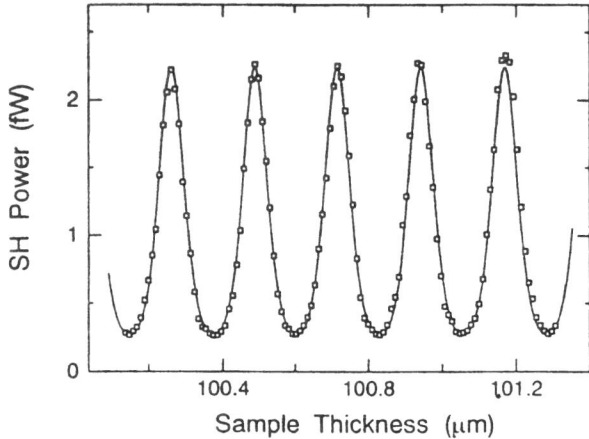

FIG. 6. Second harmonic power versus sample thickness for measurement of $\chi^{(2)}_{36}$ of GaAs by the wedge technique at 1.533 μm.

It is clear that the general trend of the previous chapters—i.e., that the smaller-bandgap semiconductors with higher dielectric constants have greater nonlinear susceptibility—is confirmed.

It is interesting to compare the experimental results of various groups with the results obtained from the full-bandstructure calculations of Moss,

TABLE I

SECOND-ORDER SUSCEPTIBILITIES OF SELECTED SEMICONDUCTORS IN pm/V*

Crystal	$\chi^{(2)}_{ij}$	Measurement $\lambda = 1.548\,\mu m$	Wavelength $\lambda = 1.064\,\mu m$
GaAs	$\chi^{(2)}_{14}$	238	340
GaP	$\chi^{(2)}_{14}$	75	140
ZnS	$\chi^{(2)}_{33}$	18	25
	$\chi^{(2)}_{31}$	10	13
	$\chi^{(2)}_{15}$	9	12
CdS	$\chi^{(2)}_{33}$	28	38
	$\chi^{(2)}_{31}$	15	20
	$\chi^{(2)}_{15}$	16	22
ZnSe	$\chi^{(2)}_{14}$	—	108
CdTe	$\chi^{(2)}_{14}$	150	220

*From Shoji et al., 1997.

TABLE II

COMPARISON OF THE SECOND-ORDER SUSCEPTIBILITIES OF SELECTED SEMICONDUCTORS
CALCULATED BY VARIOUS METHODS WITH THEIR EXPERIMENTAL VALUES
AND SOME OF THE MATERIAL PARAMETERS USED IN CALCULATIONS

Parameter	Material				
	GaP	GaAs	GaSb	InAs	InSb
Bandgap at 300 K (eV)	2.26	1.43	0.78	0.35	0.18
Lattice constant (Å)	5.45	5.66	6.10	6.05	6.47
Dielectric constant	9.1	10.9	14.4	12.3	15.1
Bond polarity α_p	0.52	0.5	0.44	0.53	0.41
Full band[a]	160	376	963	1885	3850
$\mathbf{A} \cdot \mathbf{P}$[b]	100	160	293	268	460
Hybrid orbitals	337	566	1060	1082	1273
Experimental	140	238	910	830	1250

[a] Data from Moss, Sipe, and Van Driel (1987).
[b] Data from Aspnes (1972).

Sipe, and Van Driel (1987) and the approximate hybrid-orbitals results of Eq. (56), and with the approximate results obtained with the $\mathbf{A} \cdot \mathbf{P}$ gauge by Aspnes (1972) (Eq. 31). Using the data from Moss, Sipe, and Van Driel (1987) for lattice constants, polarities, and dielectric constants, we obtained the results for the $\chi^{(2)}_{14}$ values of various III–V materials (in picometers per volt) listed in Table II.

The full-band approach of Moss, Sipe, and Van Driel (1987) definitely provides the best fit to the experimental data, with the exception of the narrow-gap materials, for which, quite surprisingly, the simple bond-orbital approach of Eq. (56) is the best. But it should be noted that the simple bond-orbital technique show perfectly well the general trend in $\chi^{(2)}$.

Ghahramani, Moss, and Sipe (1991) performed detailed bandstructure calculations on the II–VI semiconductors and achieved reasonable agreement with the experimental results of Shoji et al. (1997). Overall, the II–VI materials have higher polarities than III–V materials of comparable bandgap energies. For example, CdTe has roughly the same bandgap energy as GaAs, and yet its polarity is 0.78, compared with 0.5 for GaAs. According to Eq. (93), for the fixed average bandgap energy, $\bar{E}$, $\chi^{(2)}$ is maximum at $\alpha_p = 1/\sqrt{3} \approx 0.58$, and hence one expects the $\chi^{(2)}$ of CdTe to be less than that of GaAs by as much as 50%. This agrees well with experimental results.

2. Experimental Demonstration of Second-order Nonlinear Processes in Semiconductors

The high values of nonlinear susceptibility in the zinc-blende semiconductors coupled with the possibility of integration with laser sources should have made them natural materials of choice for a variety of nonlinear frequency conversion devices. However, all the materials of the zinc-blende structure are optically isotropic, as shown in the previous section, and thus they cannot be used for the birefringently phasematched interaction. Let us quickly review this problem (details can be obtained in Yariv, 1989, and Shen, 1984). Consider two waves (see Fig. 7):

$$\mathscr{E}_1(z, t) = \frac{1}{2}\mathscr{E}_1(z)e^{i(k_1 z - \omega_1 t)} + c.c. \tag{100}$$

and

$$\mathscr{E}_1(z, t) = \frac{1}{2}\mathscr{E}_2(z)e^{i(k_2 z - \omega_2 t)} + c.c. \tag{101}$$

propagating collinearly in the nonlinear medium with second-order susceptibility $\chi^{(2)}$. The nonlinear polarization at the sum or difference frequency is

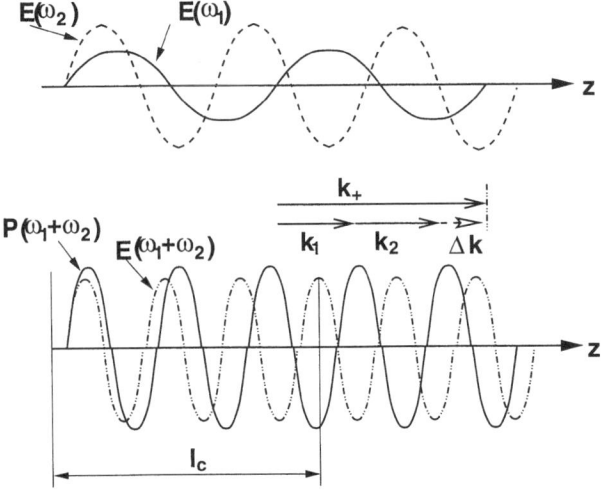

Fig. 7. Phase mismatch in sum frequency generation.

then

$$\mathcal{P}(\omega_1 \pm \omega_2) = \frac{1}{2}\varepsilon_0\chi^{(2)}(\omega_1 \pm \omega_2)\mathcal{E}_1(z)\mathcal{E}_2(z)e^{i[(k_1 \pm k_2)z - (\omega_1 \pm \omega_2)t]} \quad (102)$$

Now this nonlinear polarization gives rise to the propagating wave $\omega_1 \pm \omega_2$,

$$\mathcal{E}_\pm(z,t) = \frac{1}{2}\mathcal{E}_\pm(z)e^{i[k_\pm z - (\omega_1 \pm \omega_2)t]} + c.c \quad (103)$$

whose wavevector in the medium is $k_\pm = n_{\omega_1 \pm \omega_2}(\omega_1 \pm \omega_2)/c$. The propagation equation is

$$\frac{d^2\mathcal{E}_\pm(z,t)}{dz^2} - \frac{n^2_{\omega_1 \pm \omega_2}}{c^2}\frac{d^2\mathcal{E}_\pm(z,t)}{dt^2} = \frac{1}{\varepsilon_0 c^2}\frac{d^2\mathcal{P}(\omega_1 \pm \omega_2)}{dt^2} \quad (104)$$

Substituting Eqs. (102) and (103) and using the slow-variable amplitude approach, one obtains

$$\frac{d\mathcal{E}_\pm(z)}{dz} = i\frac{\omega_1 \pm \omega_2}{2cn_{\omega_1 \pm \omega_2}}\chi^{(2)}(\omega_1 \pm \omega_2)\mathcal{E}_1(z)\mathcal{E}_2(z)e^{i\Delta k z} \quad (105)$$

where Δk is the wavevector mismatch,

$$\Delta k = k_\pm - (k_1 \pm k_2) \quad (106)$$

It is easy to see that in the presence of mismatch the power at the sum or difference frequency oscillates as

$$P(\omega_1 \pm \omega_2) \sim |\mathcal{E}_\pm(z)|^2 = \frac{2(\omega_1 \pm \omega_2)^2}{c^2 n_{\omega_1 \pm \omega_2} n_1 n_2}|\chi^{(2)}|^2 P_1 P_2 \eta_0 \frac{\sin^2(\Delta k z/2)}{\Delta k^2} \quad (107)$$

where $\eta_0 = 377\,\Omega$, with the period of oscillation defined by the coherence length

$$l_c = \frac{\pi}{\Delta k} \quad (108)$$

Owing to large dispersion of the semiconductors in the vicinity of the bandgap, coherence length is very short — typically only a few microns, and thus the maximum power available from the nonphasematched geometry, proportional to the square of the coherence length, is miniscule.

It is therefore paramount to achieve the perfect phasematching condition

$$\Delta k = 0 \tag{109}$$

in order to make nonlinear frequency conversion practical. In most practical nonlinear crystals, the phasematching is achieved using their natural birefringence—for instance, in the case of second-harmonic generation (SHG), using one polarization for the fundamental wave and another for the second harmonic. But, as we have already mentioned, the zinc-blende structures do not exhibit natural birefringence, and thus alternative methods of phasematching have to be employed in generating nonlinear waves in these semiconductors.

One way of obtaining phasematching is to use two counterpropagating waves (Vakhshoori, Wu, and Wang, 1988) in the GaAs/AlGaAs waveguide, which creates constant second harmonic polarization along the length of the waveguide (Fig. 8). Such a "sheet of polarization" serves as a coherent source for the second harmonic wave emitted in the direction normal to the surface. The same technique was later employed by Vakhshoori and Wang (1988) to demonstrate a broadband optical correlator with the potential for integration. While the fundamental pulse propagated back and forth in a waveguide cavity, it emitted a second harmonic signal normal to the top surface of the waveguide. The variation of this second harmonic signal

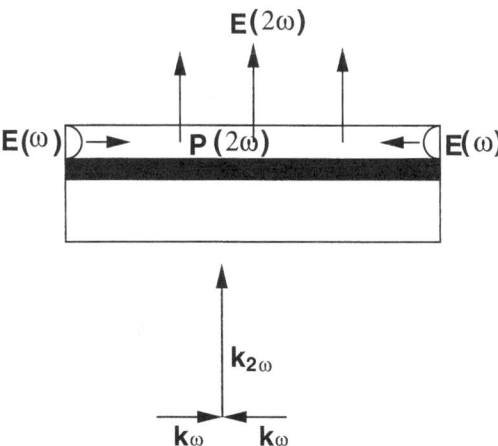

FIG. 8. Phasematching in surface-emitting geometry.

across the waveguide cavity yielded the pulse width of the fundamental beam. Using this technique, a pulse width as short as 7 ps was measured. Using the same geometry, but $\langle 111 \rangle$-grown GaAs, blue-green surface-emitting second harmonic generators were built (Vakhshoori et al., 1991). The power, however, was rather low, owing mainly to a very short propagation distance, equal to the thickness of the waveguide. For instance, Normandin et al. (1991) obtained about 2 nW of the second harmonic power using two 10-mW pumps.

One of the more promising methods of phasematching has been the "quasi-phasematching" (QPM) method introduced as early as 1962 by Armstrong et al. and later refined by other authors (Somekh and Yariv, 1972; Tang and Bey, 1973). The principle behind QPM is strikingly simple: because in every coherence length the phase between the nonlinear polarization wave, $\mathscr{P}(\omega_1 \pm \omega_2)$, and the free-running wave at the sum/difference frequency, $\mathscr{E}_\pm$, gets 180 degrees out of phase, reversing the sign of the nonlinear polarization (Fig. 9), accomplished by changing the sign of $\chi^{(2)}$, should bring them back into phase. One can think of QPM in terms of imposing the spatial modulation of spatial frequency $(2\pi\Lambda)^{-1} = \Delta k$ onto the nonlinear polarization wave:

$$\mathscr{P}(\omega_1 \pm \omega_2) \sim e^{i(k_1 \pm k_2)z} = e^{i(k_\pm - \Delta k)z} \tag{110}$$

Naturally, such modulation will result in a "sideband" with wavevector $(k_\pm - \Delta k + \Delta k = k_\pm)$ perfectly matched to the running wave $\mathscr{E}_\pm$. The highest efficiency is obtained when the sign of $\chi^{(2)}$ is reversed every coherence length.

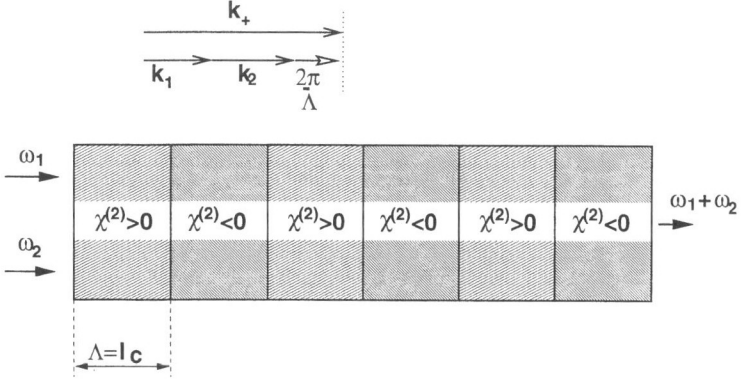

FIG. 9. Quasi-phasematching.

This is relatively easy to accomplish by periodic poling in ferroelectric materials such as $LiNbO_3$ and KTP. Unfortunately, the zinc-blende materials are not ferroelectric and thus cannot be poled, and one has to use an alternative method, essentially involving cutting and bonding.

The first demonstration of QPM doubling of the carbon dioxide laser radiation in a stack of discrete bulk GaAs plates was done as early as 1976 (Thompson, McMullen, and Anderson). The method was refined by Gordon et al. (1993), who used the diffusion-bonded stacked GaAs plates for quasi-phasematched second harmonic generation of a carbon dioxide laser. A nonlinear interaction length of 4 mm was achieved in these experiments.

More recently, a new QPM technique was developed by Yoo et al. (1995). The AlGaAs layer (Fig. 10) was epitaxially grown on a template substrate,

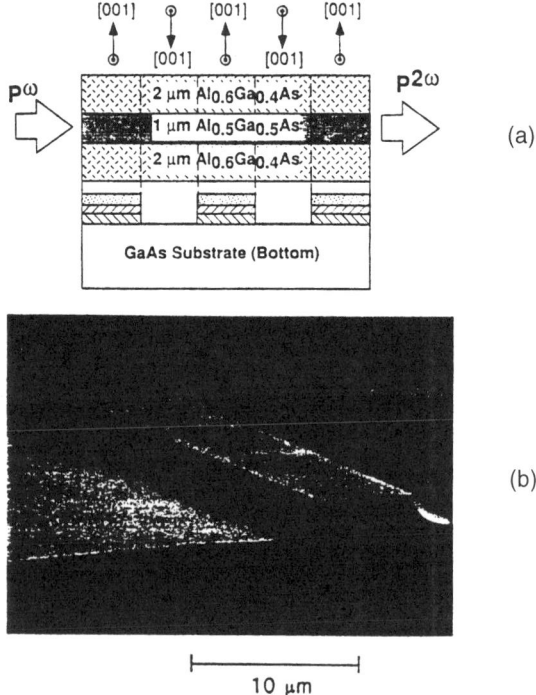

FIG. 10. Quasi-phasematching second harmonic generation (SHG) in epitaxial AlGaAs with periodic domain inversion: (a) schematic of the waveguide OMCVD grown on the template with periodic domain inversion, (b) scanning electron micrograph of a stained cross section of periodically domain inverted AlGaAs, (c) scanning electron micrograph of AlGaAs waveguide, and (d) measured second harmonic power versus the square of the fundamental power, shown with quadratic curve fit. (Data from Yoo et al., 1995.)

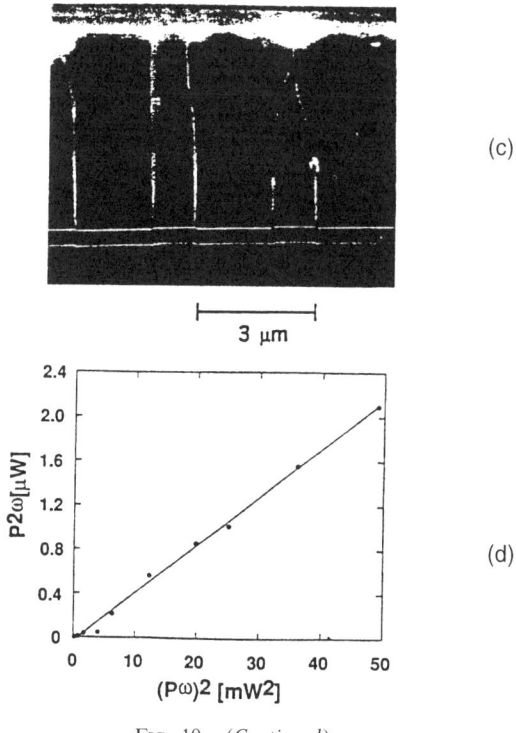

FIG. 10. (*Continued*)

where a periodic crystal domain inversion was achieved using wafer bonding, followed by selective etching and organometallic chemical vapor deposition (OMCVD). The peak conversion efficiency of 4.9%/W was achieved for the fundamental wavelength of 1466 nm. Similarly, Angell *et al.* (1994) realized a scheme for growing alternating $\langle 100 \rangle$- and $\langle 111 \rangle$-oriented regions of ZnSe and ZnTe on a GaAs substrate using an OMCVD-grown CdTe template.

Quasi-phasematching has also been demonstrated in the counter-propagating (surface-emitting) geometry. Since GaAs and AlGaAs have quite different values of $\chi^{(2)}$, by growing alternative layers of GaAs and AlGaAs one achieves modulation of $\chi^{(2)}$ in the direction normal to the surface. Obviously, owing to the nature of epitaxial growth, this is a far easier task than growing alternative layers along the plane. If the thickness of each layer is equal to the half-wavelength of the frequency-converted light, one creates a nonlinear polarization wave $\mathscr{P}(2\omega)$ (Fig. 11) with

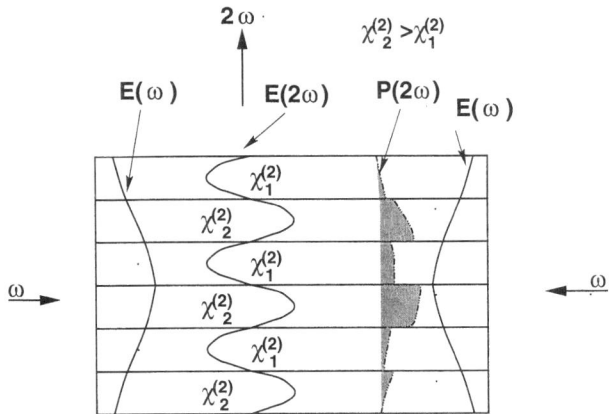

FIG. 11. Quasi-phasematching in the surface-emitting SHG geometry.

a component

$$\mathscr{P}(2\omega) \sim [\chi^{(2)}_{GaAs} - \chi^{(2)}_{AlGaAs}]\mathscr{E}^2(\omega)e^{i(\mathbf{k}_{2\omega})x} \tag{111}$$

that is perfectly matched to the nonlinear wave $\mathscr{E}_\pm$. By using AlGaAs resonant multilayers embedded in a waveguide geometry Cada et al. (1992) nd Normandin, Williams, and Chatenoud (1992) obtained a 10-million-fold enhancement for surface-emitting harmonic generation over a normal GaAs film. The experimental results agreed closely with theory, and a monolithic implementation of visible, surface-emitting, solid state diode lasers was presented for InGaAs, InP, and GaAs geometries. The surface-emitting structures have also been proposed (Normandin et al., 1991) for use in wavelength division multiplexing for optical communications.

One can also use the intermodal dispersion in the waveguides to compensate for the material dispersion and achieve phasematching. In the second harmonic generation case, for instance, one can launch the fundamental wave $\mathscr{E}(\omega)$ into the TE_0 mode, where it will propagate with effective propagation constant $\beta_\omega(TE_0)$. The second harmonic polarization wave will then propagate as

$$\mathscr{P}(2\omega) \sim \mathscr{E}^2(\omega)e^{2i\beta_\omega(TE_0)z} \tag{112}$$

It is possible to match this wave with the higher-order mode for second harmonic radiation, say TM_1, with a phasematching condition of

$$2\beta_\omega(TE_0) = \beta_{2\omega}(TM_1) \tag{113}$$

However, the efficiency of frequency conversion is also proportional to the overlap of the polarization wave and the free wave—i.e.,

$$\frac{P(2\omega)}{P(\omega)} \sim \left| \int \chi^{(2)}(x) \mathscr{E}^2_{TE_0}(x) \mathscr{E}_{TM_1} dx \right|^2 \tag{114}$$

Owing to orthogonality of modes, this overlap is very small. Dramatic improvement can be achieved, however, if one modulates the second-order susceptibility in the direction of growth, as shown in Fig. 12, essentially achieving quasi-phasematching (Ito and Inaba, 1978); Khurgin, 1988a). It has been predicted theoretically (Khurgin, 1988a) that one can achieve manifold enhancement of efficiency using bonded layers of II-VI materials.

Finally, one can achieve phasematching using the so-called form birefringence (Van der Ziel et l., 1974) in the waveguides, incorporating thin layers of materials with different indices of refraction, such as GaAs and AlGaAs. In such structures, TE and TM modes can have substantially different refractive indices—perhaps enough to compensate for the natural disper-

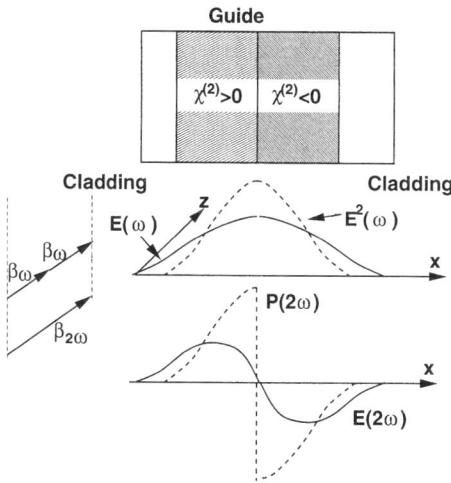

FIG. 12. Phasematching by use of intermodal dispersion in the waveguide and the domain inversion.

sion. Van der Ziel (1975) and Van der Ziel and Ilegems (1976) performed some measurements of thin GaAs/AlGaAs waveguides, and although second harmonic generation was observed, as well as parametric generation, the efficiency was low, indicating that the index contrast of two such similar materials as GaAs and AlGaAs is insufficient to provide phasematching.

Recently, however, a breakthrough was achieved by Fiore *et al.* (1997a). By selective lateral oxidation of the GaAs/AlGaAs heterostructures, thin layers of $Al_{0.7}Ga_{0.3}As$ were oxidized and converted into the amorphous layers of a very-low index ($n = 1.6$) aluminum oxide, as shown in Fig. 13. Form birefringence between *TE* and *TM* modes as high as 0.2 was measured, which was more than adequate to compensate for the dispersion for most $\chi^{(2)}$ processes. For example, tunable difference frequency generation of 4-μm radiation from 1-μm and 1.32-μm pumps was achieved, with a conversion efficiency of 0.043%/W. The tunability range was in excess of 50 cm^{-1}.

To summarize this section, although zinc-blende semiconductors have huge second-order nonlinearities, taking advantage of them requires inordinate effort and ingenious fabrication methods. At present, practical devices utilizing the bulk III–V and II–VI semiconductors are yet to materialize, but the rapid advancement of the novel epitaxial methods of growth shows promise for these materials. So far, we have considered only the use of epitaxial structures in the context of phasematching and integration with other electro-optic devices, but the use of epitaxial methods allows much more than this and may lead to engineering of large artificial nonlinearities in semiconductors, which are the subject of the next section.

IV. Second-Order Susceptibility in Semiconductor Quantum Wells and Superlattices

1. ASYMMETRIC QUANTUM WELLS

In the previous section, we established that the second-order susceptibility of a given material depends on the third order of its dipole moment, and hence it would be highly advantageous to increase the size of the structure, or to "scale it up." One way of doing this, long known to researchers, would be to use a very large molecule, typically an organic one, and indeed significant progress has been achieved in the growth and fabrication of nonlinear polymer materials and in taking advantage in the giant dipole moment that characterize such materials (Zyss, 1994). The polymers, however, are still plagued by high residual absorption, degradation, and difficulties in

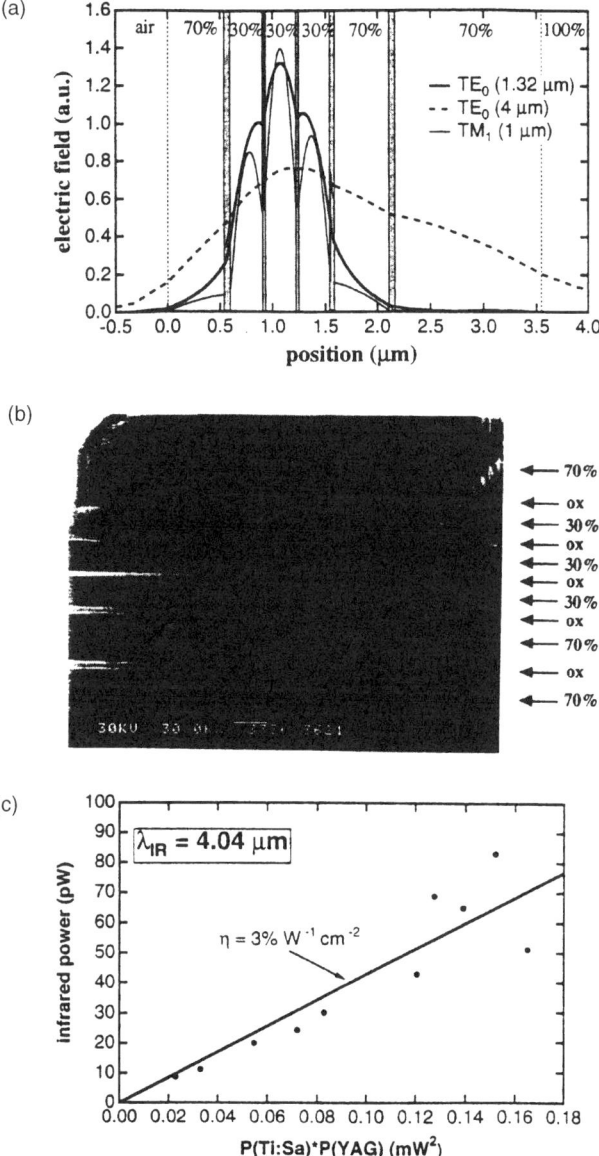

FIG. 13. Phasematched difference frequency generation using form birefringence in selectively oxidized AlGaAs/AlAs multilayers. (a) Electric fields in the oxidized waveguide: TE_0 ($\lambda = 1.32$ μm), TM_1 ($\lambda = 1$ μm), and TE_0 ($\lambda = 4$ μm). The gray zones indicate aluminum oxide. (b) Scanning electron micrograph of a stained multilayer waveguide. (c) infrared power as a function of the product of pump powers in a 1.2-mm waveguide. (Data from Fiore et al., 1997.)

phasematching. Furthermore, the polymer-based nonlinear optical devices are not integrable with more conventional optoelectronic devices. At the same time, the rapid progress in the epitaxial growth of semiconductors that has occurred since late 1970s has led to explosive developments in the low-dimensional semiconductor structures—quantum wells (QWs) and superlattices (SLs) that have unique electronic and optical properties (Weisbuch and Vinter, 1994). In low-dimensional structures, of which the most typical is the GaAs/AlGaAs symmetric QW shown in Fig. 14(a), the electron in the conduction band is confined in the low-bandgap material (GaAs) by the heterostructural barriers, preventing the electron from moving into the higher-bandgap material (AlGaAs). A similar situation exists for the holes in the valence band.

The depth of a QW can vary from a few atomic layers to as much as a few hundred angstroms and is limited only by the coherence length of the carriers, and the height of the barriers determined by the band offsets of the materials comprising the structure is typically on the order of a few hundred millielectronvolts.

The confining potential can be shaped in many ways—square, parabolic,

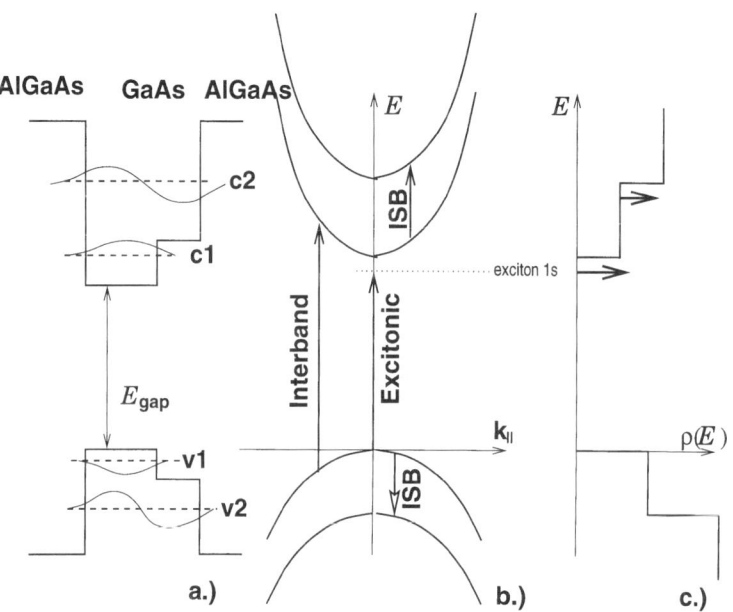

FIG. 14. Semiconductor quantum wells: (a) bandstructure, (b) in-plane dispersion and optical transitions, and (c) density of states.

symmetric or asymmetric—appearing as nothing but the "large-scale model" of the classical atomic or molecular potential described at the beginning of this section. It is therefore only natural to expect the electron placed in such potential to behave similar to the classical electron and exhibit nonlinearities proportional to the cube of the size of the QW. This was noticed by Gurnik and DeTemple as early as 1983, and the detailed theory was developed in late 1980s (Khurgin, 1987, 1988b, 1989).

The description of the physical processes in QWs that is presented here is extremely simplified in that it disregards a number of intricacies associated with band nonparabolicities, the presence of other bands (i.e., split-off valence bands and higher-order conduction bands), and other factors (Bastard, 1988; Chuang, 1995; Kane, 1957, 1966). However, the simple effective mass analysis given here has been shown to explain very well the main features of the nonlinear optical properties of the semiconductor heterostructures (Khurgin, 1988b; Fiore et al., 1995a, 1995b). Bastard (1988) was the first to introduce the concept of envelope wavefunctions, and we will follow his approach in our description of carriers in QWs.

While the carriers are free to move in the plane of growth (xy), their motion in the direction normal to the plane of growth becomes quantized, and their wavefunctions near the bandgap are best described as

$$\Psi_{c,v}(\mathbf{r}) = |c(v), n, \mathbf{k}_\parallel\rangle = u_{c(v),\mathbf{k}}(\mathbf{r})e^{i\mathbf{k}_\parallel \cdot \mathbf{r}} f_{c(v),n}(z) \qquad (115)$$

where $u_{c(v),\mathbf{k}}$ is a periodic (atomic) part of the Bloch function, described by the antibonding combinations of the S states in the conduction (c) band and bonding combinations of P orbitals for the valence (v) band, $\mathbf{k}_\parallel$ and $\mathbf{r}_\parallel$ are the wavevector and coordinate in the plane of the QWs, and $f_{c(v),n}(z)$ is the so-called envelope function of the nth subband. In the effective mass approximation, the envelope functions are the eigenfunctions of the simple effective mass Hamiltonian:

$$\left[-\frac{\hbar^2}{2m_{c(v)}}\frac{d^2}{dz^2} + V_{c(v)}(z)\right] f_{c(v),n}(z) = \mathscr{E}_{c(v),n} f_{c(v),n}(z) \qquad (116)$$

where $V_{c(v)}(z)$ is the profile of the conduction (valence) band potential, $m_{c,v}$ are the effective masses, and $\mathscr{E}_{c(v),n}$ is the energy of the bottom (top) of the nth conduction (valence) subband. The total energy of the state $|c(v), n, \mathbf{k}_\parallel\rangle$ is

$$E_c(n, \mathbf{k}_\parallel) = E_{c,n} + \frac{\hbar^2 k_\parallel^2}{2m_c} + E_{\text{gap}} \qquad (117)$$

for the conduction band, and

$$E_v(n, \mathbf{k}_\parallel) = -E_{v,n} - \frac{\hbar^2 k_\parallel^2}{2m_{v,\parallel}} \tag{118}$$

for the valence band, where $m_{v,\parallel}$ is the in-plane effective mass in the valence band. The subband dispersion is shown in Fig. 14(b).

Also, owing to the confinement, the smooth parabolic density of the bulk semiconductor changes dramatically, and acquires the stepladder-like character shown in Fig. 14(c). The absorption spectrum also changes significantly. The changes are associated primarily with the fact that the translational symmetry in the direction of growth is no longer present in QWs and hence a number of new transitions become allowed (Fig. 14b):

- *Free Carrier Interband Transitions.* These are the transitions between the state in the nth valence subband, $|v, n, \mathbf{k}_{\parallel,v}\rangle$, and the lth conduction subband, $|c, n, \mathbf{k}_{\parallel,c}\rangle$. The selection rules for these transitions are determined by the strength of the dipole matrix elements:

1. *Heavy holes:*

$$x^{nl}_{hh,c} = \frac{P}{\sqrt{2}\, m_0 \omega_{\text{gap}}} \langle f_{c,l} | f_{hh,n} \rangle \delta(\mathbf{k}_{\parallel,v}, \mathbf{k}_{\parallel,c}) \tag{119}$$

$$z^{nl}_{hh,c} = 0 \tag{120}$$

2. *Light holes:*

$$x^{nl}_{lh,c} = \frac{P}{\sqrt{6}\, m_0 \omega_{\text{gap}}} \langle f_{c,l} | f_{lh,n} \rangle \delta(\mathbf{k}_{\parallel,v}, \mathbf{k}_{\parallel,c}) \tag{121}$$

$$z^{nl}_{lh,c} = \frac{2P}{\sqrt{6}\, m_0 \omega_{\text{gap}}} \langle f_{c,l} | f_{lh,n} \rangle \delta(\mathbf{k}_{\parallel,v}, \mathbf{k}_{\parallel,c}) \tag{122}$$

where $P = \langle X | \mathbf{P} | iS \rangle$ is a Kane's matrix element of the momentum (Kane, 1957) and $\omega_{\text{gap}} = E_{\text{gap}}/\hbar$. It is worthwhile to note that for all the III–V semiconductors, the value of P is roughly the same, given, according to Shantharma et al. (1984), as

$$\frac{2P^2}{m_0} \sim 15 \div 25 \,\text{eV} \tag{123}$$

- *Excitonic Interband Transitions.* These are the enhancements of the band-to-band transitions. They are defined by the sharp peaks associated with the 1s excitons located below the edge of each allowed band-to-band transition, and with the strength of each excitonic transition (Miller, Weiner, and Chemla, 1986):

$$x_{v,c}^{nl}|_x = x_{v,c}^{nl} \cdot \Psi_{1s}^{nl}(0) \qquad (124)$$

or

$$z_{v,c}^{nl}|_x = z_{v,c}^{nl} \cdot \Psi_{1s}^{nl}(0) \qquad (125)$$

where $\Psi_{1s}^{nl}(\mathbf{r}_\parallel)$ is the 2-D excitonic wavefunction.

- *Intersubband (ISB) Transitions.* These are the transitions between the different subbands within the same conduction or valence band. For the conduction band intersubband transitions,

$$z_c^{nl} = \langle c, l, \mathbf{k}_{\parallel,l}|z|c, n, \mathbf{k}_{\parallel,n}\rangle = \langle f_{c,l}|z|f_{c,n}\rangle \delta(\mathbf{k}_{\parallel,n}, \mathbf{k}_{\parallel,l}) \qquad (126)$$

$$x_c^{nl} = \langle c, l, \mathbf{k}_{\parallel,l}|x|c, n, \mathbf{k}_{\parallel,n}\rangle = 0 \qquad (127)$$

In the effective mass approximation, only the z (TM)-polarized electromagnetic radiation can cause the intersubband transitions within the conduction band. More precise calculations show that the in-plane (TE) transitions are not strictly forbidden, but their strength is much smaller than that of TM-polarized ones.

Although there have been theoretical investigations of the nonlinearities associated with the transitions within the valence bands (Tsang and Chuang, 1995; Li and Khurgin, 1993; Qu and Ruda, 1993) showing that both TE and TM polarized transitions are allowed, there has been only one experimental observation of the $\chi^{(2)}$ associated with transitions between the valence subbands (Seto *et al.*, 1995), so we will not consider them here.

The magnitude of the dipole of the intersubband transition is of the same order as the size of the well—i.e., typically a few times larger than the interband dipole. It should be noted, however, that the "giant" size of the dipole is by no means an inherent property of intersubband transitions, but is simply related to the fact that the intersubband transition energies are typically an order of magnitude lower than the band-to-band transition energies (Khurgin, 1993). It follows directly from the oscillator sum (Eq. 94).

In the symmetric QWs, the selection rules are extremely simple: the interband transitions, proportional to the overlap of the envelope wavefunction, are allowed only between the states of the same parity, and the intersubband transitions, proportional to the intersubband dipole moment, are allowed only between the states of opposite parity. Naturally, it is impossible to find three states $|l\rangle$, $|m\rangle$, and $|n\rangle$ with the nonzero combination

$$\mu_{ln}\mu_{nm}\mu_{ml} = -e^3 \langle l|\mathbf{r}|n\rangle\langle n|\mathbf{r}|m\rangle\langle m|\mathbf{r}|l\rangle \tag{128}$$

that, according to Eq. (56), is necessary to obtain nonzero $\chi^{(2)}$. Once the symmetry of the confinement potentials $V_c(z)$ and $V_v(z)$ is broken, the selection rules are no longer so strict, and one can obtain second-order nonlinearity. The asymmetry can be incorporated either externally, by applying the bias (Fig. 15a), or internally, by growing the QWs with various

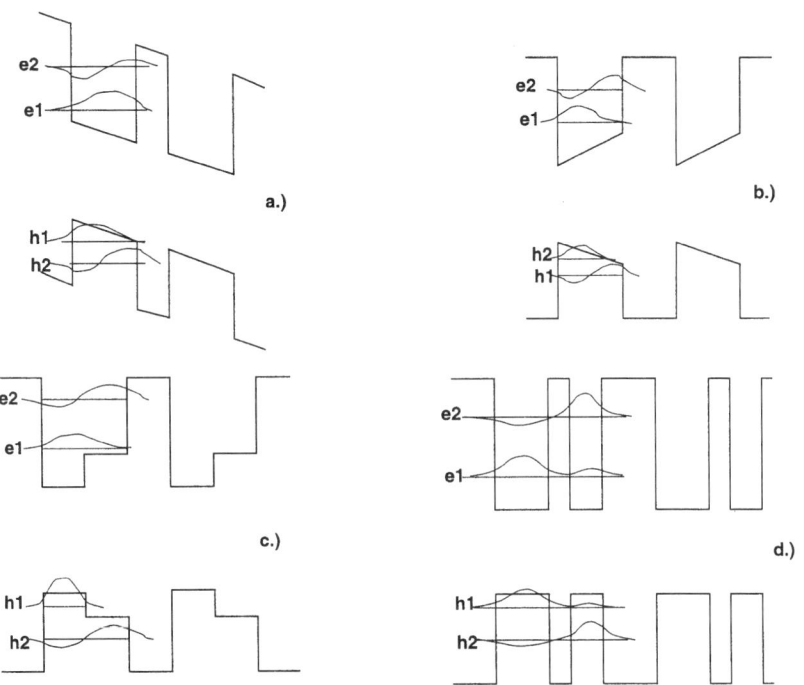

FIG. 15. Asymmetric semiconductor QWs: (a) biased QW, (b) graded bandgap QW, (c) stepped QW, and (d) ACQW.

forms of asymmetry—graded bandgap QW (Fig. 15b), stepped QW (Fig. 15c), or asymmetric coupled QW (ACQW) (Fig. 15d).

Since all the optical processes in Eq. (128) must originate and end up at the same ground state *l* within the same band, the number of the interband transitions in this expression must be even—i.e., either zero or two. The first group of processes involves three virtual transitions within the same band, and we shall refer to them as intersubband optical nonlinearities. The second group of processes involves two interband transitions and one intersubband transition and we shall refer to them as interband optical nonlinearities.

2. Intersubband Second-Order Nonlinearities: Theory

Because the intersubband dipole matrix element can be as much as an order of magnitude larger than the band-to-band elements, the second-order nonlinearity associated with the intersubband processes can be very large. This fact was first noticed by Gurnick and DeTemple (1983), who modeled the quantum well using the anharmonic potential similar to the potential U considered in Section II. Indeed, their calculations predicted second harmonic coefficients in excess of 100 nm/V. In 1989, Khurgin considered realistic asymmetric QW structures and obtained analytical expressions for all of the most important second-order nonlinear coefficient—the second harmonic coefficient, the linear electro-optic effect, and optical rectification—and investigated the dependence of these coefficients on the compositions and dimensions of QWs. Over the years, there have been a number of numerical calculations for the different special cases of $\chi^{(2)}$, including sum and difference frequency generation, for the specific structures (Tsang, Ahn, and Chuang, 1988; Yuh and Wang, 1989; Rosencher and Bois, 1991; Iconic, Milanovic, and Tjapkin, 1989). The importance of the self-consistency of these calculations has been examined. Here, following Khurgin (1989), we consider the general features of the intersubband $\chi^{(2)}$.

When the QWs are *n*-doped with the dopant density $N_d \leq 10^{18}$ cm^{-3}, then, in general, only the states in the ground subband $|c, 1, \mathbf{k}_\parallel \rangle$ are occupied by the carriers. Furthermore, in the parabolic band approximation (Eq. 117), the intersubband transition energy $E_c(n)$ does not depend on the in-plane wavevector $\mathbf{k}_\parallel$. In this case, neglecting electron-electron interactions, the electrons behave as an ensemble of identical independent electrons. Then the task of finding the nonlinear susceptibility of electrons in QWs is entirely identical to the task of finding the $\chi^{(2)}$ value of the electrons in bonding orbitals of semiconductors, as previously considered. According to Eq. (56), the expression for $\chi^{(2)}$ is

$$\chi^{(2)}_{33}(\omega_1+\omega_2,\omega_1,\omega_2) = \frac{N_d e^3}{2\varepsilon_0 \hbar^2} \sum_{m,n} \langle 1|z|m\rangle\langle m|z|n\rangle\langle n|z|1\rangle$$

$$\times \left[\frac{1}{(\omega_{n1}-\omega_1-\omega_2)(\omega_{m1}-\omega_1)} \right.$$

$$+ \frac{1}{(\omega_{n1}-\omega_1-\omega_2)(\omega_{m1}-\omega_2)}$$

$$+ \frac{1}{(\omega_{n1}+\omega_2)(\omega_{m1}-\omega_1)} \frac{1}{(\omega_{n1}+\omega_2)(\omega_{m1}-\omega_1)} \quad (129)$$

$$+ \frac{1}{(\omega_{n1}+\omega_1+\omega_2)(\omega_{m1}+\omega_1)}$$

$$\left. + \frac{1}{(\omega_{n1}+\omega_1+\omega_2)(\omega_{m1}+\omega_2)} \right]$$

Consider now the following cases:

1. *Long Wavelength Off-Resonant Limit (Fig. 16a)*. In this case, using the completeness of the envelope wavefunctions, we obtain

$$\chi^{(2)}_{33}(0) \approx \frac{3N_d e^3}{\varepsilon_0 \bar{E}} \langle 1|(z-z_{11})^3|1\rangle \qquad (130)$$

where $\bar{E}$ is the average intersubband energy—i.e., it is on the order of the conduction band offset, and

$$z_{11} = \langle 1|z|1\rangle \qquad (131)$$

For the band offset of about $0.3 \div 0.5$ eV, the QW width on the order of 100 Å, and the doping of 10^{18} cm^{-3}, one can expect to find $\chi^{(2)}$ values on the order of a few tens of nm/V—i.e., two orders of magnitude higher than in bulk GaAs. Invoking the oscillator sum rule (Eq. 94), we can rewrite Eq. (130) as

$$\chi^{(2)}_{33}(0) \approx \frac{192\pi^2 N_d \varepsilon_0}{e r_{ex}^2} \langle 1|(z-z_{11})^3|1\rangle|\langle 1|(z-z_{11})^2|1\rangle|^2 \qquad (132)$$

where r_{ex} is the excitonic radius in a given material. This form of the

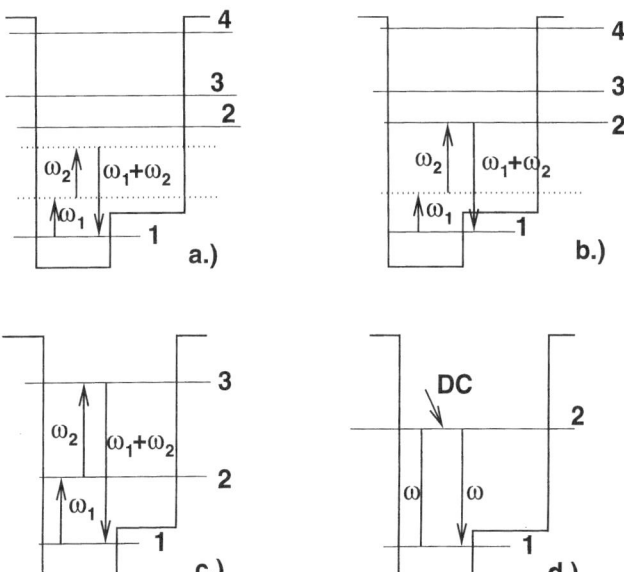

FIG. 16. Diagrams for calculations of intersubband $\chi^{(2)}$ in asymmetric QWs: (a) off-resonant, (b) single-resonant, (c) double-resonant, and (d) resonant electro-optic effect $\chi^{(2)}(\omega; 0; \omega)$. (Data from Khurgin, 1989.)

expression, quite similar to the classical form of $\chi^{(2)}$, is given here to emphasize this very strong dependence of the susceptibility on the size of QW. One should, however, be reminded once again that the main reason for high $\chi^{(2)}$ here is the fact that the intersubband transitions are occurring at low frequencies, typically in the far-IR region of the spectrum.

2. *Single-Resonant Case (Fig. 16b).* While the large magnitude of the dipole is not an inherent and unique characteristic of the intersubband transitions, these transitions still hold one great advantage—their sharp resonant character. Therefore, one can choose the sum frequency $\omega_1 + \omega_2$ to be very close to one of the transitions, typically the lowest one, $1 \to 2$. Then the resonant terms in Eq. (129) dominate the expression for $\chi^{(2)}$, and one obtains

$$\chi^{(2)}_{33}(\omega_1 + \omega_2, \omega_1, \omega_2) = \frac{N_d e^3}{2\delta\omega\varepsilon_0 \hbar^2} \sum_m \langle 1|z|m\rangle\langle m|z|2\rangle\langle 2|z|1\rangle \times \left[\frac{1}{\omega_{m1} - \omega_1} + \frac{1}{\omega_{m1} - \omega_2}\right] \quad (133)$$

where the detuning is

$$\delta\omega = \omega_{n1} - \omega_2 - \omega_1 - i\gamma \qquad (134)$$

Assuming that $\omega_{m1} \gg \omega_2, \omega_1$, we obtain, once again, using the completeness of the intersubband states,

$$\chi^{(2)}_{33}(\delta\omega, \omega_1, \omega_2) = \frac{N_d e^3}{\hbar\delta\omega\varepsilon_0 \bar{E}} \langle 2|z|1\rangle\langle 1|(z - z_{11})^2|2\rangle \qquad (135)$$

For the detuning on the order of 10 meV, the second-order susceptibility can reach hundreds of nanometers per volt.

3. *Double-Resonant Case (Fig. 16c)*. Besides other advantages, already mentioned, the biggest advantage of intersubband transitions is their flexibility: using the methods of bandgap engineering, one can create a situation where a double-resonant condition can be achieved. Then only one term in Eq. (129) remains, and the expression for $\chi^{(2)}$ becomes

$$\chi^{(2)}_{33}(\delta\omega, \omega_1, \omega_2) = \frac{N_d e^3}{(\hbar\delta\omega)^2 \varepsilon_0} \langle 1|z|3\rangle\langle 3|z|2\rangle\langle 2|z|1\rangle \qquad (136)$$

In this case, for the detuning on the order of 10 meV, the double-resonant second-order susceptibility can reach as much as 1000 nm/V. Similar resonant enhancement can be obtained in the case of the linear electro-optic effect (Fig. 16d):

$$\chi^{(2)}_{33}(\omega, 0, \omega) = \frac{N_d e^3}{(\hbar\delta\omega)^2 \varepsilon_0} |\langle 1|z|2\rangle|^2 (z_{22} - z_{11}) \qquad (137)$$

Numerous authors have analyzed the second-order susceptibility numerically and considered its optimization (Capasso, Sirtori, and Cho, 1994; Almogy and Yariv, 1995; Rosencher et al., 1996; Ikonic, Milanovic, and Tjapkin, 1989; Tomic, Milanovic, and Ikonic, 1997). Their calculations confirmed the predictions outlined in Khurgin (1989), showing that, for a given operating wavelength, $\chi^{(2)}$ can be optimized by methods of bandgap engineering, by essentially optimizing the product of three matrix elements. Typically, the ACQWs (Fig. 15d) provide a greater degree of flexibility and thus can lead to greater values of $\chi^{(2)}$. An interesting fact was observed by

Khurgin (1989): if one introduces the "polarity parameter" α_p for the ACQW as

$$\alpha_p = \int_{QW1} |f_{c,1}(z)|^2 \, dz - \int_{QW2} |f_{c,1}(z)|^2 \, dz \tag{138}$$

—i.e., as the difference between the charges in the left and right wells of the ACQW structure, $\chi^{(2)}$ reaches its maximum at $\alpha_p \sim 0.5$, or roughly at the same value of polarity at which diatomic molecules (or polar bonds in bulk crystals, as considered above) exhibit the highest values of $\chi^{(2)}$. Thus, as first noticed by Gurnick and DeTemple (1983), there exists a complete analogy between the asymmetric QWs and polar molecules and bonds.

It is important, however, to notice that the high resonant $\chi^{(2)}$ in intersubband transitions comes at a steep price—the resonant absorption of either fundamental or harmonic signals. As a result, the nonlinear wave amplitude does not increase linearly, as predicted by the simple theory, but first saturates and then decreases. The saturation occurs roughly at the distance of one absorption length, and therefore the figure of merit (FOM) for the performance of the nonlinear frequency converter (Khurgin, 1989) is

$$\text{FOM} = \left| \frac{\chi^{(2)}(\omega_2 + \omega_1, \omega_2, \omega_1)}{\Im[\chi^{(1)}(\omega_2 + \omega_1)]} \right|^2 \tag{139}$$

For the single-resonant case,

$$\text{FOM} = \left| \frac{e|\delta\omega|\langle 1|(z - z_{11})^2|2\rangle}{2\langle 1|z|2\rangle \gamma \bar{E}} \right|^2 \tag{140}$$

For the double-resonant case, we obtain

$$\text{FOM} = \left| \frac{e\langle 1|z|3\rangle\langle 3|z|2\rangle}{4\langle 1|z|2\rangle \hbar\gamma} \right|^2 \tag{141}$$

where the additional factor of 2 in the denominator reflects the fact that both fundamental and harmonic waves are being absorbed. The results (Eq. 140 and 141) indicate that one can obtain higher conversion efficiency by going away from the resonance, but it may require a greater length of active medium. As always, one must look for the compromise (Khurgin, 1989; Almogy and Yariv, 1995; Rosencher et al., 1996).

The nature of intersubband transitions within the conduction band allows only the interaction between *TM*-polarized waves, making phasematching quite difficult. As we have already mentioned, a number of authors have predicted large nonlinear susceptibilities for both *TE* and *TM* waves in *p*-doped QWs (Tsang and Chuang, 1995; Li and Khurgin, 1993; Qu and Ruda, 1993), where the transitions between light and heavy holes are *TE*-polarized. It should make phasematching by the anisotropy plausible, but the quality of the *p*-doped structures is generally far worse than the quality of the *n*-doped ones, and the range of frequencies is limited to the far-IR range. This leaves one with necessity of using quasi-phasematching, with the nonlinear susceptibility of the asymmetric well changed by the applied electric field (Myer et al., 1995), or by impurity-induced disordering (Street et al., 1997). Index changes induced by the intersubband absorption (Almogy, Seger, and Yariv, 1994) can also produce phasematching. The other promising method of phasematching—by growing periodic "quantum well domains" (Khurgin, 1988c)—has been proposed and utilized (Janz, Chatenoud, and Normandin, 1994; Fiore et al., 1997b) for the quasi-phasematching of interband nonlinearities, which is discussed later, but this concept can be used for the intersubband nonlinearities as well.

A typical "quantum well domains" structure is shown in Fig. 17. By reversing the order of growth, one can grow the mirror images of asymmetric QWs having the opposite sign of $\chi^{(2)}$; then, if one grows $\lambda/2$ multiple QW layers with opposite $\chi^{(2)}$'s one can "fit" the second harmonic polarization $\mathscr{P}(2\omega)$ into the free-running second harmonic wave $\mathscr{E}(2\omega)$ in the surface-emitting geometry, shown in Fig. 17, or in longitudinal geometry, using the selective etching method.

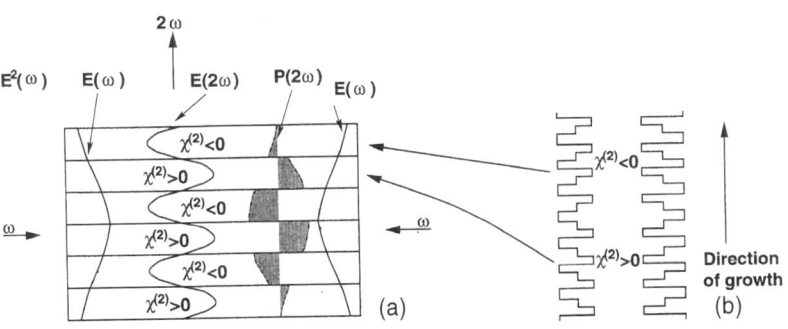

FIG. 17. Quantum well domains for surface SHG phasematching: (a) frequency doubling arrangements and (b) band profiles of domains.

3. INTERSUBBAND SECOND-ORDER NONLINEARITIES: EXPERIMENT

The first experimental observation of $\chi^{(2)}$ based on intersubband transitions in the asymmetric QW was reported by Fejer *et al.* (1989). They used biased GaAs/AlGaAs quantum wells and a single-resonance scheme to generate the second harmonic of the CO_2 laser radiation. The difficulties of their experiment were associated with the presence of a strong background second-harmonic signal resulting from the bulk $\chi^{(2)}$, but the background signal could be subtracted by measuring only the part of the nonlinear signal that depended on the bias field. The value of $\chi^{(2)}$ measured in this experiment was 28 nm/V — i.e., two orders of magnitude larger than the bulk GaAs value, exactly as predicted in Eq. (135).

Rosencher and Bois (1989) and Boucaud *et al.* (1990) improved on this result by using the double-resonant scheme with asymmetric step quantum wells. As did Fejer *et al.* (1989), they studied the second harmonic generation at 10.6 μm, and measured $\chi^{(2)}$ values as great as 720 nm/V. Their results are shown in Fig. 18. Similar results were obtained for the sum frequency generation in the stepped QWs (Liu *et al.*, 1995a, 1995b).

Capasso, Sirtori, and Cho (1994) used a different material system — InGaAs/InAlAs multiple ACQWs grown on an InP substrate (Sirtori *et al.*, 1991; Capasso, Sirtori, and Cho, 1994). They used a double-resonant scheme and measured the SHG coefficient in excess of 75 nm/V. Sirtori *et al.* (1994) also obtained a far-IR difference generation in a similar structure. By mixing two detuned CO_2 lasers in the ACQW, they obtained 60-μm radiation and measured $\chi^{(2)}(\omega_1 - \omega_2, \omega_1, \omega_2)$ as large as 1000 nm/V. These results are presented in Fig. 19.

The most often used GaAs/AlAs system has conduction band offset of less than 320 meV, which limits it to generation of radiation longer than 5 μm, while in the second most popular system — InGaAs/InAlAs grown on INP — the maximum offset is only slightly bigger — about 500 meV. In the highly strained InGaAs/AlAs heterostructures grown on GaAs, the conduction band offset can reach an astounding 1.4 eV. Obviously, one cannot observe the intersubband transitions in this range because the range of frequencies above 1 meV happens to be greater than the fundamental bandgap of the material. But Chui *et al.* (1994a, 1994b) observed intersubband transitions in excess of 500 meV.

Using this system, the double-resonant second harmonic generation of the 2-μm radiation was achieved and $\chi^{(2)}$ values of about 20 nm/V were measured. The same group (Chui, 1995) achieved successful tunable mid-IR difference frequency generation by beating the radiation from two detuned $\lambda \sim 2\,\mu$m optical parametric oscillator (OPOs), and the difference frequency

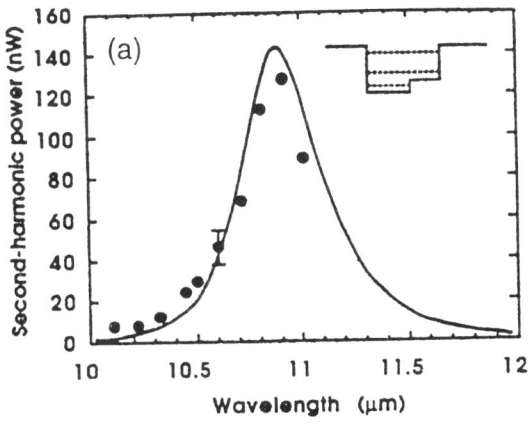

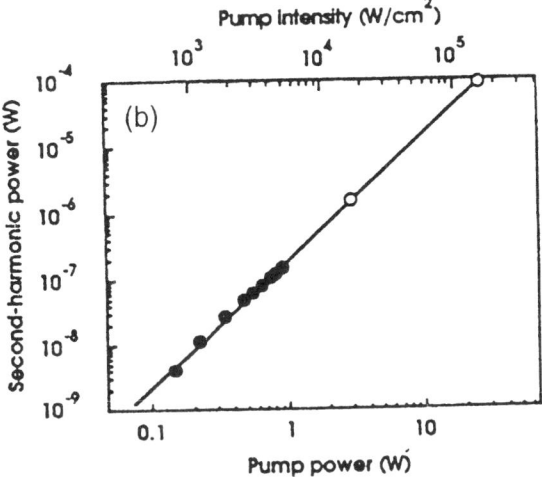

FIG. 18. Double-resonant SHG in stepped asymmetric GaAs/AlGaAs QWs. (a) SHG power versus pump CO_2 wavelength. Incident CO_2 power is 0.8 W. The conduction band profile of the asymmetric QW is shown in the inset. (b) SHG power versus input CO_2 pump power at 10.6 μm. (Data from Boucaud et al., 1990.)

generated in this scheme could be tuned from 8 to 12 μm. The maximum $\chi^{(2)}$ measured in these experiments was 12 nm/V. This demonstration and that of Sirtori et al. (1994) are very promising because they offer an exciting opportunity for development of the all-semiconductor optical parametric amplifier capable of producing tunable radiation in the mid-to-far-IR range.

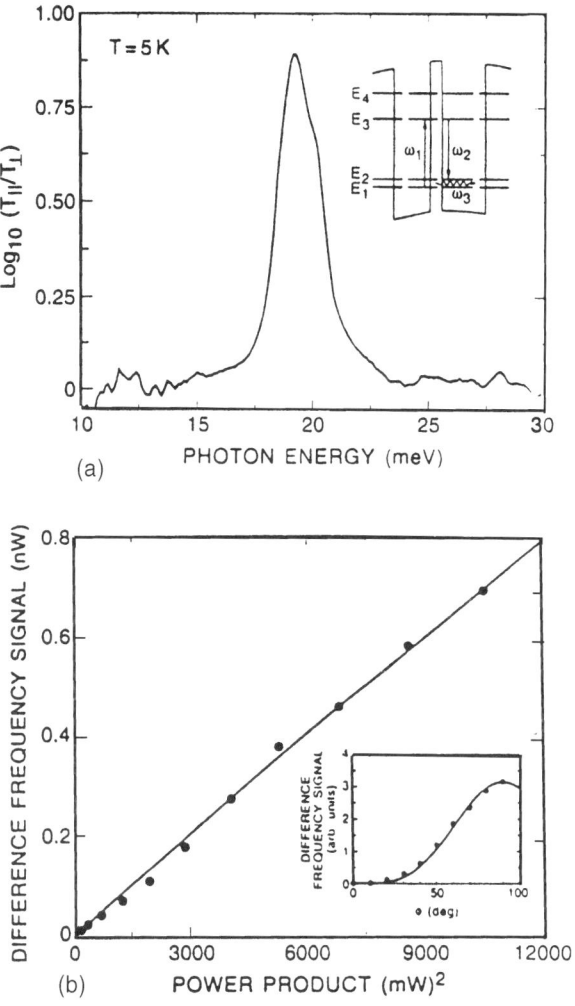

FIG. 19. Double-resonant difference frequency generation in an ACQW: (a) conduction band profile and energy levels (inset) and the absorption spectrum of the ACQW and (b) measured power of far-IR difference frequency as a function of the product of two pump power beams. (Data from Sirtori *et al.*, 1994.)

4. Interband Second-Order Nonlinearities: Theory and Experiment

We have seen that nonlinearity based exclusively on the intersubband transitions is indeed very large, but it has three drawbacks—the necessity of doping, difficulty in phasematching, and, most important, limitation to the mid-to-far-IR range by the band-offset values. As we have already mentioned, the main reason for the large intersubband nonlinearities is simply the lower frequency at which they occur (Khurgin, 1993). Taking this into account, one can say that the main advantage of asymmetric QWs lies in their flexibility—i.e., in the possibility of engineering proper dipole moments and resonant frequencies. Then, if one looks at the interband transitions in asymmetric QWs, one can see that the expected enhancement of $\chi^{(2)}$ over its bulk value cannot be as large as in intersubband transitions, since only one "giant intersubband dipole" is involved, rather than three. But, on the other hand, the flexibility of design is preserved, and the range of useful wavelengths is greatly expanded in comparison with the intersubband scheme. One can also expect to gain additional enhancement in $\chi^{(2)}$ from the excitonic transitions. From these simple observations, one must conclude that the interband nonlinearities in asymmetric QWs should be considered to be promising.

These nonlinearities first attracted attention in 1987 (Khurgin), and it was immediately realized that the situation involving interband transition was far more complicated than the one considered for the intersubband transitions. The main difference is that in the interband transitions one cannot identify just one ground state—every state in every valence subband acts as a ground state with its own envelope function, as shown in Fig. 20. Furthermore, there can be two types of processes: one involving two band-to-band transitions and one intersubband transition in the conduction band—the so-called "virtual electron process"—and one involving two band-to-band transitions and one intersubband transition in the valence band—the so-called "virtual hole process." These two processes, both illustrated in Fig. 20, have opposite signs of $\chi^{(2)}$ and tend to compensate each other. Finally, the interband transitions have different energies as functions of $\mathbf{k}_\parallel$, so that one cannot use the simple "N identical systems" model appropriate for the intersubband processes.

These complications were noted and dealt with in the first complete calculations of the interband $\chi^{(2)}$ in asymmetric QWs (Khurgin, 1987, 1988b). It was shown that in the long-wavelength limit the $\chi^{(2)}$ of the system actually vanishes. Indeed, assuming that all the transition frequencies are roughly identical, we obtain, limiting our attention to holes of one type—for instance, the heavy holes—

1 Second-Order Nonlinearities and Optical Rectification

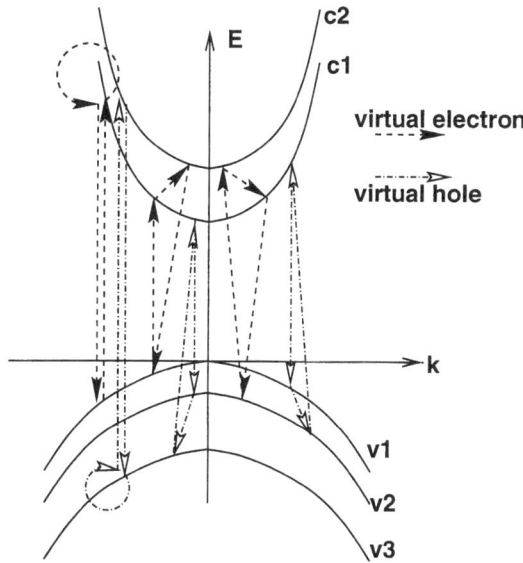

FIG. 20. Transitions involved in the interband $\chi^{(2)}$ in QWs.

$$\chi^{(2)}_{xzx}(0) \sim \frac{3e^3 P^2}{2m_0^2 \omega_{\text{gap}}^2 \varepsilon_0 \bar{E}^2} \rho_{2D} L_z^{-1}$$

$$\times \sum_{nlm} [\langle f_{hh,l}|f_{c,m}\rangle\langle f_{c,m}|z|f_{c,n}\rangle\langle f_{c,n}|f_{hh,l}\rangle$$

$$- \langle f_{c,l}|f_{hh,m}\rangle\langle f_{hh,m}|z|f_{hh,n}\rangle\langle f_{hh,n}|f_{c,l}\rangle] \quad (142)$$

$$\sim \chi^{(1)}(0) \frac{e}{\bar{E}} \sum_{nlm} [\langle f_{hh,l}|f_{c,m}\rangle\langle f_{c,m}|z|f_{c,n}\rangle\langle f_{c,n}|f_{hh,l}\rangle$$

$$- \langle f_{c,l}|f_{hh,m}\rangle\langle f_{hh,m}|z|f_{hh,n}\rangle\langle f_{hh,n}|f_{c,l}\rangle]$$

where L_z is the period of multiple QW structure and $\rho_{2D} = m_r/\pi\hbar_2$ is the reduced two-dimensional density of state (m_r is the reduced effective mass). The first term in the square brackets corresponds to the electronic $\chi^{(2)}$, and the second one to the heavy-hole $\chi^{(2)}$. Now, assuming that the envelope states in each band make up a complete set of eigenfunctions, we obtain

$$\chi^{(2)}_{xzx}(0) \approx \chi^{(1)}(0) \frac{e}{\bar{E}} \sum_l [\langle f_{hh,l}|z|f_{hh,l}\rangle - \langle f_{c,l}|z|f_{c,l}\rangle] = 0 \qquad (143)$$

since the sum of the dipoles over the complete set of envelope functions is zero (or our material would have been a segneto-electric material). Evidently, there are many terms that cancel each other in the sum over subbands and **k**. Only when one of the frequencies approaches a resonance with one of the subbands, one term dominates all other terms, and the cancellation becomes incomplete. Khurgin (1988b) performed careful analysis of the cancellation problem and derived the expressions for $\chi^{(2)}$ for the cases of second harmonic generation, the linear electro-optic effect, and optical rectification. Similar results were later obtained by Fiore *et al.* (1995a) for the case of biased QWs.

In these works it was established that the dominant component of the second-order susceptibility is indeed $\chi^{(2)}_{xzx}$, and its values, depending on the QW geometry and detuning, are between 5 and 20 pm/V. The other two nonzero components allowed by the symmetry, $\chi^{(2)}_{zxx}$ and $\chi^{(2)}_{zzz}$, are much smaller, primarily because of the absence of resonance near the bandgap. It should be noted, however, that in strained QWs (Chuang, 1995) the lowest-energy transition is the one between the light hole band and the conduction band, and so one can expect to find a reasonably large $\chi^{(2)}_{zzz}$ there.

In fact, large $\chi^{(2)}_{zzz}$ values were calculated by Atanasov, Bassani, and Agranovich (1994), Kelaidis, Hutchings, and Arnold (1994), and Qu *et al.* (1994a), but these authors did not use the sum used by Khurgin (1988b) and by Fiore *et al.* (1995a), and without using such rules, one cannot easily truncate the summation over the bands and has to rely on numerical analysis only. Subsequent experiments did not find the component to be appreciable. Other authors (Harshman and Wang, 1992; Shimizu, Kuwata-Gonokami, and Sakaki, 1992) greatly overestimated the value of $\chi^{(2)}$: by not considering all the terms, they obtained values as high as 10^4 pm/V near the excitonic peak. On the other hand, more involved calculations by Tsang, Chuang, and Lee (1990) and by Tsang and Chuang (1990) that considered the excitonic enhancement confirmed the results of Khurgin (1988b), giving $\chi^{(2)}_{zxx}$ around 20 pm/V.

The first experimental demonstration of the interband $\chi^{(2)}$ was achieved by Janz, Chatenoud, and Normandin (1994) and by Qu *et al.* (1994b) in asymmetric stepped GaAs/AlGaAs QWs in the reflection mode. They measured $\chi^{(2)}_{xzx} = 13$ nm/V, which is in complete agreement with the analysis of Khurgin (1988b) but is also nearly identical to $\chi^{(2)}_{zxx} = 11$ nm/V, which could not be explained by the theory and could perhaps be traced to the interface effect.

Fiore *et al.* (1995b) performed careful and unambiguous measurements of $\chi^{(2)}$ in the biased QWs. Their experimental apparatus is shown in Fig. 21(a). Using the dependence of second harmonic power on the applied bias allowed the symmetric QW nonlinearity to be completely separated from the bulk and surface contributions. The results of the experiment (Fig. 21b) show that for QW samples, the efficiency of SHG increases with the bias, as

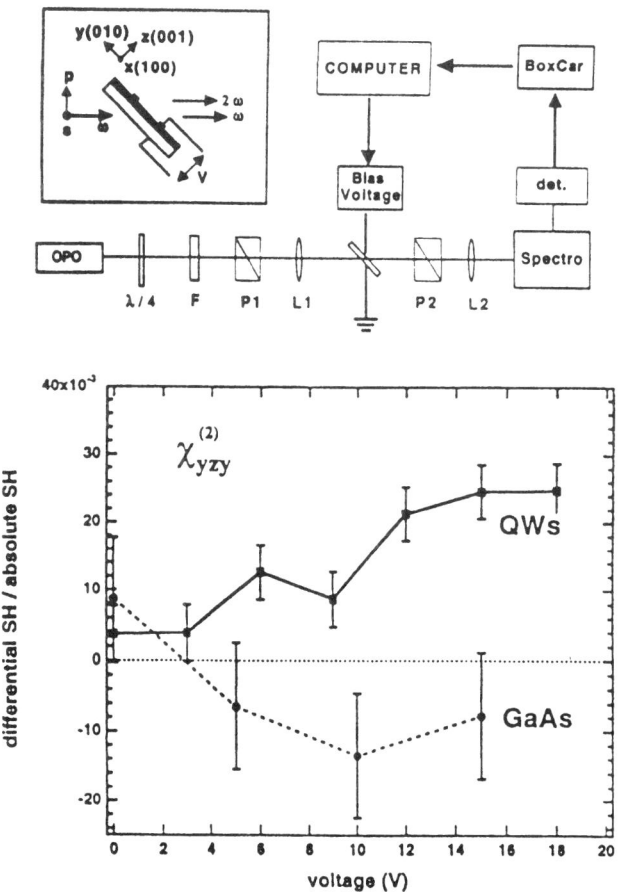

FIG. 21. Interband SHG in DC-biased QWs: (a) experimental apparatus and (b) ratio of differential second harmonic power to bulk second harmonic power as a function of applied bias. (Data from Fiore *et al.*, 1995b.)

expected from the theory (Khurgin, 1988b; Fiore et al., 1995a), while for bulk GaAs samples, used for reference, there is no increase in SHG efficiency with bias, because, as we know from the previous section, the bulk $\chi^{(2)}$ depends only on the intrinsic asymmetry of the bond. The results, which are in complete agreement with the theoretical predictions of Fiore et al. (1995a) and Khurgin (1988b), are $\chi^{(2)}_{xzx} = 25$ nm/V and $\chi^{(2)}_{zxx} = 11$ nm/V, which is too small to be measured.

Most recently, Fiore et al. (1997b) achieved quasi-phasematching in the surface-emitting geometry using the so-called quantum well domains (Khurgin, 1988c) with periodic reversal of the asymmetry (Fig. 22). A substantial increase in the efficiency of the second harmonic generation was achieved, and $\chi^{(2)}_{xzx} = 20$ nm/V was measured. There have been other observations of the second-order nonlinearities in asymmetric QWs (Xie et al., 1991; Lue and Ma, 1993), but the bulk and surface effects in these works were not separated from the QW nonlinearities.

It was also stressed by Khurgin (1988b) that $\chi^{(2)}$ should be increased substantially by using the material with higher band offsets, in which the mutual cancellation effects of different subbands can be reduced substantially. It has been suggested that the II–VI materials be used for this purpose (Wong, 1994). Indeed, $\chi^{(2)}$ has been observed in ZnSe/ZnCdSe asymmetric QWs, and blue light has been generated (Pellegrini et al., 1995).

One should note that, although the magnitude of the interband $\chi^{(2)}$ is only tens of picometers per volt — i.e., three to four orders of magnitude less than

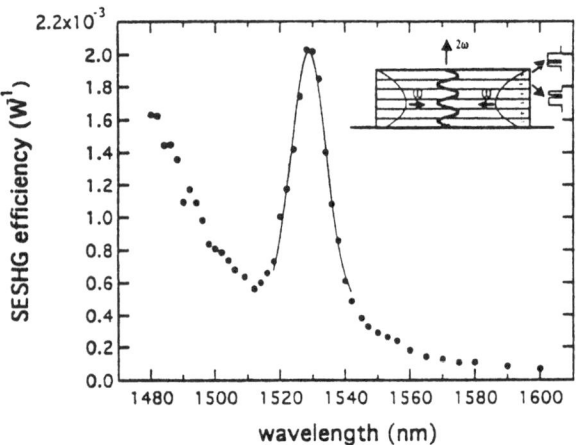

FIG. 22. Surface-emitting SHG efficiency as a function of pump wavelength in waveguide with ACQWs. (Data from Fiore et al., 1997b.)

the intersubband $\chi^{(2)}$—it is still quite comparable with the standard nonlinear materials—LiNbO$_3$, KTP, and others—and, unlike them, the semiconductor structures can be integrated monolithically with lasers and other electro-optic semiconductor devices. It is therefore not overly optimistic to expect that these asymmetric structures will develop into practical devices.

V. Optical Rectification and Terahertz Emission in Semiconductors

1. Brief Review of the Optical Rectification Terms in $\chi^{(2)}$

When we considered the semiclassical derivation of the second-order susceptibility, we had already noted that one can also use the expression for the off-resonant susceptibilities (Khurgin, 1988b) for the resonant case by inserting the decay (dephasing) rate of the off-diagonal elements of the density matrix $i\gamma$ into the resonant denominators. There are, however, terms that also include the diagonal elements of density matrix, corresponding to the population density ρ_{nn}. These matrix elements decay with a different rate, $T_n^{-1} < 2\gamma$, and, as we have already shown, ρ_{nn} can be very efficiently excited when both interacting waves are nearly resonant with the transition $|1\rangle \to |n\rangle$. This is the case of near-resonant far-IR difference generation (its ultimate limit being optical rectification), described by Eq. (60), which we reproduce here with all the nonresonant states disregarded and the resonant state designated as $|2\rangle$:

$$\chi^{(2)}_{i,j,k}(\Delta\omega;\omega_1,\omega_2) = \frac{N}{2\varepsilon_0\hbar^2} \frac{\omega_1 - \omega_2 + 2i\gamma}{\omega_1 - \omega_2 + iT_s^{-1}} \frac{\Delta\mu^i_{21}\mu^j_{21}\mu^k_{12}}{(\omega_{21} - \omega_1 - i\gamma)(\omega_{21} - \omega_2 + i\gamma)} \quad (144)$$

where $\Delta\omega = \omega_1 - \omega_2$, and where we use T_s for the relaxation rate of the level $|2\rangle$ to avoid confusion with the common designation of the dephasing time T_2. Now, T_s in this expression indicates that there are real carriers transferred from level 1 to level 2 that are responsible for the process of generating the low-frequency $\Delta\omega$. Therefore, this seems to be a "real process," limited by the scattering time T_s. On the other hand, optical rectification is often referred to as a "virtual process" that is instantaneous in time. In order to clarify this situation, it is best to consider the situation in the time domain.

Consider the scalar situation, with the z-axis being the direction of the asymmetry—i.e., one of the tetragonal bonds for the bulk semiconductor, or the growth axis for the QW. Consider the levels being broadened into the bands, with the reduced density of states $\rho_r(E_{21})$, where $E_{21} = \hbar\omega_{21}$ is the transition energy. Then, when both ω_2 and ω_1 are above the absorption edge,

$$\hbar\omega_1 - E_{\text{gap}} \gg \hbar\gamma \tag{145}$$

we obtain, from Eq. (144),

$$\chi^{(2)}(\Delta\omega) = \frac{1}{2\varepsilon_0} \frac{\Delta\omega + 2i\gamma}{\Delta\omega + iT_s^{-1}} \int_{Eg}^{\infty} \frac{\Delta\mu_{21}|\mu_{21}|^2}{(E_{21} - \hbar\omega_1 - i\hbar\gamma)(E_{21} - \hbar\omega_1 + \hbar\Delta\omega + i\hbar\gamma)} \rho_r(E_{21}) dE_{21} \tag{146}$$

Now, because of Eq. (145), we can take the lower limit in the integral in Eq. (146) as $-\infty$ and obtain, using contour integration,

$$\chi^{(2)}_{\text{res}}(\Delta\omega) = \frac{2\pi}{2\varepsilon_0} \frac{\Delta\omega + 2i\gamma}{\Delta\omega + iT_s^{-}} \frac{\Delta\mu_{21}|\mu_{21}|^2 \rho_r(\hbar\bar{\omega})}{\hbar(\Delta\omega + 2i\gamma)} = \frac{\Delta\mu_{21}}{\hbar(\Delta\omega + iT_s^{-1})} \mathfrak{F}[\chi^{(1)}(\bar{\omega})] \tag{147}$$

where $\bar{\omega} = \frac{1}{2}(\omega_2 + \omega_1)$. Equation (147) allows a very simple physical interpretation, if we introduce the absorption coefficient

$$\alpha(\bar{\omega}) = \frac{\bar{\omega}}{nc} \mathfrak{F}[\chi^{(1)}(\bar{\omega})] \tag{148}$$

and consider the oscillations of the intensity of the combined wave:

$$I(t) = \frac{n}{\eta_0} \mathcal{E}^2(t) = \frac{n}{2\eta_0} E_1^2 + \frac{n}{2\eta_0} E_2^2 + \frac{n}{\eta_0} E_1 E_2 \cos(\Delta\omega t) \\ = I_1 + I_2 + 2\sqrt{I_1 I_2} \cos(\Delta\omega t) \tag{149}$$

Then we can obtain the expression for the polarization at the beat frequency:

$$\mathcal{P}_{\text{res}}(\Delta\omega) = \varepsilon_0 \chi^{(2)}_{\text{res}}(\Delta\omega) E_1 E_2 = \frac{\Delta\mu_{21} T_s}{1 + i\Delta\omega T_s} \cdot \frac{\alpha(\bar{\omega}) 2\sqrt{I_1 I_2}}{\hbar\bar{\omega}} \tag{150}$$

One can see from Eq. (150) the physical process at play here: each time a photon gets observed, the electron is transferred from state $|1\rangle$ to state $|2\rangle$ and the dipole $\Delta\mu_{21}$ is generated. The dipole lifetime is T_s. If the incoming light is modulated with frequency $\Delta\omega$, the absorption-induced dipole will follow the oscillation, but at higher frequencies, above the 3-db frequency $\omega_{3db} = T_s^{-1}$, the dipole cannot follow. The frequency characteristic of $\chi_{res}^{(2)}(\Delta\omega)$ is shown in Fig. 23(a).

Now consider the off-resonant case, in which

$$E_{gap} - \hbar\omega_1 \gg \hbar\gamma \qquad (151)$$

Now we can replace all the denominators in Eq. (146) by some average detuning $\bar{\delta}$ and obtain

$$\chi_{eff}^{(2)}(\Delta\omega) = \frac{N}{2\varepsilon_0} \frac{\Delta\omega + 2i\gamma}{\Delta\omega + iT_s^{-1}} \frac{\Delta\mu_{21}|\mu_{21}|}{\hbar^2(\bar{\delta})^2} = \frac{\Delta\mu_{21}}{2\hbar\gamma} \frac{\Delta\omega + 2i\gamma}{\Delta\omega + iT_s^{-1}} \mathfrak{F}[\chi^{(1)}(\bar{\omega})] \qquad (152)$$

The off-resonant susceptibility has strikingly different character depending on the beat frequency $\Delta\omega$. For $\Delta\omega \ll \gamma$, Eq. (152) becomes identical to Eq. (147)—i.e., we are dealing simply with the off-resonant absorption, generating some real carriers in state $|2\rangle$. When, however, the frequency increases way above broadening, $\Delta\omega \gg \gamma$, we obtain

$$\chi_{off}^{(2)}(\Delta\omega) = \frac{\Delta\mu_{21}}{2\hbar\gamma} \mathfrak{F}[\chi^{(1)}(\bar{\omega})] \qquad (153)$$

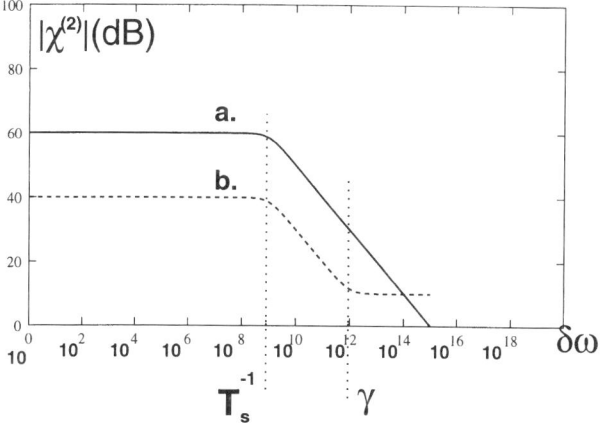

FIG. 23. Optical rectification and THz generation—frequency response of $\chi^{(2)}(\delta\omega; \omega_1, \omega_2)$: (a) resonant and (b) off-resonant.

The frequency response of the system becomes flat (Fig. 23b), with no relaxation time involved: this is a clear characteristic of an "instantaneous" or "virtual" process. It means that in the case of fast off-resonant pulse, no real carriers will be generated—with population following the pulse adiabatically, without scattering. More precise analysis, for arbitrary detuning, will show that the process is limited in time by the inverse of detuning—i.e., it is indeed nearly instant.

Let us go to the time domain and assume that we are dealing with the short pulse of frequency close to $\bar{\omega}$:

$$\mathscr{E}(t) = \frac{1}{2}\int_0^\infty E(\omega)e^{-i\omega t}\,d\omega + c.c. \tag{154}$$

Instantaneous power is

$$I(t) = \langle \mathscr{E}^2(t)\rangle \frac{n}{\eta_0} = \frac{n}{4\eta_0}\int_0^\infty\int_0^\infty E_1(\omega_1)E_2^*(\omega_2)e^{-i(\omega_1-\omega_2)t}\,d\omega_2\,d\omega_1 + c.c. \tag{155}$$

where averaging is done over the time longer than the optical oscillations period $\bar{\omega}^{-1}$. Now, each combination of frequencies ω_2, ω_1 will produce the response

$$\mathscr{P}(\omega_1-\omega_2;\omega_1,\omega_2) = \varepsilon_0\chi^{(2)}(\Delta\omega)E_1(\omega_1)E_2^*(\omega_2)e^{-i(\omega_1-\omega_2)t}$$

$$= \varepsilon_0 E_1(\omega_1)E_2^*(\omega_2)\int_0^\infty \chi^{(2)}(\tau)e^{-i(\omega_2-\omega_1)(t-\tau)}\,d\tau + c.c. \tag{156}$$

where

$$\chi^{(2)}(\tau) = \frac{1}{2\pi}\int \chi^{(2)}(\Delta\omega)e^{-i\Delta\omega\tau}\,d\Delta\omega \tag{157}$$

is the susceptibility in the time domain, or the response function. Integrating Eq. (156) over ω_2, ω_1, we obtain

$$\mathscr{P}(t) = \int_0^\infty\int_0^\infty \mathscr{P}(\omega_1-\omega_2;\omega_1,\omega_2)\,d\omega_2\,d\omega_1 = \frac{2}{nc}\int_0^\infty I(t-\tau)\chi^{(2)}(\tau)\,d\tau \tag{158}$$

Applying Eq. (157) to Eq. (147), one can obtain the pulse response for the resonant case

$$\chi^{(2)}_{\text{res}}(\tau) \sim e^{-\tau/T_s} \tag{159}$$

which is simply an exponential decay with time constant T_s, while for the off-resonant case, from Eq. (152),

$$\chi^{(2)}_{\text{off}}(\tau) \sim \delta(\tau) + (2\gamma - T_s^{-1})e^{-\tau/T_s} \tag{160}$$

meaning that there are both instantaneous and slow components in the polarization at difference frequencies. In the limiting case when there is no dephasing and the lifetime of the matrix element ρ_{21} is determined by its relaxation — i.e., when $\gamma = T_{21}^{-1} = T_s^{-1}/2$ — the slow component disappears, but of course, in the semiconductors at room temperature, dephasing is very large and $T_{21} \ll T_s$, so one can expect to see both fast and slow components.

We can now apply the results of this simple analysis to the actual practical cases of generating low-frequency radiation in bulk semiconductors and quantum wells.

2. Optical Rectification and Coherent Terahertz Emission in Bulk Semiconductors

The first experimental observations of optical rectification in bulk crystals were made in the 1960s (Bass et al., 1962; Ward, 1966), and in these experiments the "instantaneous" response was reported for the excitation well below the absorption edge of the dielectric crystals. Researchers turned their attention to the semiconductor crystals much later, in the late 1980s, when it became obvious that, in the absence of other sources of coherent emission in the THz range ($\lambda = 30 \div 300\,\mu m$), one could use optical frequency mixing, or the transient of the sub-picosecond optical rectification signal, to produce such THz radiation, which could be used in the study of ultrafast semiconductor circuits, in spectral analysis, and many other applications.

According to our derivation, and common wisdom, in order to achieve the THz pulsations one must use nonresonant, virtual processes, and it was there, below the bandgap of GaAs, where the coherent THz pulsations produced by the laser beam were first observed (Hu, Zhang, and Austin, 1991). However, when the researchers of the same group moved their excitation energy above the bandgap (Zhang et al., 1992), the THz signal not only failed to disappear but, on the contrary, was enhanced by two orders of magnitude (Fig. 24).

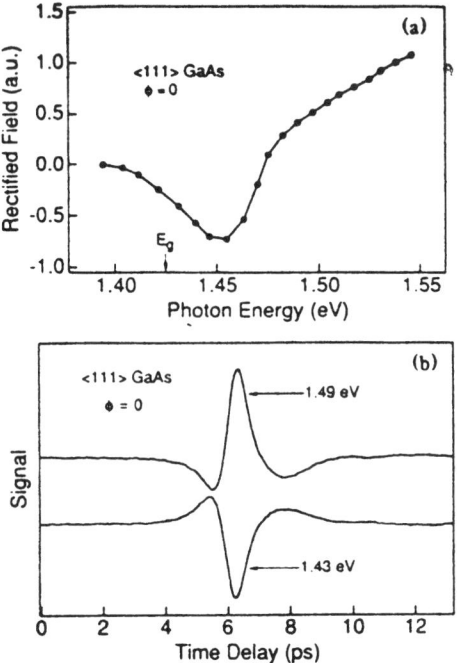

FIG. 24. Terahertz radiation from ⟨111⟩ GaAs: (a) peak value of the rectified signal versus laser energy and (b) temporal waveforms of the radiated field. (Data from Zhang et al., 1992.)

Similar effects were observed in InP samples of different orientations (Greene et al., 1991; Chuang et al., 1992). The original explanation of the above-the-gap emission by the current surge in the depletion field near the semiconductor surface (Zhang et al., 1990; Zhang and Austin, 1992; Planken et al., 1992a) could not explain the strong dependence of the intensity of the emitted THz radiation on the polarization of the incoming optical pulse relative to the crystal axes, and it was suggested by both Chuang et al. (1992) and Zhang et al. (1992) that bulk optical rectification plays an important role in generation of the THz signals. Moreover, careful studies by Saeta, Greene, and Chuang (1993), who had studied the dependence of the THz signal on crystal orientation and polarization of the laser pulse, revealed that 65 to 95% of the THz signal follows the angular dependence consistent with the bulk $\chi^{(2)}$ and not with the asymmetry introduced by the depletion field.

But it follows from the just-derived expression for the above-the-gap $\chi^{(2)}$ (Eq. 159) that $\chi^{(2)}_{\text{res}}$, associated with the excitation of the real carriers, should have the same lifetime T_s as the carriers themselves—i.e., the recombination time τ_r. This time, in undoped semiconductors, is measured in hundreds of picoseconds, and therefore one cannot expect a truly fast response to have been measured in Zhang *et al.* (1992) and Greene *et al.* (1991).

Khurgin (1994) has shown that, when applied to the case of bulk III–V semiconductors, this reasoning is too simplistic because in deriving Eq. (147) we assumed that we had N ground states $|1, \mathbf{k}\rangle$, all with the same built-in permanent dipole moment(PDM) $\bar{\mu}_{11}$, and N excited states $|2, \mathbf{k}\rangle$, all with the same built-in PDM $\mathbf{\mu}_{22}$. In this model, the only way the photoexcited dipole $\Delta\mu\rho_{22}$ can relax is by means of the relaxation from the excited to the ground state—i.e., recombination. In the III–V semiconductors, this is not true, because the states there belong to four distinct bonding orbitals (Fig. 25a), where we have chosen the system of coordinates associated with $[1\bar{1}0]$, $[110]$, and $[001]$ axes. Each of the four bonding orbitals $|b_i\rangle$ has a dipole moment

$$\mathbf{\mu}_{b,i} = -e\langle b_i|\mathbf{r}|b_i\rangle = \mu_b \hat{\mathbf{b}}_i \approx \frac{1}{2}\alpha_p l_b \hat{\mathbf{b}}_i \tag{161}$$

where $\hat{\mathbf{b}}_i$ is a unit vector in the direction of the *i*th bond. Each of the four

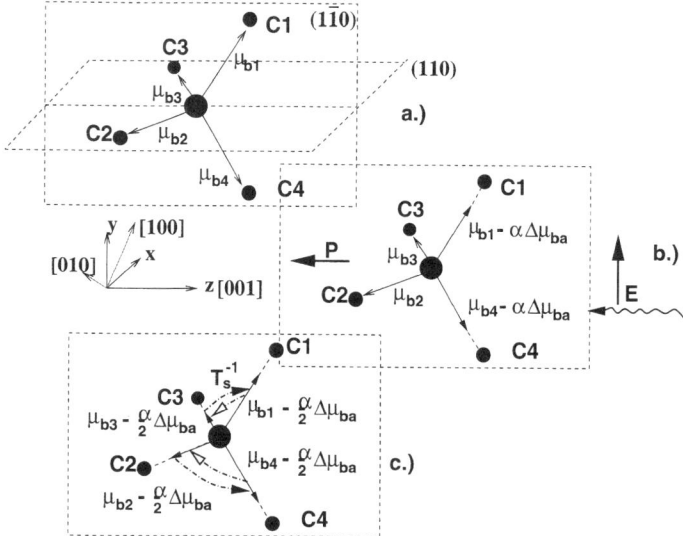

FIG. 25. Optical rectification in bulk zinc-blende semiconductors—PDMs of four bonds: (a) before arrival of short optical pulse $E(t)$, (b) during the pulse, and (c) after the pulse.

antibonding orbitals $|a_i\rangle$ has a dipole moment

$$\boldsymbol{\mu}_{a,i} = -e\langle a_i|\mathbf{r}|a_i\rangle = \mu_a \hat{\mathbf{b}}_i \approx -\frac{1}{2}\alpha_p l_b \hat{\mathbf{b}}_i \qquad (162)$$

The value of μ_b was calculated from the experimental data for most of the III–V and II–VI semiconductors (Hass and Henvis, 1962) and was found to vary from 0.8 atomic units in GaSb ($\alpha_p = 0.45$) to 0.91 atomic units in GaAs ($\alpha_p = 0.48$) to 1.5 atomic units in a more ionic InP ($\alpha_p = 0.58$). Note that in the absence of photoexcitation the bonding dipoles cancel each other—i.e., there is no permanent polarization in the material:

$$\sum_{i=1}^{4} \boldsymbol{\mu}_{b,i} = \mu_b \sum_{i=1}^{4} \hat{\mathbf{b}}_i = 0 \qquad (163)$$

Thus, when the electron is absorbed within a bond, it is transferred from the bonding orbital (valence band) to the antibonding band (conduction orbital), whereas in real space it simply corresponds to the movement of the center of the electron charge from the anion toward the cation. Thus, the process of absorption of a photon is accompanied by the generation of a dipole:

$$\Delta\boldsymbol{\mu}_{ba,i} = \boldsymbol{\mu}_{a,i} - \boldsymbol{\mu}_{b,i} \qquad (164)$$

Whether the photogenerated dipoles add up to create DC polarization depends on the polarization of the incoming light. If the light is polarized along one of the principal axes of the crystal, the probabilities of the photon being absorbed by each bond are identical, and four photogenerated dipoles will cancel each other. If, on the other hand, as shown in Fig. 25(b), the incoming short pulse of light is polarized along the [110] direction—i.e., in the ($1\bar{1}0$) plane—only two bonds that lie in that plane—i.e., b_1 and b_4—can absorb the light. As a result, once the electrons are transferred to the antibonding orbitals, the dipole moments of bonds b_1 and b_4 are reduced by some amount $\alpha\Delta\mu_{ba}$, where α is proportional to the pulse energy—i.e., $E(\bar{\omega})E^*(\bar{\omega})\Delta t_p$. The dipoles of four bonds do not compensate each other anymore, and there is a DC dipole directed along the $z = [001]$ axis.

Because all four antibonding orbitals belong to the conduction band, and all four valence orbitals belong to the valence band, the *intraband* relaxation processes will redistribute the photogenerated electrons and holes equally among all four orbitals (Fig. 25c), equalizing their dipole moments once again, and bringing the DC dipole back to zero. Therefore, the lifetime of the permanent polarization T_s in all the equations derived above is not the band-to-band recombination time, but the time that it takes to equalize the electron and hole populations between four bonding and four antibonding orbitals within each band, and is much shorter than the recombination time, being typically on the order of a few hundred femtoseconds.

This simple theory explains why, even when the semiconductor is excited above the bandgap, the time response can be of sub-picosecond duration. In fact, one can estimate the second-order susceptibility as

$$\chi^{(2)}_{\text{res}}(\Delta\omega) = \frac{\Delta\mu_{ba} T_s}{\sqrt{3}\,\hbar(1 + i\Delta\omega T_s)} \frac{n\alpha(\bar{\omega})\lambda}{2\pi} \quad (165)$$

where $\alpha(\bar{\omega})$ is the absorption coefficient. But these considerations fail to take into account the fact that in the vicinity of the fundamental bandgap, the states in the conduction and valence bands are not the bond orbital states, but their wavevector-dependent mixtures. Khurgin (1994) used the theory of Kane (1957) to evaluate the wavevector-dependent PDMs of the states in valence $\boldsymbol{\mu}_{vv}(\mathbf{k})$ bands and conduction $\boldsymbol{\mu}_{cc}(\mathbf{k})$ bands. First of all, since the lowest Γ_6^c conduction band is described by the antibonding combination of the S-type functions of anion and cation, it includes all four antibonding orbitals, and thus its permanent dipole is zero:

$$\boldsymbol{\mu}_{cc}(\mathbf{k}) = 0 \quad (166)$$

This is indeed a very important result, showing that the low-frequency, optically excited polarization can be attributed entirely to the photoexcited holes and not to the electrons. Therefore, both the magnitude and speed of $\chi^{(2)}_{\text{res}}$ are determined by scattering processes in the valence band. The PDM in the valence band depends only on the direction of the wavevector, and has opposite signs for light and heavy holes:

$$\boldsymbol{\mu}_{hh}(\mathbf{k}) = -\boldsymbol{\mu}_{lh}(\mathbf{k}) = \frac{\mu_b}{2\sqrt{3}}(-\sin 2\theta \cdot \cos\phi \hat{\mathbf{x}} + \sin 2\theta \cdot \sin\phi \hat{\mathbf{y}} + \sin^2\theta \cdot \sin 2\phi \hat{\mathbf{z}})$$

$$(167)$$

where ϕ and θ are the azimuthal and polar angles in the coordinate system associated with principal crystal axes $\hat{\mathbf{x}} = [100]$, $\hat{\mathbf{y}} = [001]$, and $\hat{\mathbf{z}} = [001]$. The magnitude of the dipole moment is shown in Fig. 26, and one can clearly see the manifestation of the tetragonal symmetry of the zinc-blende lattice — the magnitude of the PDM reaches its maximum along four $\langle 111 \rangle$

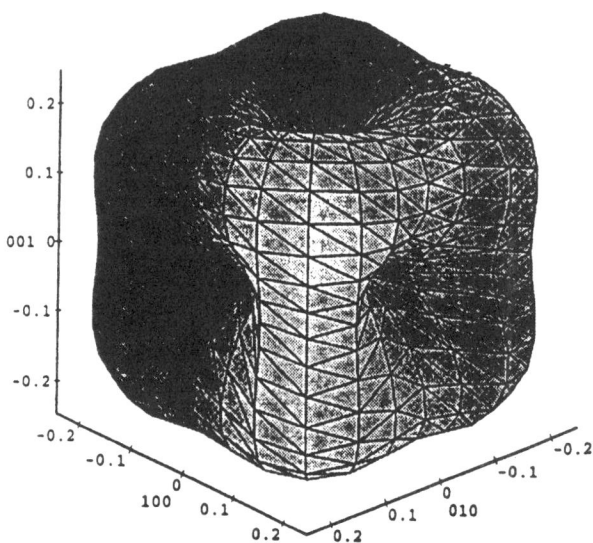

FIG. 26. Magnitude of the PDM of the hole state as a function of the direction of the wavevector **k**. (Data from Khurgin, 1994.)

directions. Also, one can see from Eq. (167) that the sign of the PDM projections changes when the wavevector is rotated 90 degrees. Therefore, it is this type of 90-degree scattering within each valence band as well as heavy-to-light hole scattering that are most responsible for the speed of the PDM decay, and thus for the speed of the THz generation process. According to Singh (1993) and Ridley (1982), these processes occur on a sub-picosecond scale, rendering the THz by the above-the-bandgap excitation feasible. Performing integration over **k**, one obtains the value of $\chi^{(2)}_{\text{res}}$:

$$\chi^{(2)}_{\text{res}}(\Delta\omega) = \frac{1}{10} \frac{\mu_b T_s}{\sqrt{3}\hbar(1 + i\Delta\omega T_s)} \frac{n\alpha(\bar{\omega})\lambda}{2\pi} \quad (168)$$

which is an order of magnitude less than Eq. (165) predicts by the simple bond orbital picture, pointing to the importance of the mixing of bond orbitals near the Γ point. Figure 27 shows the dispersion of the optical rectification coefficient:

$$r_{14} = \frac{1}{2}\chi^{(2)}_{xyz}(0) \quad (169)$$

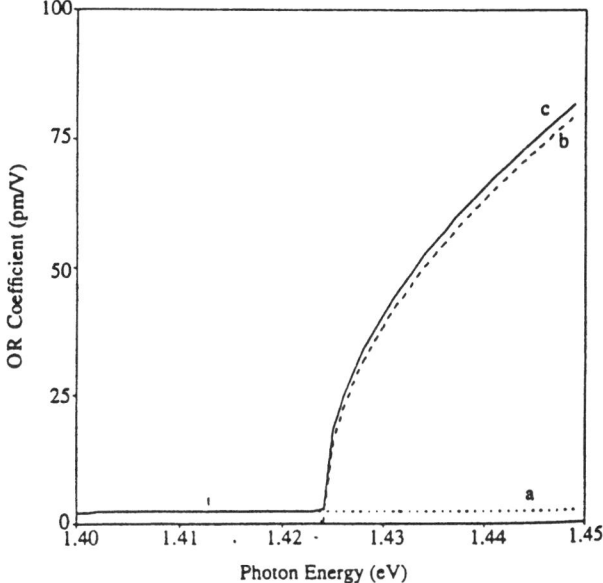

FIG. 27. Dispersion of the optical rectification coefficient $\frac{1}{2}\chi^{(2)}(0;\omega,\omega)$ near the absorption edge of GaAs: (a) "fast" component, (b) "slow" component, and (c) combined. (Data from Khurgin, 1994.)

It includes the "slow" or "real" component (Eq. 168) and an additional "virtual" (below-the-gap) component of much smaller magnitude. One can see that as the exciting radiation frequency exceeds the absorption edge, $\chi^{(2)}$ increases many fold, but it must be kept in mind that the figure of merit — i.e., the ratio of $\chi^{(2)}$ to absorption — obviously stays constant.

One can also mention that the treatment of the PDMs of bond orbitals as the sources of $\chi^{(2)}$ has recently been successfully expanded to the process that is its reverse — i.e., the linear electro-optic effect (Khurgin and Voisin, 1997).

3. OPTICAL RECTIFICATION AND TERAHERTZ EMISSION IN SEMICONDUCTOR QUANTUM WELLS

Because researchers have been so successful in generating terahertz responses in the bulk semiconductors, it would be only natural to attempt to take advantage of the large dipole moments associated with transition between two envelope states in QWs to increase the THz power. The situation with QWs is, however, quite different from the bulk situation, since in both conduction bands and valence bands (Fig. 28a) there is only one envelope state. Therefore, if one excites a real electron-hole pair, a very large PDM (measuring tens of angstroms) associated with this pair

$$\Delta\mu_{vc} = \mu_{cc,1} - \mu_{vv,1} = -e[\langle f_{c,1}^2(z)\rangle - \langle f_{v,1}^2(z)\rangle] \quad (170)$$

will appear, but with no other envelope states, the only way for it to relax would be through recombination—i.e., on a time scale of hundreds of picoseconds. Thus, the resonant $\chi^{(2)}$ (Eq. 147) would be very slow in such a two-level system, and the only means of obtaining a THz signal in this two-level scheme would be below-the-gap excitation—i.e., the virtual process (Eq. 153). The first proposals for this scheme were made in 1987 (Yamanishi; Chemla, Miller, and Schmitt-Rink), and it was given a detailed treatment by Khurgin (1988b). This effect was first demonstrated in 1990 (Sakata *et al.*). Unfortunately, because the absorption edge in QWs is not sharp, the virtual effect always gets obscured by the very strong real process associated with the resonant absorption. Moreover, the long-lived space charge tends to screen the asymmetry and reduce the value of $\chi^{(2)}$.

When one considers the intersubband transitions (Fig. 28b), the photo-induced intersubband dipole $\Delta\mu_{21}$ has a much shorter lifetime than the interband one, and one can achieve fast response, on the order of a few hundred femtoseconds. The first experimental demonstrations of optical

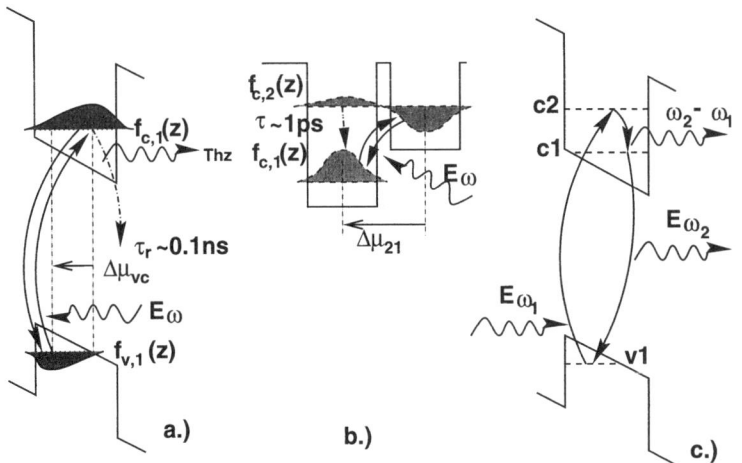

FIG. 28. Optical rectification and difference frequency generation in asymmetric QWs: (a) two-level interband, (b) two-level intersubband, and (c) three-level interband.

rectification in intersubband transitions (Rosencher *et al.*, 1989; Bois *et al.*, 1990) indeed confirmed large optical rectification coefficients of tens of nanometer per volt. It was shown in Unterrainer *et al.* (1995) that one can obtain optical rectification of far-IR pulses in the doped QWs, with transition energies as low as 10 meV, meaning that one can build photovoltaic detectors of far-IR light using optical rectification. Intersubband photovoltaic detection of IR light was demonstrated in Levine, Gunapala, and Hong (1991), Schneider (1993), and Schneider *et al.* (1997). Optical rectification is also possible in biased parabolic QWs (Guo and Gu, 1993). One of the advantages of intersubband transition in ACQWs, such as the one shown in Fig. 28b is that by varying the barrier thickness one can optimize the strength of the response versus the speed — i.e., one can design application-specific structures.

One can still get a very strong THz response near the bandgap if one considers the situation in which two different subbands are simultaneously excited in one of the bands (Fig. 28c). For the first time, this has been observed in ACQW structures with two closely spaced levels in the conduction band (Leo *et al.*, 1991; Roskos *et al.*, 1992). Later on, Planken *et al.* (1992b) observed THz oscillations with simultaneous excitation of light and heavy hole excitons in the valence band. Coherent control of the charge oscillations in the coupled QW structures by using phase-locked series of optical pulses was achieved by Plankten *et al.* (1993) and Brener *et al.* (1993), and additional enhancement of oscillations was achieved by Weiner (1994), who used the technique of femtosecond pulse shaping. Good theoretical models of this effect, including excitonic and many-body models, have evolved (Luo *et al.*, 1993, 1994; Chansungsan, Tsang, and Chuang, 1994). Most authors describe the observed phenomena as coherent charge oscillations between two subbands, but it is easy to show that this effect can be considered a transient of $\chi^{(2)}_{\text{res}}$. Indeed, using the same density-matrix technique discussed elsewhere in this chapter, we can obtain the expression

$$\chi^{(2)}_{14,\text{res}}(\Delta\omega) = \frac{2\pi}{2\varepsilon_0 \hbar} \frac{\mu_{v1c1}\mu_{c1c2}\mu_{c2v1}}{\omega_{c2 \to c1} - \Delta\omega - i\gamma} \rho_r$$

$$\approx \frac{\mu_{c1c2}}{\hbar(\omega_{c2 \to c1} - \Delta\omega - i\gamma)} \sqrt{\mathfrak{F}[\chi^{(1)}(\omega_1)]\mathfrak{F}[\chi^{(1)}(\omega_2)]}$$

(171)

where ρ_r is a reduced two-dimensional density of states and $\omega_{c2 \to c1} = (E_{c2} - E_{c1})/\hbar$ is the intersubband transition frequency. So, the terahertz response is indeed proportional to the simultaneous absorption in both subbands, and in the time domain it is

$$\chi^{(2)}_{14,\text{res}}(\tau) \sim e^{-i\omega_{c2 \to c1}\tau - \gamma\tau} \tag{172}$$

—i.e., the oscillations occur at the intersubband transition frequency and decay with the coherence time γ^{-1}. This is indeed what was observed by Planken et al. (1992b), as shown in Fig. 29.

It is also worthwhile to note that, in addition to creating $\chi^{(2)}$ associated with envelope transitions, quantum confinement also changes the bulk $\chi^{(2)}$ by changing the character of bond-orbital mixing near the Γ point (Khurgin, 1996). This effect occurs even in symmetric QWs and does not require an external field. It can drastically enhance the bulk rectification coefficient for multiple QWs grown in the $\langle 110 \rangle$ or $\langle 111 \rangle$ direction, but at the same time it makes the effect slower.

We conclude this section by observing that although, historically, the main thrust in development of second-order nonlinear optics has been the up-conversion of radiation, for which the semiconductors studied here are not very competitive as a result of high absorption at short wavelengths, the more recent trend has been toward down-conversion—parametric generation, frequency shifting for WDM networks, and generation of THz radiation—for which the semiconductors, both bulk and structured, have several advantages associated with huge dipole elements involved in the transitions. The case of down-conversion into the far-IR THz domain considered here is sufficiently distinct from the more general nonresonant nonlinear processes to justify this special treatment, and in our opinion this is an area where nonlinear optics of semiconductors will find a number of practical applications.

VI. Conclusions

In this chapter, we have given a brief description of the nature of second-order nonlinear processes in semiconductors. We have developed two alternative methods of estimating the magnitude of $\chi^{(2)}$—using $\mathbf{A} \cdot \mathbf{P}$ and $\mathbf{E} \cdot \mathbf{r}$ gauges—and have obtained order-of-magnitude agreement with the experimental results, which we have also reviewed.

Basically, the fact that $\chi^{(2)}$ in zinc-blende semiconductors is typically an order of magnitude higher than $\chi^{(2)}$ in more conventional materials can be traced to a very simple observation: *each bond in a zinc-blende semiconductor is asymmetric and contributes to $\chi^{(2)}$ with a minimum amount of cancellation.* In other materials, such as $LiNbO_3$ or $KTiPO_4$, for instance, only one bond in each unit cell contributes to $\chi^{(2)}$. Unfortunately, the same tetragonal arrangement of bonds that is so beneficial for $\chi^{(2)}$ renders semiconductors

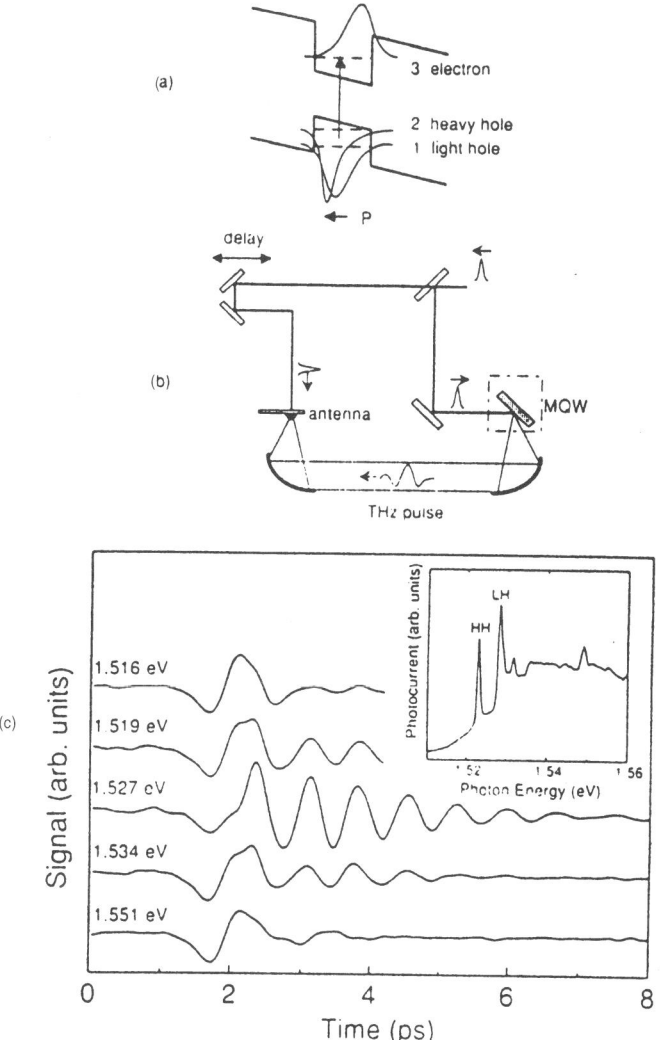

FIG. 29. Terahertz pulse generation using coherent excitation of light and heavy excitons in QWs: (a) wavefunction envelopes in a biased QW, (b) experimental setup for THz generation measurements, and (c) measured THz waveforms for several excitation wavelengths. Inset in (c): the measured photocurrent spectrum. (Data from Planken *et al.*, 1992b.)

isotropic and makes phasematching extremely difficult, requiring the advanced growth and fabrication techniques described in this chapter.

It is this rapid advance in growth techniques that has led to the development of an entirely new field — nonlinearity in asymmetric semiconductor quantum wells. Significant progress has been achieved in this area in the course of the last few years, and giant second-order nonlinearities have been measured.

Special attention is given in this chapter to optical rectification and THz pulse generation in bulk semiconductors and quantum wells. A simple theoretical treatment is backed up by several of the most recent experimental results.

There are a few important topics that could have been included in this chapter but have been left out, because we have tried to concentrate on a few areas that we deem to be of the most interest to readers. For example, this chapter is entirely devoted to III–V and II–VI semiconductors, because they are the ones with which readers are most familiar, and also because they are already in wide use in optoelectronics in such devices as lasers, photodetectors, modulators, and numerous passive optical elements. As a result, such important infrared nonlinear materials as III–VI materials, such as GaSe and chalcopyrites ($ZnGeP_2$, for example), have been left out. These crystals have $\chi^{(2)}$ values on the order of 100 pm/V or more and exhibit anisotropy sufficient for birefringence phasematching (Vodopyanov, 1993; Bhar, Das, and Vodopyanov, 1995). Another area not touched on here is generation of THz pulses in photoconductive antennas (Verghese, McIntosh, and Brown, 1997), which can, of course, be thought of as a $\chi^{(2)}$ process. Finally, since the surface of any material, even a cubic one such as Si, has no inversion symmetry, one can obtain all the $\chi^{(2)}$ phenomena on the semiconductor surface and use it as an excellent spectroscopic probe of the surface. A good review of the latest results is given in McGilp (1996). One can also use $\chi^{(2)}$ in semiconductors to obtain the third-order nonlinear effects by cascading (Kelaidis, Hutchings, and Arnold, 1994; Lee, Khurgin, and Ding, 1996). But to include all the new fields would be to divert the reader's attention from the main idea of the chapter — the idea that semiconductors and their heterostructures have good potential for serving as second-order nonlinear components in integrated optoelectronic devices and systems.

List of Abbreviations and Acronyms

A	anion
ACQW	asymmetric coupled quantum well
C	cation
FOM	figure of merit

IR	infrared
ISB	intersubband
OMCVD	organometallic chemical vapor deposition
OPO	optical parametric oscillator
PDM	permanent dipole moment
QPM	quasi-phasematching
QW	quantum well
SHG	second harmonic generation
SL	superlattice

References

Almogy, G., Segev, M., and Yariv, A. (1994). *Opt. Lett.* **19**, 1192.
Almogy, G. and Yariv, A. (1995). *Nonlinear Opt. Phys. Mat.* **4**, 401.
Angell, M. J., Emerson, R. M., Hoyt, J. L., Gibbons, J. F., Eyres, L. A., Bortz, M. L., and Fejer, M. M. (1994). *Appl. Phys. Lett.* **64**, 3107.
Armstrong, J. A., Bloembergen, N., Ducuing, J., and Pershan, P. S. (1962). *Phys. Rev.* (1962). **127**, 1918.
Aspnes, D. (1972). *Phys. Rev. B* **6**, 4648.
Atanasov, R., Bassani, F., and Agranovich, V. M. (1994). *Phys. Rev. B* **50**, 7809.
Bass, M. A., Franken, P. A., Ward, J. F., and Weinreich, G. (1962). *Phys. Rev. Lett.* **9**, 446.
Bastard, G. (1988). *Wave Mechanics Applied to Semiconductor Heterostructures*. Academic Press, New York.
Bell, M. I. (1971). *Phys. Rev. B* **6**, 516.
Bhar, G. C., Das, S., and Vodopyanov, K. L. (1995). *Appl. Phys. B* **61**, 187.
Bois, P., Rosencher, E., Nagle, J., Marinet, E., Boucaud, P., and Julien, F. H. (1990). *Superlattices and Microstructures* **8** 369.
Boucaud, P., Julien, F. H., Yang, D. D., Lourtioz, J.-M., Rosencher, E., Bois, P., and Nagle, J. (1990) *Appl. Phys. Lett.* **57**, 215.
Boyd, R. W. (1992). *Nonlinear Optics*. Academic Press, San Diego.
Brener, I., Planken, P. C. M., Nuss, M. C., Pfeifer, L., Leaird, D. E., and Weiner, A. M. (1993). *Appl. Phys. Lett.* **63**, 2213.
Butcher, P. N. and Cotter, D. (1990). *The Elements of Nonlinear Optics*. Cambridge University Press, Cambridge.
Butcher, P. N. and McLean, T. P. (1963). *Proc. Phys. Soc.* **81**, 219.
Butcher, P. N. and McLean, T. P. (1964). *Proc. Phys. Soc.* **83**, 579.
Cada, M., Svilans, M., Janz, S., Bierman, R., Normandin, R., and Glinski, J. (1992). *Appl. Phys. Lett.* **61**, 2090.
Capasso, F., Sirtori, C., Cho, A. Y. (1994). *IEEE J. Quantum Electron.* **30**, 1313.
Chang, R. K., Ducuing, J., and Bloembergen, N. (1965). *Phys. Rev. Lett.* **15**, 415.
Chansungsan, C., Tsang, L., and Chuang, S. L. (1994). *J. Opt. Soc. Am. B* **11**, 2508.
Chemla, D. S., Miller, D. A. B., and Schmitt-Rink, S. (1987). *Phys. Rev. Lett.* **59**, 1018.
Choy, M. M. and Byer, R. L. (1976). *Phys. Rev. B* **14**, 1693.
Chuang, S. L. (1995). *Physics of Optoelectronic Devices*. J. Wiley & Sons, New York.
Chuang, S. L., Schmitt-Rink, S., Greene, B. I., Saeta, P. N., and Levi, A. F. J. (1992). *Phys. Rev. Lett.* **68**, 102.
Chui, H. C., Martinet, E. L., Fejer, M. M., and Harris, J. S. (1994a). *Appl. Phys. Lett.* **64**, 736.
Chui, H. C., Martinet, E. L., Woods, G. L., Fejer, M. M., and Harris, J. S. (1994b). *Appl. Phys. Lett.* **64**, 3365.

Chui, H. C., Woods, G. L., Fejer, M. M., Martinet, E. L., and Harris, J. S. (1995). *Appl. Phys. Lett.* **66**, 3365.
Cohen, M. L. and Chelikowsky, J. R. (1988). *Electronic Structure and Optical Properties of Semiconductors*. Springer-Verlag, Berlin.
Fejer, M. M., Yoo, S. J. B., Byer, R. L., Harwit, A., and Harris, J. S. (1989). *Phys. Rev. Lett.* **62**, 1041.
Fiore, A., Beauliew, Y., Janz, S., McCaffrey, J. P., Wasilewsky, Z. R., and Xu, D. X. (1997a). *Appl. Phys. Lett.* **70**, 2655.
Fiore, A., Berger, V., Rosencher, E., Bravetti, P., Laurent, N., and Nagle, J. (1997b). *Appl. Phys. Lett.* **71**, 3622.
Fiore, A., Rosencher, E., Berger, V., and Nagle, J. (1995a). *Appl. Phys. Lett.* **67**, 3765.
Fiore, A., Rosencher, E., Vinter, B., Weill, D., and Berger, V. (1995b). *Phys. Rev. B* **51**, 13192.
Flytzanis, C. and Ducuing, J. (1969). *Phys. Rev.* **178**, 1218.
Fong, C. Y. and Shen, Y. R. (1975). *Phys. Rev. B* **12**, 2325.
Ghahramani, E., Moss, D. J., and Sipe, J. E. (1991). *Phys. Rev. B* **43**, 9700.
Gordon, L., Woods, G. L., Eckardt, R. C., Route, R. R., Feigelson, R. S., Fejer, M. M., and Byer, R. (1993). *Electr. Lett.* **29**, 1942.
Greene, B. I., Federici, J. F., Dykaar, D. R., Levi, A. F. J., and Pfeifer, L. N. (1991). *Opt. Lett.* **16**, 48.
Guo, K.-X. and Gu, S.-W. (1993). *Phys. Rev. B* **47**, 16322.
Gurnick, M. K. and DeTemple, T. A. (1983). *IEEE J. Quantum Electron.* **19**, 791.
Harshman, P. J. and Wang, S. (1992). *Appl. Phys. Lett.* **60**, 1992.
Hass, M. and Henvis, B. W. (1962). *J. Phys. Chem. Solids* **23**, 1099.
Hase, Y., Kumata, K., Kano, S. S., Ohashi, M., Kondo, T., Ito, R., and Shiraki, Y. (1992). *Appl. Phys. Lett.* **61**, 145.
Hu, B. B., Zhang, X.-C., and Austin, D. H. (1991). *Phys. Rev. Lett.* **67**, 2709.
Ikonic, Z., Milanovic, V., and Tjapkin, D. (1989). *IEEE J. Quantum Electron.* **25**, 54.
Ito, H. F. and Inaba, H. (1978). *Opt. Lett.* **2**, 139.
Janz, S., Chatenoud, F., and Normandin, R. (1994). *Opt. Lett.* **19**, 622.
Jha, S. S. and Bloembergen, N. (1968). *Phys. Rev.* **171**, 891.
Johnston, W. D. Jr. and Kaminow, I. P. (1969). *Phys. Rev.* **188**, 1209.
Kane, E. O. (1957). *J. Phys. Chem. Solids* **1**, 249.
Kane, E. O. (1966). In *Semiconductors and Semimetals*, Vol. 1 (R. K. Willardson and A. C. Beer, eds.). Academic Press, New York.
Kelaidis, C., Hutchings, D. C., and Arnold, J. M. (1994). *IEEE J. Quantum Electron.* **30**, 2998.
Khurgin, J. B. (1987). *Appl. Phys. Lett.* **51**, 2100.
Khurgin, J. B. (1988a). *J. Appl. Phys.* **64**, 5026.
Khurgin, J. B. (1988b). *Opt. Lett.* **13**, 603.
Khurgin, J. B. (1988c). *Phys. Rev. B* **38**, 4062.
Khurgin, J. B. (1989). *J. Opt. Soc. Am. B* **6**, 1673.
Khurgin, J. B. (1993). *Appl. Phys. Lett.* **62**, 1390.
Khurgin, J. B. (1994). *J. Opt. Soc. Am. B* **11**, 2492.
Khurgin, J. B. (1996). *J. Opt. Soc. Am. B* **13**, 2129.
Khurgin, J. B. and Voisin, P. (1997). *Semicond. Sci. Technol.* **12**, 1378.
Lee, S.-J., Khurgin, J. B., and Ding, Y. J. (1996). *Opt. J. Quantum Electron.* **28**, 1617.
Leman, G. and Friedel, G. (1962). *J. Appl. Phys.* **33S**, 281.
Leo, K., Shah, J. S., Gobel, E. O., Damen, T., Schmitt-Rink, S., Schaefer, W., and Kohler, K. (1991). *Phys. Rev. Lett.* **66**, 201.
Levine, B. F. (1973). *Phys. Rev. B* **7**, 2600.
Levine, Z. H. and Allan, D. C. (1991a). *Phys. Rev. Lett.* **66**, 41.
Levine, B. F. and Bethea, C. G. (1972). *Appl. Phys. Lett.* **20**, 272.

Levine, B. F., Gunapala, S. D., and Hong, M. (1991b). *Appl. Phys. Lett.* **59**, 1969.
Li, S. and Khurgin, J. B. (1993). *Appl. Phys. Lett.* **62**, 1727.
Liu, H. C., Jianmeng, L., Costard, E., Rosencher, E., and Nagle, J. (1995a). *IEEE J. Quantum Electron.* **31**, 1659.
Liu, H. C., Jianmeng, L., Costard, E., Rosencher, E., and Nagle, J. (1995b). *Solid-State Electronics* **40**, 567–570.
Lue, J. T. and Ma, K. W. (1993). *IEEE Photon. Technol. Lett.* **5**, 37.
Luo, M. S. C., Chuang, S. L., Planken, P. C. M., Brener, I., and Nuss, M. C. (1993). *Phys. Rev. B* **48**, 11043.
Luo, M. S. C., Chuang, S. L., Planken, P. C. M., Brener, I., Roskos, H. G., and Nuss, M. C. (1994). *IEE J. Quantum Electron.* **30**, 1478.
McGilp, J. F. (1996). *J. Phys. D* **29**, 1812.
Meyer, J. R., Hoffman, C. A., Bartoli, F. J., Ram-Mohan, L. R. (1995). *Appl. Phys. Lett.* **67**, 608.
Miller, R. C. (1964). *Appl. Phys. Lett.* **5**, 17.
Miller, D. A. B., Weiner, J. S., and Chemla, D. S. (1986). *IEEE J. Quantum Electron.* **22**, 1816.
Moss, D. J., Sipe, J. E., and Van Driel, H. M. (1987). *Phys. Rev. B* **36**, 9708.
Normandin, R. S., Letourneau, S., Chatenoud, F., and Williams, R. L. (1991). *IEEE J. Quantum Electron.* **27**, 1520.
Normandin, S. R., Williams, R. L., and Chatenoud, F. (1992). *Electr. Lett.* **26**, 2088.
Pellegrini, V. A., Parlangeli, A., Borger, M., Atanasov, R. D., Beltram, F., Vanzetti, L., and Franciosi, A. (1995). *Phys. Rev. B* **52**, 527.
Phillips, J. C. (1968). *Phys. Rev.* **166**, 832.
Planken, P. C. M., Brener, I., Nuss, M. C., Luo, M. S. C., and Chuang, S. L. (1993). *Phys. Rev. B* **48**, 4903.
Planken, P. C. M., Nuss, M. C., Brener, I., Goosen, K. W., Luo, M. S. C., Chuang, S. L., and Pfeifer, L. (1992a). *Phys. Rev. Lett.* **69**, 3800.
Planken, P. C. M., Nuss, M. C., Knox, W. H., Miller, D. A. B., and Goosen, K. W. (1992b). *Appl. Phys. Lett.* **61**, 2009.
Qu, X. H., Bottomley, D. J., Ruda, H., and Springthorpe, A. J. (1994a). *Phys. Rev. B* **50**, 5703.
Qu, X. H. and Ruda, H. (1993). *Appl. Phys. Lett.* **62**, 1296.
Qu, X. H., Ruda, H., Janz, S., and Springthorpe, A. J. (1994b). *Appl. Phys. Lett.* **65**, 3176.
Ridley, B. K. (1982). *Quantum Processes in Semiconductors*. Clarendon, Oxford.
Robinson, F. N. H. (1967). *Bell System Tech. J.* **46**, 913.
Rosencher, E. and Bois, P. (1989a). *Electr. Lett.* **25**, 1063.
Rosencher, E. and Bois, P. (1991). *Phys. Rev. B* **44**, 11315.
Rosencher, P., Bois, P., Nagle, J., Costard, E., and Delaitre, S. (1989b). *Appl. Phys. Lett.* **55**, 1597.
Rosencher, E., Fiore, A., Vinter, B., Berger, V., Bois, P., and Nagle, J. (1996). *Science* **271**, 168.
Roskos, H. G., Nuss, M. C., Shah, J. S., Leo, K., and Miller, D. A. B. (1992). *Phys. Rev. Lett.* **68**, 2216.
Saeta, P. N., Greene, B. I., and Chuang, S. L. (1993). *Appl. Phys. Lett.* **63**, 3482.
Sakata, Y., Yamanishi, M., Yamaoka, Y., Kodama, S., Kan, Y., and Suemune, I. (1990). *Jap. J. Appl. Phys.*, Part 2 **29**, L1973.
Schneider, H. (1993). *J. Appl. Phys.* **74**, 4789.
Schneider, H., Schonbein, C., Walther, M., Schwarz, K., Fleissner, J., and Koidl, P. (1997). *Appl. Phys. Lett.* **71**, 246.
Seto, M., Helm, M., Moussa, Z., Boucaud, Z., Julien, P., Lourtioz, F. H., Nutzel, J.-M., and Abstreiter, G. (1995). *Appl. Phys. Lett.* **65**, 2969.
Shantharma, L. G., Adams, A. R., Ahmad, C. N., and Nicholas, R. J. (1984). *J. Phys. C. Solid State Phys.* **17**, 4429.
Shen, Y. R. (1984). *The Principles of Nonlinear Optics*. J. Wiley & Sons, New York.

Shimizu, A., Kuwata-Gonokami, M., and Sakaki, H. (1992). *Appl. Phys. Lett.* **61**, 399.
Shoji, A., Kondo, I. T., Kitamoto, A., Shirani, M., and Ito, R. (1997). *J. Opt. Soc. Am. B* **14**, 2268.
Singh, J. (1993). *Physics of Semiconductors and Their Heterostructures*. McGraw-Hill, New York, Ch. 10–12 and references therein.
Sipe, J. E. and Ghahramani, E. (1993). *Phys. Rev. B* **48**, 11705.
Sirtori, C., Capasso, F., Faist, J., Pfeifer, L. N., and West, K. W. (1994). *Appl. Phys. Lett.* **65**, 445.
Sirtori, C., Capasso, F., Sivko, D. L., Chu, S. N. G., and Cho, A. Y. (1991). *Appl. Phys. Lett.* **59**, 2302.
Somekh, S. and Yariv, A. (1972). *Appl. Phys. Lett.* **21**, 140.
Soref, R. A. and Moos, H. W. (1964). *J. Appl. Phys.* **35**, 2152.
Street, M. W., Whitbread, N. D., Hutchings, D. C., Arnold, J. M., Marsh, J. H., Aitchison, J. S., Kennedy, G. T., and Sibbett, W. (1997). *Opt. Lett.* **22**, 1600.
Tang, C. L. and Bey, P. P. (1973). *IEEE J. Quantum Electron.* **9**, 9.
Thompson, D. E., McMullen, J. D., and Anderson, D. B. (1976). *Appl. Phys. Lett.* **29**, 113.
Tomic, S., Milanovic, V., and Ikonic, Z. (1997). *Phys. Rev. B* **56**, 1033.
Tsang, L., Ahn, D., and Chuang, S. L. (1988). *Appl. Phys. Lett.* **52**, 697.
Tsang, L. and Chuang, S. L. (1990a). *Phys. Rev. B* **42**, 5229.
Tsang, L. and Chuang, S. L. (1995). *Appl. Phys. Lett.* **60**, 2543.
Tsang, L., Chuang, S. L., and Lee, S. M. (1990b). *Phys. Rev. B* **41**, 5942.
Unterrainer, K., Heyman, J. N., Craig, K., Galdrikian, B., and Sherwin, M. S. (1995). *Superlattices and Microstructures* **17**, 159.
Vakhshoori, D., Fischer, R. J., Hong, M., Sivco, D. L., Zydik, G. J., and Cho, A. Y. (1991). *Appl. Phys. Lett.* **59**, 896.
Vakhshoori, D. and Wang, S. (1988a). *Appl. Phys. Lett.* **53**, 347.
Vakhshoori, D., Wu, M. C., and Wang, S. (1988b). *Appl. Phys. Lett.* **52**, 424.
van der Ziel, J. P. (1975). *Appl. Phys. Lett.* **26**, 60.
van der Ziel, J. P. and Ilegems, M. (1976). *Appl. Phys. Lett.* **29**, 200.
van der Ziel, J. P., Miller, R. C., Logan, R. A., Nordland, W. A., and Mikulyak, R. M. (1974). *Appl. Phys. Lett.* **25**, 238.
van Vechten, J. A. (1969). *Phys. Rev.* **182**, 891.
Verghese, S., McIntosh, K. A., and Brown, E. R. (1997). *Appl. Phys. Lett.* **71**, 2743.
Vodopyanov, K. L. (1993). *J. Opt. Soc. Am. B* **10**, 1723.
Ward, J. F. (1965). *Rev. Mod. Phys.* **37**, 1.
Ward, J. F. (1966). *Phys. Rev.* **143**, 569.
Weiner, A. M. (1994). *J. Opt. Soc. Am. B* **11**, 2480.
Weisbuch, C. and Vinter, B. (1994). *Quantum Semiconductor Structures: Fundamentals and Applications*. Academic Press, Boston.
Wong, K. B. and Jaros, M. (1994). *Phys. Rev. B* **50**, 17238.
Wynne, J. J. and Bloembergen, N. (1969). *Phys. Rev.* **188**, 1211.
Xie, Y. L., Chen, Z. H., Cui, D. F., Pan, S. W., Deng, D. Q., and Zhou, Y. L. (1991). *Phys. Rev. B* **43**, 12477.
Yamanishi, M. (1987). *Phys. Rev. Lett.* **59**, 1014.
Yariv, A. (1989). *Quantum Electonics*. J. Wiley & Sons, New York.
Yoo, S. J., Bhat, R., Caneau, C., and Koza, M. A. (1995). *Appl. Phys. Lett.* **66**, 3410.
Yuh, P. F. and Wang, K. L. (1989). *Appl. Phys.* **65**, 4377.
Zhang, X.-C. and Austin, D. H. (1992a). *J. Appl. Phys.* **71**.
Zhang, X.-C., Hu, B. B., Darrow, J. T., and Austin, D. H. (1990). *Appl. Phys. Lett.* **56**, 1011.
Zhang, X.-C., Jin, Y., Yang, K., and Scholwater, L. J. (1992b). *Phys. Rev. Lett.* **69**, 2303.
Zyss, J. (1994). *Molecular Nonlinear Optics: Materials, Devices and Physics*. Academic Press, Boston.

CHAPTER 2

Nonlinearities in Active Media

Katherine L. Hall

ADVANCED NETWORKS GROUP
MASSACHUSETTS INSTITUTE OF TECHNOLOGY
LINCOLN LABORATORY
LEXINGTON, MASSACHUSETTS

Erik R. Thoen and Erich P. Ippen

DEPARTMENT OF ELECTRICAL ENGINEERING AND COMPUTER SCIENCE
MASSACHUSETTS INSTITUTE OF TECHNOLOGY
CAMBRIDGE, MASSACHUSETTS

	LIST OF ACRONYMS	83
I.	INTRODUCTION	84
II.	ACTIVE SEMICONDUCTOR MEDIA	85
	1. *Linear Gain in Active Materials*	86
	2. *Carrier Confinement and Waveguiding*	91
	3. *Nonlinear Gain in Active Media*	94
III.	MEASUREMENT TECHNIQUES	100
	1. *Orthogonally-Polarized Collinear Pump-Probe Technique*	101
	2. *A Heterodyne Pump-Probe Technique*	106
IV.	NONLINEAR INDEX OF REFRACTION IN ACTIVE MEDIA	112
V.	DATA ANALYSIS AND INTERPRETATION	119
VI.	SHAPING AND SATURATION OF SHORT PULSES IN ACTIVE WAVEGUIDES	134
VII.	FOUR-WAVE MIXING	141
VIII.	APPLICATIONS	145
	1. *Wavelength Converters*	146
	2. *All-Optical Switching*	148
IX.	SUMMARY	153
	REFERENCES	155

List of Acronyms

AOM   acousto-optic modulator
APM   additive pulse modelocked

FCA	free carrier absorption
FWM	four-wave mixing
KK	Kramers-Kronig
MQW	multiple quantum well
PSD	polarization sensitive delay
PZT	piezoelectric transducer
SHB	spectral hole burning
SLMQW	strained layer multiple quantum well
SOA	semiconductor optical amplifier
TE	transverse electric
TM	transverse magnetic
TPA	two-photon absorption
UNI	ultrafast nonlinear interferometer
WDM	wavelength division multiplexed
XGM	cross-gain modulation
XPM	cross-phase modulation

I. Introduction

Semiconductor lasers and optical amplifiers are of interest for possible applications to broadband optical communications and switching systems. Recently, InGaAsP-based devices, operating in the optical communications bands of 1300 and 1500 nm, have received particular attention. However, there has also been intensive investigation of GaAs-based devices emitting near 800 nm and strained InGaAs-based devices emitting near 980 nm. While much is known about the linear and small signal characteristics of these devices, the physical mechanisms responsible for their nonlinear and dynamic properties are still being studied. The nonlinear responses of these active semiconductor waveguides are of interest for several reasons. They influence the modulation response and mode stability of diode lasers, they limit the speed of and produce crosstalk between multiplexed signals in optical amplifiers, and they may be useful for the design of nonlinear waveguide modulators and switches. Dynamic changes in carrier density, limited by the speed of the carrier lifetime, are responsible for many of the observed effects. However, experiments have shown that nonequilibrium changes in the carrier energy distributions are also important.

In this chapter, we review the experimental and theoretical investigations into the origin of the nonlinear responses of active semiconductor waveguides. This chapter is organized in the following manner. In Section II, we provide the theoretical background necessary to understand linear gain in active media. In addition, because most experimental work has been performed on semiconductor lasers and optical amplifiers, we discuss the electrical and optical structures of these devices. Finally we describe the physical mechanisms responsible for nonlinear gain in active media. In Section III we review the most common experimental techniques used to measure the nonlinear responses of nonlinear materials and waveguides, and we discuss traditional as well as newly developed pump-probe techniques. In Section IV we discuss the mechanisms responsible for refractive index nonlinearities and discuss the coupling between the index and the gain in active semiconductor media. In Section V we review the models for the nonlinear responses of semiconductor media, discuss data interpretation, and highlight the behavior and time recovery of various optically induced gain and refractive index changes. In Section VI we review the impact of the nonlinear response on the saturation and propagation of short optical pulses in these waveguides. Section VII is devoted to a discussion of frequency domain measurements (four-wave mixing) and interpretation of the nonlinear responses of active media. In Section VIII, we highlight some current application areas in which the nonlinear responses of active waveguides are being exploited. Finally, in Section IX, we summarize the chapter.

II. Active Semiconductor Media

Active semiconductor materials are distinguished from passive semiconductor materials in that they are pumped, electrically or optically, to create nonequilibrium carrier populations and can provide gain to photons with energies exceeding the material bandgap energy, E_{gap}. In an oversimplified but physically useful picture, the semiconductor material can be considered as an ensemble of two-level systems, with upper levels in the conduction band and lower levels in the valence band. The photon energy $\hbar\omega$ of the incident light defines the upper-lower energy level difference over which a transition can be made. Here, $\hbar = h/2\pi$, where h is Planck's constant and ω is the optical carrier frequency. The probabilities of absorption and emission between lower and upper states are governed by the relative carrier densities. If the density of carriers in some lower-energy state E_2 exceeds the density of carriers in some upper state $E_1 = E_2 + \hbar\omega$, the probability that a photon of energy $\hbar\omega$ will stimulate a transition from the lower state to the upper state, is greater than the probability that it will stimulate a transition from the upper state to the lower state. Stated another way, absorption is more probable than stimulated emission. In this case, the intensity of an optical beam decreases as it passes through the material because, on average, photons have been absorbed. On the other hand, if the density of carriers in the upper state exceeds the density of carriers in the lower state (population inversion), stimulated emission is more probable than stimulated absorption and the intensity of the optical beam (on average) increases as it passes through the material. This increase in the intensity of the optical beam is called gain. Changing the carrier densities of the upper and lower states by electrical or optical carrier injection, changes the photon energies at which gain and absorption occur. In reality, the gain in semiconductor materials is not adequately described by this simple picture. There are continua of energy levels called bands. The energy distribution of carriers within these bands is determined by interactions between carriers and by the statistical properties of the carriers. Transition probabilities are affected by wave-vector selection and polarization selection rules. Nevertheless, the simple picture provides a reasonable basis for understanding what follows.

1. LINEAR GAIN IN ACTIVE MATERIALS

Linear gain in active materials has been treated extensively in the literature (e.g., Casey and Panish, 1978; Agrawal and Dutta, 1986; Thompson, 1985; Dutta, 1980; Dutta, 1981), but we review it here for the sake of completeness and because we will refer to these simple calculations of linear gain to explain the nonlinear gain. In a semiconductor material, with three-dimensional freedom of motion for electrons, the number of electrons per unit volume, N, allowed in a given region of k-space is given by

$$N = \frac{k^3}{3\pi^2} \tag{1}$$

where k is the magnitude of the electron wave-vector (Ashcroft and Mermin, 1976). To determine the density of electronic levels, as a function of energy in the conduction band, $\rho_c(E)$, we use the dispersion relation for the electrons to relate k to E. For a parabolic band with $E = \hbar^2 k^2 / 2m^*$,

$$\rho_c(E) = \frac{dN}{dE} = \frac{1}{2\pi^2} \left(\frac{2m_e^*}{\hbar^2}\right)^{3/2} E^{1/2} \tag{2}$$

where m_e^* is the effective mass of the electrons. A similar relationship is attained for the density of states of holes in the valence band, $\rho_v(E)$. The difference in parabolicity between valence and conduction bands is simply reflected in the effective mass. Figure 1 shows a schematic of a conduction band and a valence band in a direct band semiconductor. In general, the band-structure of a semiconductor is quite complicated and depends on many parameters, including the injected carrier density, the concentration

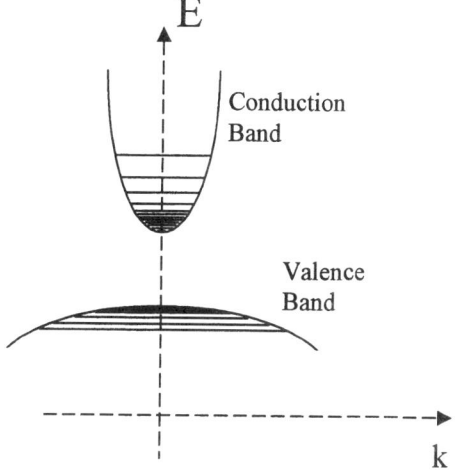

FIG. 1. Schematic of the parabolic band model for a direct band semiconductor material.

of impurities, and the temperature. This parabolic band model is the simplest case, but is adequate for describing the physical origins of linear and nonlinear gain in these active materials.

The probability that a given energy state in the conduction band is actually occupied by an electron is given by the Fermi function

$$f_c(E) = \frac{1}{[1 + e^{(E-E_{Fc})/k_B T_c}]}, \tag{3}$$

where E_{Fc} is the quasi-Fermi energy for electrons in the conduction band, k_B is Boltzmann's constant, and T_c is the temperature of the conduction-band electron distribution. Note that $f(E_{Fc}) = \frac{1}{2}$ and that the probability that a particular energy state is occupied depends on the temperature of the carrier distribution, T_c. The density of carriers is given by

$$N = \int f_c(E)\rho_c(E)dE \tag{4}$$

for electrons in the conduction band, and by

$$P = \int [1 - f_v(E)]\rho_v(E)dE \tag{5}$$

for holes in the valence band.

Remember that in a simple two-level energy model, the transition probability is related to these relative carrier densities. If we consider the transition between two energy levels, E_1 in the conduction band and E_2 in the valence band, the rate of stimulated absorption is given by

$$r_{21} = B' f_v(E_2)[1 - f_c(E_1)] S(E_{21}) \tag{6}$$

where $S(E_{21})$ is the photon flux density at the energy difference $E_1 - E_2$, and B' is the transition probability (Casey and Panish, 1978) given by

$$B' = \frac{\pi q^2 h}{m_c^* m_v^* \varepsilon_0 n_g^2 E_{21}} |M|^2 \tag{7}$$

where $|M|^2$ is the momentum matrix element, ε_0 is the permittivity of free space, n_g is the group velocity index of refraction, m_v^* is the effective mass of holes in the valence band, and q is the electronic charge. Physically, stimulated absorption takes an electron from an occupied state in the valence band, $f_v(E_2)$, to an empty state in the conduction band, $[1 - f_c(E_1)]$. The rate of stimulated emission is given by

$$r_{12} = B f_c(E_1)[1 - f_v(E_2)] S(E_{21}) \tag{8}$$

Because B is the same as in the equation for stimulated emission and absorption ($B = B'$), the net direction of stimulated transition is determined by the relative carrier concentrations in the two energy levels, E_1 and E_2.

An incident photon flux density, $S(E_{21})$, experiences gain when the rate of stimulated emission exceeds the rate of absorption, $r_{12} > r_{21}$. This condition is known as the Bernard-Duraffourg condition (Bernard and Duraffourg, 1961). The gain can be related to the net stimulated emission-rate by

$$g(E_{21}) = \frac{r_{12} - r_{21}}{S(E_{21}) v_g} = \frac{B[f_c(E_1) - f_v(E_2)]}{c/n_g} \tag{9}$$

where $v_g = c/n_g$ is the group velocity. Note that for gain to exist, $f_c(E_1) > f_v(E_2)$. This condition is not related to the total concentration of electrons or holes; rather, for gain to exist, the occupational probability of E_1 must exceed the occupational probability of E_2. For the quasi-neutrality condition to hold, the total concentrations of electrons and holes must be equal.

Because there are many combinations of energy levels in the conduction and valence bands that satisfy the condition $E = E_1 - E_2$, to find the total gain for photons with energy E, we must sum Eq. (9) over all energy levels separated by E. Then, the gain in the semiconductor material is given by

$$g(E) = \frac{\pi q^2 \hbar}{\varepsilon_0 m_e^* m_v^* c n_g E} \int_{-\infty}^{\infty} |M|^2 \rho_v(E_2) \rho_c(E_2 + E) [f_c(E_2 + E) - f_v(E_2)] dE_2 \quad (10)$$

where E_2 is the energy state in the valence band and $E_1 = E_2 + E$ is the energy of the state in the conduction band. Figure 2 shows the gain curve calculated for bulk InGaAsP material, at room temperature, with a carrier density of $N = 1.8 \times 10^{18}$ cm^{-3}. For these calculations, we have assumed that the k-selection rule is not obeyed (Casey and Panish, 1978). This assumption, which is reasonably valid near the band edge and for high carrier densities, is sufficient for our calculations, because we are trying to demonstrate the dependence of the gain on such parameters as carrier density and temperature and are not looking for exact quantitative values for the gain coefficients. Other parameters used to calculate the gain curve are given in Table I.

Single photons with energies less than the bandgap or equal to the electron-hole quasi-Fermi energy separation cannot stimulate transitions. However, photons with energies less than the quasi-Fermi energy separation for electrons and holes (but greater than the bandgap energy), on average, stimulate emission and experience gain. Photons with energies larger than the quasi-Fermi energy separation stimulate absorption. For the gain curve

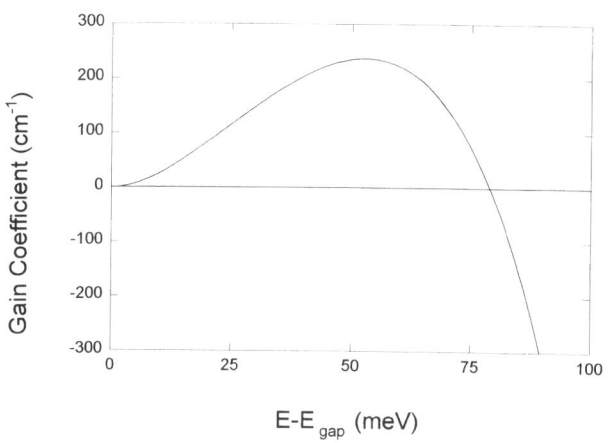

FIG. 2. Typical gain curve for an InGaAsP bulk laser.

TABLE I

TABLE OF PARAMETERS USED IN THE CALCULATION OF THE LINEAR GAIN FOR A BULK INGAASP WAVEGUIDE

Name	Symbol	Value
Index of refraction	n_g	3.6
Bandgap energy	E_g	750 meV
Split-off band energy	Δ	312 meV
Free electron mass	m_0	5.7×10^{-13} meV·sec^2/cm^2
Conduction band mass	m_e^*	0.041 m_0
Heavy hole mass	m_h^*	0.620 m_0
Light hole mass	m_l^*	0.050 m_0
Valence band mass	m_v^*	0.630 m_0

shown in Fig. 2, the gain region extends from E_{gap} to $E_{gap} + 80$ meV. At the transparency point, $E_{trans} = E_{gap} + 80$ meV, stimulated emission and stimulated absorption are equally probable and the gain coefficient is zero. This transparency energy is equal to the separation between the quasi-Fermi energies for electrons and holes, $E_{trans} = E_{Fc} - E_{Fv}$. The absorption (or linear loss regime) includes all energies greater than E_{trans}. In this regime the gain coefficient is negative.

Note that the gain curve in Fig. 2 depends on many parameters. For example, it depends on the carrier (free electrons and holes) density. The steady state injected carrier density is used to determine the quasi-Fermi energy for the carriers, which determines the occupational probability that is used to calculate the gain. The carrier density can be changed by current injection or by optical excitation. Figure 3 shows a series of gain curves calculated for different carrier densities ranging from 1.6×10^{18} cm^{-3} to 2.4×10^{18} cm^{-3}. The carrier temperatures were fixed at room temperature, $T = 300$ K.

It is worth noting that the gain coefficients calculated in this chapter are determined using a simple model in which many complicating effects are ignored. For example, we have not considered the effect of carrier-concentration-dependent bandgap energy reduction (Agrawal and Dutta, 1986). The bandgap renormalization should not affect the value of the peak gain, but it may cause a slight shift in the band edge, and possibly in the peak-gain wavelength for the device. Also, we have assumed that the active region of the device is almost intrinsic and have not considered the effects of band-tail states. Finally, we have assumed that the valence and conduction bands are purely parabolic, an approximation that breaks down for high-lying energy states. Despite these many approximations, we find that

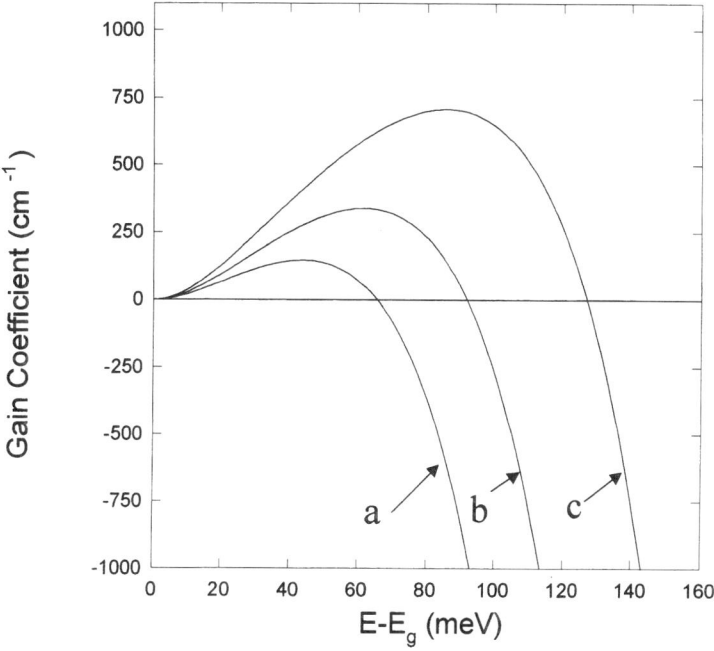

FIG. 3. Calculated gain curves as a function of excess energy above the bandgap. The carrier densities are (a) $1.6 \times 10^{18}\,\mathrm{cm}^{-3}$, (b) $2.0 \times 10^{18}\,\mathrm{cm}^{-3}$, and (c) $2.4 \times 10^{18}\,\mathrm{cm}^{-3}$.

our calculations are in good agreement with more complete calculations reported in the literature (Dutta, 1980). In addition, we can use the simplified models to give insight to the physical mechanisms responsible for the gain and refractive index changes induced in active media in the presence of intense optical fields.

2. CARRIER CONFINEMENT AND WAVEGUIDING

It was recognized very soon after the advent of the laser that these active semiconductors could be used as gain elements for compact lasers (Basov *et al.*, 1961). Soon after this prediction, groups at a number of research facilities demonstrated lasing in GaAs diodes (Hall *et al.*, 1962; Nathan *et al.*, 1962; Quist *et al.*, 1962; Holonyak and Bevacqua, 1962). In these demonstrations, an electric current was passed through a simple GaAs *p-n* junction chip with lateral cleaved facets. Passing current through the device caused minority carriers to be "injected" across the *p-n* junction, where they could recombine with majority carriers through spontaneous or stimulated emission. The large index discontinuity between the semiconductor material and the surrounding air allowed cleaved facets to serve as reflectors that formed a self-contained resonator cavity in the plane of the *p-n* junction. When the "injected" carrier density was higher than some threshold, laser oscillation occurred and light was emitted from the device. In that simple device, the "active region," or the width of the region that provided gain, was simply defined by the diffusion length of the carriers in the material. The threshold currents for those original semiconductor diode lasers were high, in part because there was no good way to confine the light in the cavity to the plane where the carrier inversion had been created and because the inversion (or active) region was large.

Several important changes have been made in the standard active semiconductor device since those days. A simplified picture of a semiconductor laser diode is shown in Fig. 4. The diode consists of a semiconductor waveguide layer, sandwiched in the middle of a *p-n* junction. Waveguiding is achieved by placing a high-refractive-index core or active-region material between two lower-refractive-index cladding materials. This design is fortuitous because there is (almost always) an inverse relationship between the bandgap and the refractive index of the semiconductor material used in these devices so that the low-index cladding layers that provide waveguiding also create potential barriers at the interface and help confine the injected carriers to the active region of the device. In this way, the optical field and the injected carriers are confined in the same spatial region of the device, the so-called active region. The optical field and injected carriers can be confined in the orthogonal transverse dimension by selectively applying a bias current to a narrow metal electrode along the top of the device, and/or by etching grooves in the material to form a ridge. In a typical single-

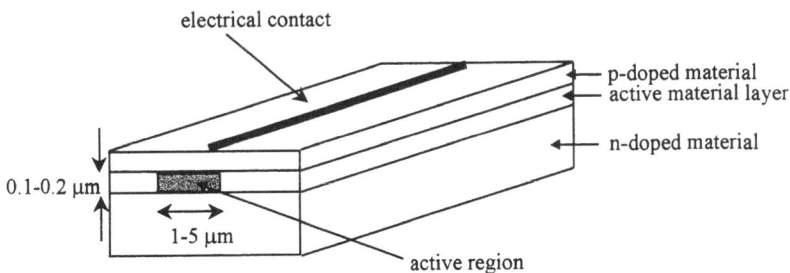

FIG. 4. Schematic diagram of a semiconductor laser diode.

transverse-mode laser diode, the active region is between 1 and 5µm wide and the active layer is between 0.1 and 0.2 µm thick. The end mirrors for these devices are formed by cleaving the facets of the chip normal to the waveguide.

The waveguiding properties and the carrier-confinement properties of these semiconductor structures can be optimized for different applications. For example, the material growth and device fabrication for low-threshold-current lasers differ significantly from those for high-gain devices. Also, the active region of a semiconductor diode laser can be much more complicated than a single material layer. Alternating layers of higher-bandgap and lower-bandgap materials can create so-called quantum well devices (see Chapter 1). In these structures, the carriers can be confined in the quantum wells and the optical field is confined by the separate confinement layer. Carriers that are injected into this structure must diffuse across the confinement layer and be captured by the quantum wells, where they can provide gain to an optical field. For these devices, the carriers are confined to a two-dimensional space within the active region, and the gain is determined by the two-dimensional density of states

$$\rho(E) = \frac{dN}{dE} = \frac{m^*}{\pi h} \tag{11}$$

(Ashcroft and Mermin, 1976), rather than by the three-dimensional density of states (Eq. 2) described above. For our general discussion of the physical mechanisms responsible for nonlinear gain, the dimensionality of the active region is not important. However, we will discuss some effects that are unique to quantum well diodes and discuss the advantages and limitations of two-dimensional and three-dimensional active regions.

In many cases, the nonlinearities in active media devices are investigated using semiconductor optical amplifiers rather than diode lasers. In an amplifier, the facet reflectivity is reduced, either through the use of an antireflection coating or by angling the optical waveguide with respect to the cleaved facets. Both methods spoil the cavity and inhibit lasing. Optical amplifiers are useful to study because the carrier density of the device can be changed with no associated cavity effects to complicate the interpretation of the data. Also, there are no standing wave effects in the cavity and no interference effects because of multiple trips of the optical probing beams through the active semiconductor cavity. Finally, very high carrier densities can be achieved, and band-filling effects are easily observed because the carrier density is not clamped by laser action.

3. Nonlinear Gain in Active Media

With an understanding of the physical mechanisms responsible for linear gain in these semiconductor diodes, and a simple picture of the device structure, we can consider the proposed sources for the nonlinear gain and absorption. Previously, we discussed the carrier dynamics resulting from interband effects, intraband effects, and diffusion effects (Hall et al., 1994). Interband effects are attributed to changes in carrier density caused by stimulated absorption, stimulated emission, and spontaneous emission. The simplest description of these carrier density dynamics is given phenomenologically by the rate equation

$$\frac{dN}{dt} = \frac{I}{qV} - \frac{N}{\tau_s} - g_d(N - N_t)S \tag{12}$$

where I is the injection (bias current), V is the volume of the active region, q is the electronic charge, τ_s is the upper state spontaneous emission lifetime, N_t is the carrier density that provides transparency for the incident photons, g_d is the differential gain (dg/dN), and S is the photon density. In words, the carrier density is increased by passing a current through the diode and generating electron-hole pairs. The carrier density is decreased by spontaneous emission, which means that in the absence of impinging photons, the electron-hole pairs recombine with a characteristic time constant τ_s. The carrier density is either increased or decreased by stimulated absorption or emission, depending on whether the total carrier density is lower or higher than the transparency carrier density N_t.

In the most general case, carrier recombination can be caused by nonradiative as well as radiative processes. The total spontaneous recombination rate is given by

$$R(N) = A_{nr}N + B(N)N^2 + CN^3 \tag{13}$$

where A_{nr} is the inverse of the nonradiative recombination time resulting from defects and surface recombination in the laser, $B(N)$, is the radiative recombination coefficient, and C is the coefficient of the higher-order, nonradiative Auger process. Most active semiconductor devices are fabricated from high-quality epitaxial materials, and there is no significant recombination caused by defects and surface effects. Then $A_{nr} = 0$, and the carrier lifetime is expressed as

$$\frac{1}{\tau_s} = B(N)N + CN^2 \tag{14}$$

Note that both the radiative and Auger recombination rates depend on carrier density. Experimentally, it has been shown that

$$B(N) = B_0 - B_1 N \tag{15}$$

where $B_0 \simeq 0.6 \times 10^{-10}\,\text{cm}^3/\text{s}$ and $B_1/B_0 \simeq 2 \times 10^{-19}\,\text{cm}^3$ for long-wavelength diode lasers at room temperature (Uji et al., 1983; Wintner and Ippen, 1984; Olshansky et al., 1984). There are many interactions of four-particle states that may contribute to the Auger recombination in these diodes. Experimentally, Auger recombination is distinguished by the $I^{-2/3}$ dependence of carrier lifetime on carrier density (Wiesenfeld et al., 1994b):

$$\tau_s = \frac{1}{3}\left(\frac{qV}{\sqrt{C}}\right)^{2/3} I^{-2/3} \tag{16}$$

For InGaAsP materials, C is on the order of $5 \times 10^{-29}\,\text{cm}^6/\text{s}$ (Su et al., 1982; Sermage et al., 1983; Henry et al., 1983; Mozer et al., 1983; Thompson, 1983; Uji et al., 1983; Wintner and Ippen, 1984). This nonradiative recombination can become especially important in optical amplifiers because the unclamped carrier densities can exceed $5 \times 10^{-18}\,\text{cm}^{-3}$. In long-wavelength semiconductor amplifiers, more than 50% of the spontaneous recombination effects can be attributed to Auger recombination (Wiesenfeld, 1996). Researchers have demonstrated upper state lifetimes of 20 ps (Manning et al., 1994b) at very high bias currents.

The upper state lifetime can be further reduced in the presence of an optical holding beam (Mikkelsen et al., 1993; Wiesenfeld et al., 1993; Manning et al., 1994b; Manning and Davies, 1994). The optical holding beam provides an additional recombination mechanism — namely, stimulated emission — and modifies the free carrier lifetime. This effect is important when the rate of stimulated emission exceeds the rate of spontaneous recombination. In this case, the effective carrier lifetime τ_e becomes

$$\frac{1}{\tau_e} = \frac{1}{\tau_s} + \frac{P}{E_{\text{sat}}} \tag{17}$$

where P is the power of the optical holding beam and E_{sat} is the saturation energy (Wiesenfeld, 1996), described in Section III of this chapter.

a. Diffusion Effects

As discussed above, an induced change in carrier density will recover to its equilibrium density by means of spontaneous emission, with τ_s in the range of 200 ps to 1 ns, depending on the injected carrier density (Agrawal and Dutta, 1986). However, some diode structures show a bias-independent recovery time (Eisenstein et al., 1989) that has been explained by spatial carrier gradients. In these structures, bias current creates carrier densities in regions outside the active region of the diode. When an optical signal impinges on this diode, it couples to carriers in the active region of the device but not to those in the carrier storage regions. A gradient of carriers between the storage regions and the active region is created, and carriers flow from the area of higher concentration to the area of lower concentration. The time it takes for carrier diffusion to replenish the carrier distribution in the active region depends on a number of parameters, such as the size of the active region and the mode confinement factor. For standard devices, the carrier density gradient extends approximately 1 μm into the cladding regions, yielding a diffusion time of approximately 100 ps. Bias-independent recovery times as short as 100 ps have been observed.

Besides such spatial hole burning effects associated with carrier storage regions or "pumped" cladding regions of the devices, there are spatial hole burning effects on a smaller spatial scale owing to the longitudinal standing waves set up in the laser resonator. The standing wave pattern causes a sinusoidal variation in the carrier density whereby the distance from the maximum carrier density to the minimum carrier density is $\lambda/4n$, or approximately 0.1 μm. Longitudinal spatial hole burning is not a factor in experiments on semiconductor amplifiers, where antireflection coatings on the facets spoil the cavity and eliminate standing wave effects.

Longer-range diffusion of carriers across a confinement layer in a multiple quantum well laser is another effect that has been identified as a contributor to nonlinear gain in active semiconductor devices (Eisenstein et al., 1991; Weiss et al., 1992; Wiesenfeld et al., 1994b). This so-called "well-barrier hole burning" (Rideout et al., 1991) arises when carriers in the optical confinement layer diffuse across this layer to "drop into" quantum wells that have been depleted of carriers by optical excitation. The time constant associated with this effect is structure-dependent. For typical separate confinement layer widths of 100 to 300 nm, diffusion (and capture) times of 2 to 10 ps have been measured. Carrier diffusion affects the modulation response of multiple quantum well (MQW) lasers (Nagarajan et al., 1992; Kan et al., 1992; Tessler and Eisenstein, 1993) and the chirp of directly modulated lasers (Nagarajan and Bowers, 1993).

b. *Carrier Scattering Effects*

Other contributors to the nonlinear gain (and absorption) of these active semiconductor materials are fundamental to the material itself, and are not so strongly dependent on device structure. One such nonlinearity is spectral

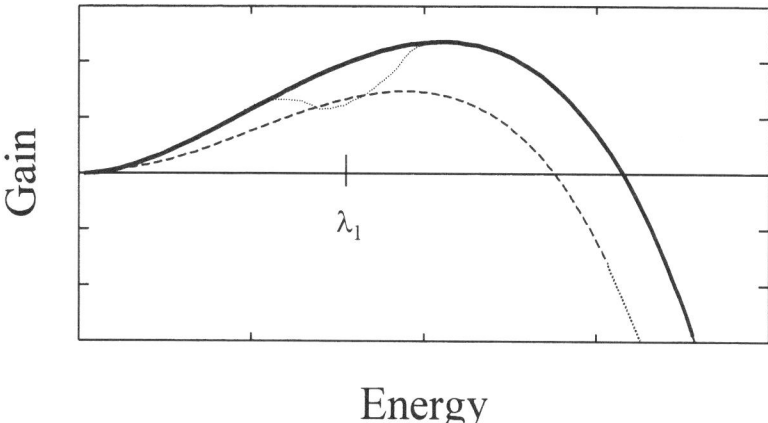

FIG. 5. Pictorial representation of the effect of spectral hole burning. In this picture, a spectral hole (dotted line) is burned in the equilibrium gain curve shown by the solid line. Carriers scatter to fill in the hole, but the gain curve recovers to an overall lower value (dashed line) because the carrier concentration has been reduced by stimulated emission.

hole burning (SHB). In an inhomogeneously broadened gain medium, light at a specific wavelength causes stimulated transitions only between specific energy levels, not across the entire gain or absorption spectrum. In a semiconductor material, this implies that stimulated transitions tend to distort—i.e., burn a hole in—the Fermi distribution. The width of the spectral hole depends on k-vector selectivity and the homogeneous dephasing rate, thought to be extremely fast (<10 fs) at high carrier densities (Knox et al., 1986; Knox et al., 1988). Figure 5 shows a pictorial representation of this behavior. In the gain region, an optical beam with center wavelength λ_1 will stimulate emission, but will tend to deplete the carrier density over only a limited range. In the absorption regime, too, an optical beam generates carriers over only a limited energy range, bleaching a hole in the absorption. At the transparency point, where there are no net stimulated transitions, there is no spectral hole burning. These localized nonuniformities or holes in the carrier distributions persist, while carrier-carrier scattering redistributes the total carrier density into a new Fermi distribution that contains greater or fewer total carriers depending on whether carriers were added or removed by the stimulated transitions.

c. *Carrier Heating Effects*

Another nonlinearity that is fundamental to the semiconductor gain medium results from nonequilibrium carrier heating (Stix et al., 1986;

Kessler and Ippen, 1987; Hall et al., 1990b). Even when the carriers have achieved a Fermi distribution, the temperature of this distribution may differ from that of the lattice. When electron and hole distributions are heated relative to the lattice, the gain of the active semiconductor medium is reduced. This effect, which arises from the temperature dependence of the Fermi function, is illustrated in Fig. 6. For these calculated gain spectra, the carrier density is held constant, but the electron and hole temperatures are varied. The solid line in Fig. 6 gives a calculated peak gain coefficient commonly quoted in the literature (Dutta, 1980), where the hole and electron distributions are described by a temperature $T_c = T_v = 300\,\text{K}$. The dashed and dotted lines show the effects of heating the carrier distributions to 320 and 340 K, respectively. Heating the carrier distribution compresses the gain and shifts the gain peak to longer wavelengths (lower energy). These calculations suggest that significant changes in the gain coefficient are possible for moderate changes in the electron and hole distribution temperatures. For example, heating the distributions 40 K above room temperature reduces the diode gain by a factor of 2.

The carrier distribution can be heated by several mechanisms. Free carrier absorption (FCA) can create highly energetic carriers in both the conduction and valence bands (Hall et al., 1992b; Hultgren et al., 1992). Through carrier-carrier scattering, these "hot" carriers share their energy with the rest of the distribution. As a result, the carrier distribution heats up. This distribution cools back to the lattice temperature on a slower timescale by

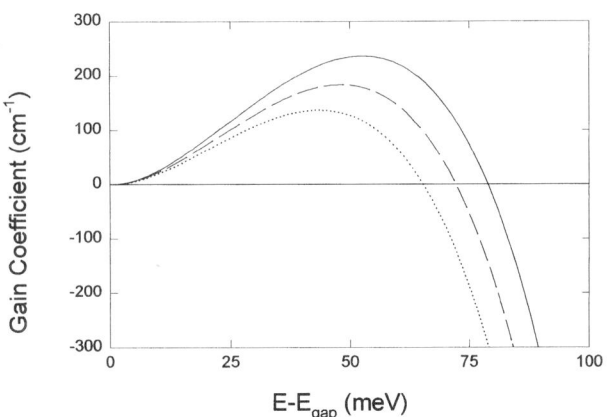

FIG. 6. Calculated gain curves versus excess energy above the bandgap for carrier temperatures of 300 K (solid line), 320 K (dashed line), and 340 K (dotted line).

coupling to the lattice vibrations (phonon emission). Notice in Fig. 6 that heating the carrier distribution reduces the gain across the entire range of energies indicated. This effect occurs because the electron quasi-Fermi level lies at higher energy, further in the absorption regime. The carrier distributions may also be heated by higher-order effects, such as two-photon absorption (TPA) (Mørk et al., 1994), but this effect is much smaller than other heating effects in active semiconductor materials biased in the gain regime.

Besides FCA, the carrier distribution also can be heated by energy changes resulting from stimulated transitions. Consider the gain curve shown in Fig. 7(a). The dashed line shows the average energy $\langle E \rangle$ of the carrier distribution, and the dotted line shows the electron-hole quasi-Fermi energy difference (E_{trans}). Carriers occupying energy levels below the average

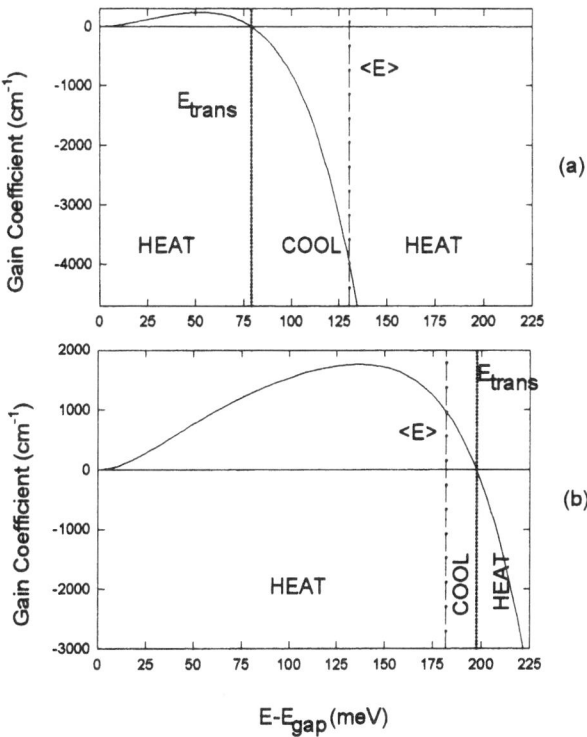

FIG. 7. Calculated gain curve versus excess photon energy along with the position of the average energy (dashed line) of the distributions, $\langle E \rangle$, for (a) $N = 1.8 \times 10^{18}$ cm^{-3} and (b) $N = 4 \times 10^{18}$ cm^{-3}. The dotted line shows the position of E_{trans}.

energy are "cold," whereas those occupying levels above the average energy are "hot." Photons with excess energies between 0 and E_{trans}, incident on this material system, stimulate emission, removing cold electrons and holes from the distribution. Removing cold carriers effectively heats the distribution. Photons with energies between E_{trans} and $\langle E \rangle$ stimulate absorption, creating cold carriers. Photons with energies greater than $\langle E \rangle$ stimulate absorption and create hot carriers. This process heats the distributions. Note that the relative positions of the average energy $\langle E \rangle$, calculated according to

$$\langle E \rangle = \frac{\int f(E)\rho(E)E\,dE}{N} \qquad (18)$$

and the electron-hole quasi-Fermi energy separation E_{trans} depend on the carrier density. However, even when the average energy is greater than the transparency energy, there is a set of photon energies for which stimulated transitions cool the distributions. In Fig. 7(b), we show a distribution in which $\langle E \rangle$ is less than E_{trans}. In this case, photon energies between 0 and $\langle E \rangle$ stimulate emission and heat the distributions. Excess photon energies between $\langle E \rangle$ and E_{trans} stimulate emission and remove hot carriers. In this case, stimulated emission effectively cools the distributions (Sun et al., 1993). Again, in the highly absorbing regime (excess photon energies greater than E_{trans}), stimulated absorption creates hot carriers and heats the distributions.

It is not clear a priori whether the dominant heating effect is FCA or stimulated transitions. Typically, the stimulated transition probability is orders of magnitude higher than the FCA probability. However, each FCA event changes the total energy of the distribution by almost 1 eV, the single photon energy. Each stimulated transition changes the total energy by only a few meV. To investigate the relative importance of these two effects, we would like to find a way to turn off one of the effects. In the next section, we will discuss how these nonlinearities are measured and identified. For now it will suffice to say that the relative importance of heating and cooling as a result of stimulated transitions and FCA depends on the way the device is biased and on the photon energies of the optical beams that are perturbing the device.

III. Measurement Techniques

In the preceding section, we discussed some of the physical mechanisms responsible for nonlinear gain in active media. In this section, we will discuss how these nonlinearities are measured and how the measurements can be interpreted. We will concentrate on time domain measurement techniques because they yield direct measurements of the nonlinear response of a material. However, we will discuss frequency domain techniques as well.

The most common technique for characterizing ultrafast dynamics of materials in the time domain is the so-called "pump-probe," or "excite and probe," technique (Ippen and Shank, 1977). Pump-probe experiments have found a wide range of applications because the technique itself is rather simple and requires no high-speed detectors or amplifiers. Instead, accurate relative timing between ultrashort pump and probe pulses is required. As the delay between pump and probe pulses is varied, the average value for the probe transmission or phase shift at each delay setting is measured. Plotting the measured probe transmission (or phase shift) as a function of pump-probe delay reveals how the gain (or refractive index) changes induced by the pump recover on an ultrafast timescale.

1. ORTHOGONALLY-POLARIZED COLLINEAR PUMP-PROBE TECHNIQUE

Figure 8 is a schematic diagram of a pump-probe arrangement. Optical pulses with short durations are used as the pump and probe pulses. In a simple implementation, the pump and probe pulses have the same center wavelength, are orthogonally polarized, and can be distinguished from each other in a polarizer. Also, a subsequent set of pump-probe pulses does not reach the device under test until all the dynamics induced by the previous

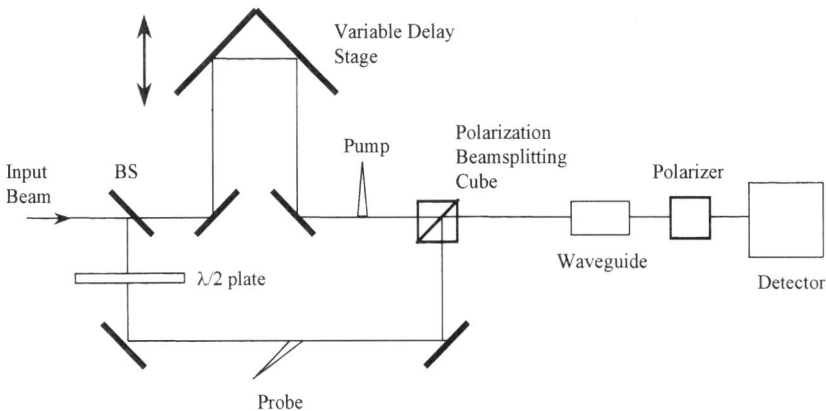

FIG. 8. Orthogonally polarized pump-probe setup for experiments on waveguides. BS, beamsplitter.

pump pulse have completely recovered. The intensity of the pump pulse is high enough to induce a nonlinear response in the device under test but low enough to remain in the small-signal regime. That is, if the intensity of the pump pulse in the experiment is reduced, the overall signature of the induced probe transmission change remains the same, except that it is scaled linearly. The presence of the pump pulse in the device under test may cause a number of effects. The carrier density may be changed as a result of stimulated transitions. The carrier distribution may be heated or the pump may burn a spatial or spectral hole. All of these effects change the probe transmission. At the output of the device under test, a polarizer is used to extinguish the pump pulses and the average probe transmission at each pump-probe delay is measured. The temporal extent of the observed results depends on the optical pulse widths of the pump and probe pulses and on the interesting dynamic responses of the material. To observe the carrier lifetime τ_s, pump-probe measurements are performed with pulse separations of hundreds of picoseconds (Eisenstein *et al.*, 1989; Manning *et al.*, 1994a). Because the lifetime effects are relatively slow, pulses of 10 to 25 ps are sufficiently narrow to observe the interesting dynamics. However, to observe effects produced by carrier scattering and carrier heating, shorter pulses, typically with optical pulsewidths less than or equal to 150 fs and shorter overall time spans, on the order of 10 ps pump-probe delay, are used.

With an idea of how nonlinearities are measured in a semiconductor device, we can describe the experimental signatures of carrier density changes, carrier heating, and spectral hole burning. Imagine, first of all, that the diode amplifier is a polarization-insensitive device and that it is biased in the absorption regime. If the probe pulse precedes the pump pulse through the device, the average transmission of the probe is unaffected by the lagging pump pulse. When the probe pulse leads the pump pulse through the device, the pump-probe time delay is negative. However, if the probe pulse overlaps the pump pulse, (zero time delay) or follows it (positive time delay), through the device, its transmission will be changed by the presence of the pump. As we discussed previously, in the absorption regime (short wavelengths, low bias currents), absorption of the pump increases the carrier density, thereby increasing the probe transmission for pump-probe delays greater than zero. This increase in carrier density persists until the carriers recombine through spontaneous emission or nonradiative recombination, or are removed by spatial diffusion. The exponential recovery of the carrier density back to its equilibrium level has a time constant of a few hundred picoseconds for modest applied bias currents. If the device is biased in the gain regime (long wavelengths, high bias currents), stimulated emission induced by the pump reduces the carrier density and the probe transmission. At the transparency point, where stimulated emission and stimulated absorption are equally probable, there is no net pump-induced change in carrier density and no associated change in probe transmission. Because the carrier density recovery time is long in comparison with the optical pulses used in the pump-probe experiments, this observed change in probe transmission (ΔT_{probe}) is proportional to the integral of the pump pulse. Experimentally, we confine our studies to pump-induced changes in probe transmission occurring on a timescale of a few picoseconds. On this timescale, the carrier density recovery is not resolved and the induced changes in probe transmission appear as "step" changes in transmission, increasing the probe transmission in the absorption regime and decreasing the transmission in the gain regime. The "step" change we refer to here is a mathematical function $u(t)$, the integral of an impulse function $\delta(t)$, given by

$$u(t - \tau) = \int_{-\infty}^{t} \delta(t' - \tau)dt' = \begin{Bmatrix} 0 \text{ if } t < \tau \\ 1 \text{ otherwise} \end{Bmatrix} \tag{19}$$

Pump-probe experiments are performed in all three regimes of operation, so that nonlinearities owing to carrier density changes can be distinguished from those owing to other processes.

For example, we have described spectral hole burning and pointed out that it occurs when stimulated transitions preferentially deplete or create carriers in a given energy range. Therefore, the transient change in gain resulting from spectral hole burning should have the same sign as the step change in gain resulting from stimulated emission and absorption. That is, it should increase the probe transmission in the absorption regime, decrease the probe transmission in the gain regime, and not change the probe transmission at the transparency point. Besides changing sign at the transparency point, the nonlinear change in gain owing to spectral hole burning should recover quickly, as the new thermalized carrier distribution is achieved by means of carrier-carrier scattering. At high carrier densities, the carrier-carrier scattering time constant is believed to be 100 fs or less (Knox et al., 1986).

Induced temperature changes in a carrier distribution have a different signature. Experiments have shown that, except in regimes of extreme absorption (Kesler, 1988; Sun et al., 1993; Mørk et al., 1994a; Willatzen et al., 1994), the average carrier distribution temperature is increased by the combined effects of stimulated transitions and FCA. Heating of the carrier distribution smears out the Fermi distribution and reduces the gain over a very large energy range. In pump-probe experiments, the carrier heating effect should reduce the probe transmission in all three regimes of operation: gain, transparency, and absorption. As mentioned previously, the carrier distribution will thermalize through carrier-carrier scattering and the heated distribution will cool back to the lattice temperature by means of phonon emission. The carrier heating recovery time constant depends on a variety of factors, including amplifier device structure and whether the heated carriers are electrons or holes. Experiments have yielded carrier heating recovery times ranging from 500 fs to 1.5 ps, in a variety of material systems (Stix *et al.*, 1986; Kessler and Ippen, 1987; Hall *et al.*, 1989; Sun *et al.*, 1993; Tatum *et al.*, 1996).

Figure 9 shows examples of the first results of pump-probe studies of gain nonlinearities in active InGaAsP devices in the three regimes of operation: absorption, transparency, and gain using sub-picosecond pulses (Hall *et al.*, 1990b). These devices were bulk, V-groove laser diodes (Wilt *et al.*, 1984) whose facets had been antireflection coated (Eisenstein *et al.*, 1988) and the

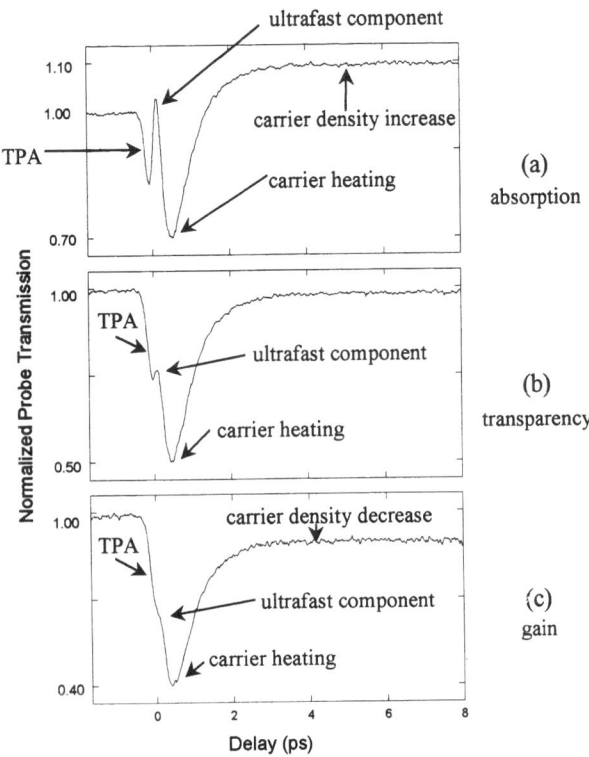

FIG. 9. Change in probe transmission as a function of orthogonally polarized pump-probe delay when the bulk SOA is biased (a) in the absorption regime, (b) at the transparency point, and (c) in the gain regime.

pump-probe pulsewidth was 150 fs. A more detailed description of the pump-probe experimental setup used to obtain these data can be found in the paper by Hall and coworkers (Hall et al., 1990a). The sequence of traces can be discussed first by comparing the long-lasting (step) components in the traces. These are the step changes in probe transmission caused by pump-stimulated changes in carrier density. As we described previously, in the absorption regime (Fig. 9a), stimulated absorption of the pump increases the carrier density and increases the probe transmission for pump-probe delays greater than zero. In the gain regime (Fig. 9c), pump-stimulated emission reduces the carrier density and the probe transmission. At the transparency point (Fig. 9b), where stimulated emission and stimulated absorption are equally probable, there is no net change in carrier density and no long-lived change in carrier density. To obtain these data in the three regimes of operation, we varied the amplifier bias current and kept the pump-probe wavelength constant. Qualitatively similar results have been obtained by holding the amplifier bias constant and varying the pump-probe wavelength (Hall et al., 1989). Also, qualitatively similar results have been measured using the semiconductor optical amplifier as the detector as well as the device under test (Hall et al., 1990c).

Besides the long-lasting changes in probe transmission resulting from carrier density changes, these data exhibit a transient decrease in transmission, recovering with a 600-fs time constant, in all three regimes of operation. Because this transient causes a reduction in probe transmission in the gain and absorption regimes, and is not negligible at the transparency point, it cannot be explained by interband transition effects alone, as can spectral hole burning. Rather, this transient decrease in transmission is caused by the heating of the carrier distribution by stimulated transitions and FCA. Heating the distribution reduces the gain in all regimes of diode operation. Another component in the diode's response is observed clearly in the absorption regime. This transient appears as an absorption bleaching (transient increase in gain) and recovers with a time constant of 100 to 250 fs. This component has the expected signature of spectral hole burning. That is, it represents a short-lived transient decrease in absorption. A remnant of this signature, however, is still apparent at the transparency point, where spectral hole burning contributions to the overall response should be zero. Thus, other mechanisms for this signature must be involved. These other mechanisms will be described later in this chapter. For now, we will simply refer to this component of the response as the ultrafast component. Finally, in all three regimes of operation, we observe an instantaneous decrease in probe transmission. We attribute this transient decrease in probe transmission to TPA because it has no measurable time constant and is only weakly dependent on carrier density and pump-probe wavelength.

2. HETERODYNE PUMP-PROBE TECHNIQUE

While the orthogonally polarized pump-probe technique is simple and yields some insights into the physical mechanisms responsible for nonlinear gain in active materials, it is limited in several ways. First, the time resolution of a semiconductor waveguide measurement can be limited by pump-probe pulse walk-off owing to polarization dispersion in the device. A group velocity mismatch resulting in 100 to 200 fs of walk-off between transverse electric (TE) polarized and transverse magnetic (TM) polarized pulses has been observed in 1-mm-long InGaAsP devices (Hall *et al.*, 1992c). Second, restricting the pump and probe pulses to orthogonal polarizations limits the ability of the pump-probe technique to characterize devices, such as unstrained multiple quantum well (MQW) lasers and compressively strained MQWs, whose gain is fundamentally polarization-sensitive. For these reasons, we devised a pump-probe setup, called the heterodyne pump-probe technique, whereby pump and probe pulses could be parallel-polarized (co-polarized), spatially overlapped, and nominally at the same wavelength, but still be distinguishable at a detector.

An experimental technique demonstrating the heterodyne pump-probe scheme is shown in Fig. 10. The source could be any short-pulse laser whose

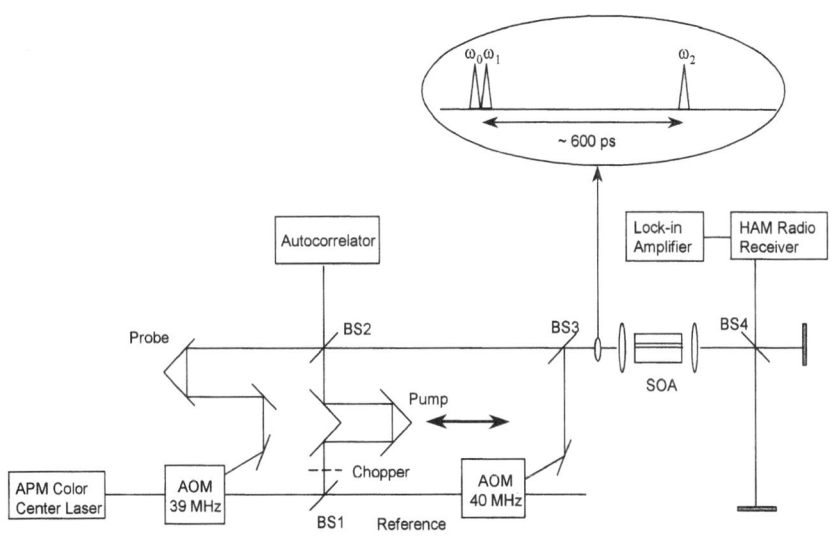

FIG. 10. Setup for the heterodyne pump-probe technique. BS, beamsplitter; AOM, acousto-optic modulator; SOA, semiconductor optical amplifier. The ham radio receiver is detecting a 1-MHz beat signal.

repetition rate is slower than the recovery of the long-lived nonlinearities in the device being measured. In our first demonstration of this technique, we used an additive pulse modelocked (APM) color center laser (Mark et al., 1989) with a repetition rate of 100 MHz. The output from the source is split into three beams—pump, probe, and reference—and we use acousto-optic modulators (AOMs) to shift the spectra of the probe and reference pulses (Hall et al., 1992a). The AOM that deflects the probe beam is driven at 39 MHz, and the probe pulse spectral components are shifted by an equivalent amount ($\omega_1 = \omega_0 + 39\,\text{MHz}$). The pump beam is chopped at a few kilohertz and is passed straight through the AOM. The spectrum of the pump pulses is not shifted ($\omega_{\text{pump}} = \omega_0$). The reference beam is deflected at a second AOM driven at 40 MHz ($\omega_2 = \omega_0 + 40\,\text{MHz}$). The absolute frequency of the RF drive to the AOM's is not important as long as neither frequency is equal to a harmonic of the source repetition rate and no other beat frequency is the same as that between the probe and the reference. The three beams are recombined collinearly with the reference pulse leading the pump-probe pair by a few hundred picoseconds so that any changes induced by the reference will have died away by the time the pump-probe pair arrives.

This single beam of pulses is coupled through the diode amplifier and into an imbalanced Michelson interferometer where the probe and reference pulses are overlapped in time and space at the detector. The probe and reference pulses will interfere and beat at their difference frequency, 1 MHz. Again, the difference frequency imposed by the two RF drives for the AOMs is not crucial, but we chose 1 MHz because we could use a high-frequency amateur radio (ham radio) receiver to detect the 1-MHz beat. The advantages of the ham radio receiver are that it has AM (amplitude modulation) and FM (frequency modulation) reception in the same frequency bands and it has a narrow-band filter that eliminates some noise on the detected signal. In this setup, for nonlinear gain measurements, the radio operates as an AM receiver, and pump-induced changes in the probe transmission appear as amplitude modulation (at the chop frequency) on the probe reference beat. This setup can be modified to perform the measurements we described previously, where the pump and probe pulses were orthogonally polarized, by placing a half-wave plate in the pump arm and rotating the polarization by 90 degrees. Note that this modification does not affect the power levels incident on the diode, nor does it affect the heterodyne detection because we are detecting the beat between reference and probe pulses. Figure 11 shows the measured change in probe transmission as a function of pump-probe delay for cross-polarized (solid line), as well as for co-polarized (dashed line) pump-probe pulses in the (a) absorption, (b) transparency, and (c) gain regimes. Note that the heterodyne detection technique produced the same

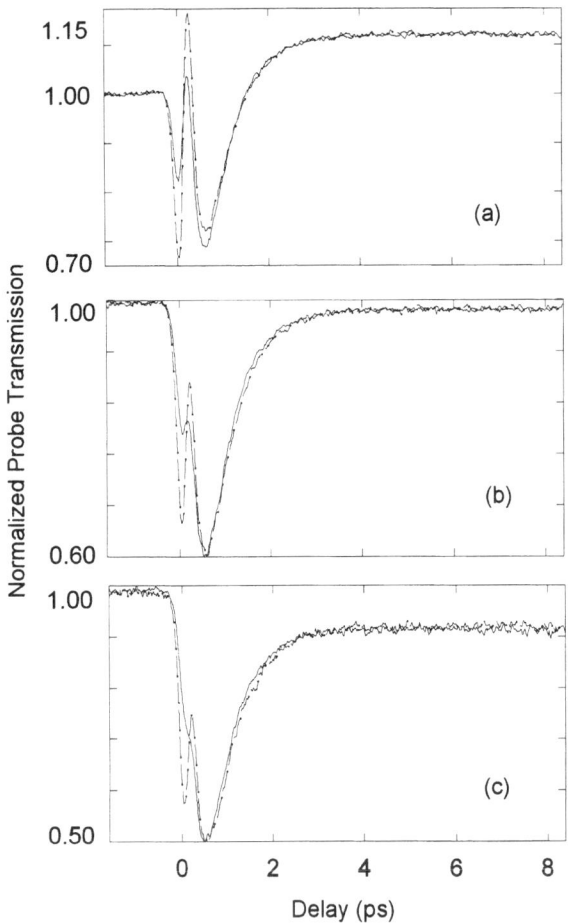

FIG. 11. Change in probe transmission as a function of cross-polarized (solid line) and co-polarized (dashed line) pump-probe delay when the bulk amplifier device is biased (a) in the absorption regime, (b) at the transparency point, and (c) in the gain regime.

results for orthogonally polarized pump-probe beams as those obtained by simply placing a polarizer after the diode amplifier and selecting the probe beam for detection (Fig. 8). Note also that for time delays greater than 2 ps, the measured responses for the co- and cross-polarized pump-probe beams are essentially the same. However, the co-polarized data are much sharper than the cross-polarized data for short time delays. There are several possible explanations for the blurring of the response function near zero time delay in the cross-polarized data. First, owing to the group velocity mismatch for orthogonal polarizations that we referred to earlier (Hall et al., 1992c), the cross-polarized pump and probe pulses are walking off from each other as they travel through the diode. This walk-off limits the time resolution of the experiment and washes out interesting fast dynamics near zero time delay. Second, as we will discuss in Section V, we expect differences in the amplitude of the coherent coupling term for co-polarized and cross-polarized beams (Ippen and Shank, 1977; Vardeny and Tauc, 1981). Finally, the difference between the two results may be attributable to an anisotropy in the waveguide nonlinearity.

The heterodyne technique allows a much more thorough characterization of the nonlinear gain of active media. In devices such as multiple quantum well (MQW) laser amplifiers (see Chapter 1), where the heavy hole–light hole degeneracy at the Brillouin zone center ($k = 0$) is broken by the well confinement, it allows us to probe the electron–heavy hole transition separately from the electron–light hole transition. In compressively strained MQW amplifiers, where there is a large energy separation between the light hole and heavy hole levels, interesting dynamics can be observed. For example, we used the heterodyne technique to study a strained layer (SL) device with an active region consisting of four 2.5-nm-thick, 1.53% compressively strained wells sandwiched between 9-nm-thick barriers. The measured emission peaks were 1.53 μm for the TE mode and 1.38 μm for the TM mode (Koren et al., 1990). The emission peaks are polarization-dependent because of the polarization selection rules (Bastard, 1988). At the Brillouin zone center, heavy hole–electron (hh-e) and light hole–electron (lh-e) transitions are allowed for TE-polarized light (parallel to the quantum layers), whereas hh-e transitions are forbidden for TM-polarized light (orthogonal to the layers). The quantum well energy levels and the allowed transitions are shown schematically in Fig. 12. These polarization selection rules are strictly true only at the Brillouin zone center ($k = 0$), but we have verified experimentally that they hold for modest applied bias currents.

Figure 13 shows the measured change in probe transmission as a function of co-polarized pump-probe delay for the strained layer (SL) MQW diode biased in the three regimes of operation. The results for the TE-polarized pump and probe are shown by the solid lines, and the results for the TM-polarized pump and probe are shown by the dashed lines. The TM-polarized results show only the instantaneous absorption owing to TPA. No other dynamics are observed, because the pump-probe wavelength ($\lambda = 1.51\,\mu$m) is below-band for the lh-e transition ($\lambda = 1.38\,\mu$m) and the hh-e transitions are forbidden. This result confirms that the effects of band mixing at bias currents below 20 mA are weak.

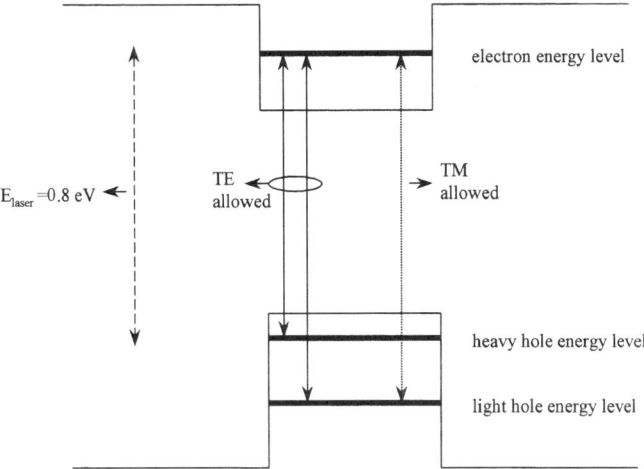

FIG. 12. Schematic of the energy levels and allowed transitions in a strained quantum well. Note that the pump-probe pulse energy is below-band for the lh-e transition.

The TE-polarized results for the SLMQW amplifier show similar dynamics to those observed previously in bulk (Fig. 11) and MQW diodes (Hall et al., 1990b; Hall et al., 1990a). The three regimes are distinguished by the sign of the step change in probe transmission resulting from stimulated transitions. All three regimes show a transient decrease in transmission that recovers with a time constant of approximately 1 ps in this case. This gain compression is a result of carrier heating. There is also an ultrafast component, appearing as a transient absorption bleaching (probe transmission increase) in the absorption regime. This component recovers with a time constant of 100 to 200 fs, and is also present, although less visible, in the gain regime and at the transparency point. In addition to these other gain changes, there is an instantaneous decrease in gain, in all three regimes, attributed to TPA.

Notice that the TPA component of the response is a larger portion of the total response for the SLMQW diode than for the bulk V-groove diode. We believe that this difference can be explained by the difference between the mode confinement factors for the bulk diode ($\Gamma = 0.3$) (Wilt et al., 1984) and the SLMQW diode ($\Gamma = 0.05$) (Agrawal and Dutta, 1986). TPA occurs across the entire spatial mode (cladding and active region) of the pump-probe electric field. However, gain dynamics—such as carrier heating, spectral hole burning, and carrier density changes—are expected to occur only in the active layer of the diode. A smaller mode confinement factor

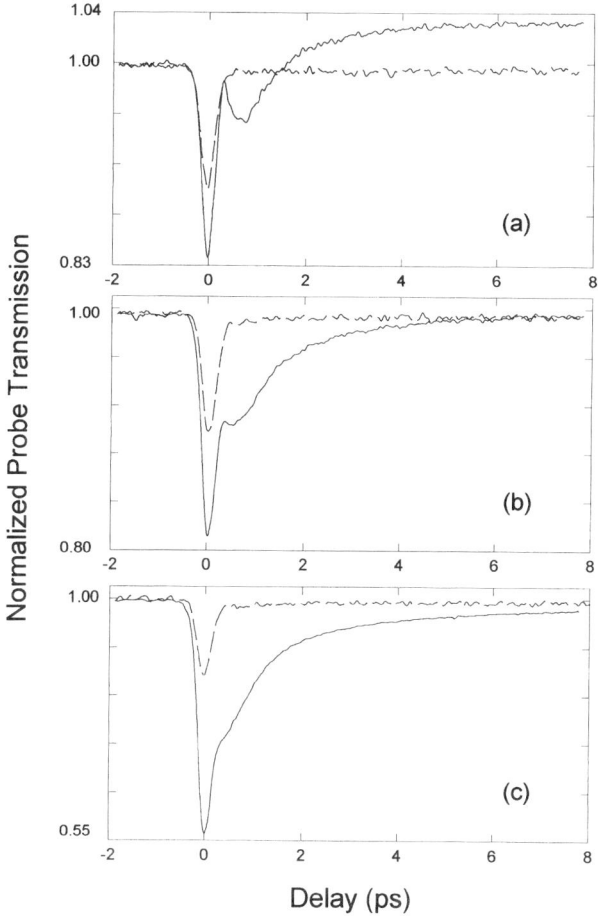

FIG. 13. Change in probe transmission as a function of co-polarized pump-probe delay in an SLMQW diode. The solid lines are for TE-polarized pump-probe pulses and the dashed lines are for TM-polarized pump-probe pulses. The bias current is (a) 10 mA in the absorption regime, (b) 12 mA at the transparency point, and (c) 18 mA in the gain regime.

means that the ratio of the signal generated in the active layer to the signal generated across the entire spatial mode of the pulses is smaller than in a diode with a larger confinement factor. Therefore, relative to the active layer signal, TPA is a larger portion of the total response in MQW diodes than in bulk diodes. The derived TPA coefficient, $\beta \sim 20$ cm/GW, is comparable in both types of diodes.

Using orthogonally polarized or cross-polarized pump and probe pulses, we take further advantage of the large valence subband energy separation and the polarization selection rules to pump below-band for one transition while probing above-band for the other. When the pump is TM-polarized, it is "below-band" for the only allowed (lh-e) transition and is able to stimulate only intraband transitions. However, the TE-polarized probe, at the same wavelength, can monitor the carrier distribution by means of the interband hh-e transition. Note that since the pump cannot stimulate interband transitions, there should be no observed change in probe transmission associated with pump-stimulated transitions. Also, nonlinearities associated with stimulated transitions, such as spectral hole burning, should not be observed in this case. Figure 14 shows the pump-probe results for the diode under the same bias conditions as those shown in Figure 13, but for cross-polarized pump (TM) and probe (TE). Note that, as predicted, there is essentially no change in gain as a result of carrier density changes because the "below-band" pump cannot stimulate transitions. Note however, that even in the absence of stimulated transitions, there is a significant reduction in transmission owing to carrier heating. These carriers are heated by means of FCA and TPA. Also note that the ultrafast component, recovering with a time constant of 100 to 200 fs and causing a "cusp" in the response curve, is still present in the cross-polarized data.

The heterodyne pump-probe technique may also be used to investigate polarization anisotropy in more complicated devices such as polarization insensitive amplifiers with alternating tensile and compressive strained quantum wells. The small-signal gain in these amplifiers is polarization insensitive because in compressively strained quantum wells, the lowest energy radiative transitions are between the conduction band and the heavy hole band; whereas in a sufficiently tensile-strained well, they are between the conduction band and the light hole band. Interestingly, the unique structure of these types of devices allows direct measurement of the interwell coupling using various polarization combinations in a heterodyne pump-probe measurement. For example, one polarization may be used to selectively excite the compressively strained well while the orthogonal polarization is used to probe the other type of well. Using this technique, Lenz et al. (1996) observed a dynamic nonlinear anisotropy in these devices that showed a 7 ps time constant that they attributed to interwell coupling dynamics.

IV. Nonlinear Index of Refraction in Active Media

One of the great strengths of the heterodyne pump-probe technique is that it can be used, with essentially no changes in the experimental setup, to

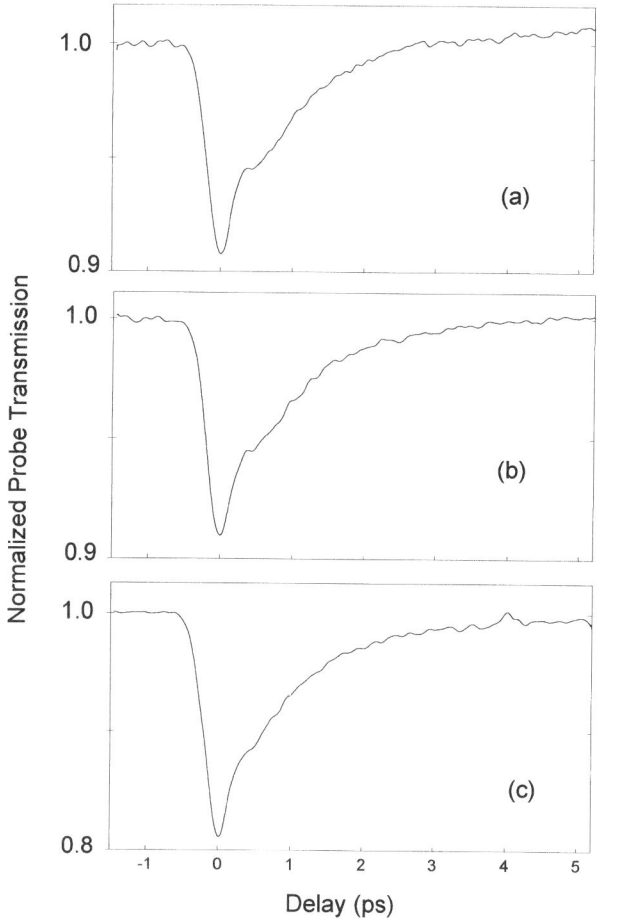

FIG. 14. Change in probe transmission as a function of cross-polarized pump-probe delay for the SLMQW diode biased in the (a) absorption ($I = 10\,\text{mA}$), (b) transparency ($I = 12\,\text{mA}$), and (c) gain ($I = 18\,\text{mA}$) regimes. The pump is TM-polarized and the probe is TE-polarized. Note that the total time delay in this plot is less than 7 ps, whereas it is 10 ps in the other data displayed in this chapter.

measure pump-induced refractive index nonlinearities. The set-up is the same as shown in Fig. 10 except that the ham radio receiver is set as an FM receiver. In these experiments, when the probe pulse precedes the pump pulse, the probe phase is not modulated and the FM receiver output is zero. However, when the probe pulse lags behind the pump pulse, its phase is modulated (chopped). The radio receiver detects this modulation in phase and generates an output signal that is proportional to its derivative (because $f \sim d\phi/dt$). The lock-in amplifier detects this signal, integrates it, and yields an output voltage that is proportional to the pump-induced phase change. This measured phase change is then calibrated with respect to a known optical phase change. In our case, one of the mirrors in the interferometer arm is modulated by a piezoelectric transducer (PZT) and the phase shift is obtained by observing fringes. Changes in refractive index are related to the measured probe phase shifts by the simple relation

$$\Delta n = \frac{\lambda}{2\pi L} \Delta\phi \qquad (20)$$

where L is the length of the active waveguide region and $\Delta\phi$ is the measured probe phase shift.

Ultrafast refractive index changes are important because they affect the chirp and the spectrum of short pulses propagating in waveguides (Koch and Linke, 1986; Agrawal and Olsson, 1989b; Grant and Sibbett, 1991; Tang et al., 1996). Refractive index nonlinearities may also be exploited in interferometric all-optical switches (Patel et al., 1998). The ultrafast refractive index nonlinearities in active media are related to the ultrafast gain nonlinearities by the Kramers-Kronig (KK) relation. If g_0 is the linear gain coefficient and n is the refractive index, the KK transformation (Hutchings et al., 1992) yields

$$n(\omega) = 1 + \frac{c}{\pi} \int \frac{g_0(\omega')}{\omega'^2 - \omega^2} d\omega' \qquad (21)$$

A perturbational change in gain or absorption, Δg_0, will have an associated change in index (Sheik-Bahae et al., 1991) given by

$$\Delta n(\omega) = \frac{c}{\pi} \int \frac{\Delta g_0(\omega')}{\omega'^2 - \omega^2} d\omega' \qquad (22)$$

Using this formalism, we can discuss the index changes associated with the gain changes we described in the preceding section.

For example, the change in gain resulting from changes in carrier population can be calculated as

$$\Delta g_0 = g_0(E, N_1) - g_0(E, N_2) \qquad (23)$$

where $g_0(E, N_i)$ is the gain curve derived for a carrier density of N_i. This change in gain can be substituted into Eq. (22) to determine the change in refractive index owing to carrier population changes. Figure 15 shows the computed change in refractive index (a) for a given change in gain (b) as a result of carrier density changes. The curves are calculated for carrier density changes of $\Delta N = +0.02 \times 10^{18}\,\mathrm{cm}^{-3}$ (solid line), $-0.02 \times 10^{18}\,\mathrm{cm}^{-3}$ (dashed line), $-0.05 \times 10^{18}\,\mathrm{cm}^{-3}$ (dotted line), and $-0.07 \times 10^{18}\,\mathrm{cm}^{-3}$ (dashed-dotted line) around $N = 1.8 \times 10^{18}\,\mathrm{cm}^{-3}$. Note that for excess photon energies near the bandgap, increasing the carrier density decreases the refractive index, and vice versa (Henry et al., 1981; Manning et al., 1983; Bennett et al., 1990). Also note that the index changes extend below the band edge.

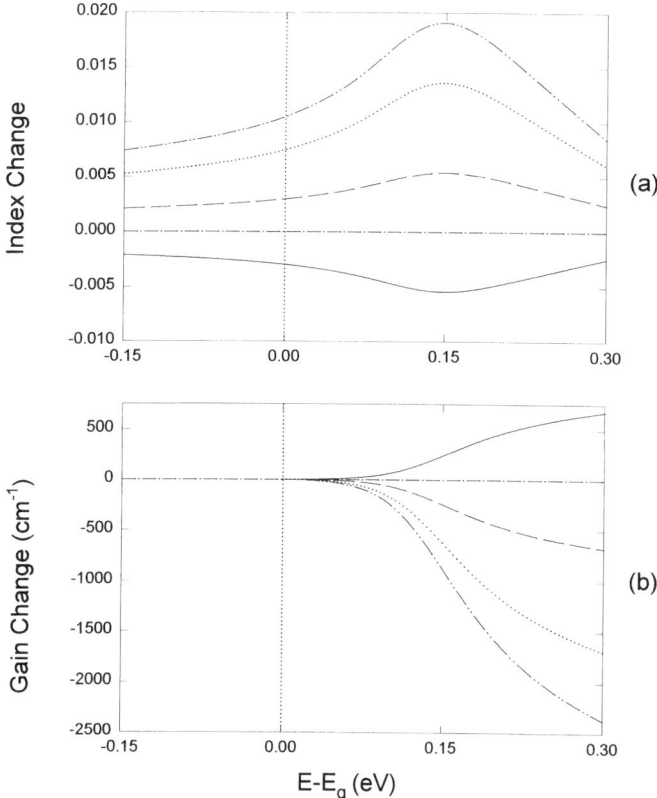

FIG. 15. Computed change in (a) refractive index and (b) gain for carrier density changes of $\Delta N = +0.02 \times 10^{18}\,\mathrm{cm}^{-3}$ (solid line), $-0.02 \times 10^{18}\,\mathrm{cm}^{-3}$ (dashed line), $-0.05 \times 10^{18}\,\mathrm{cm}^{-3}$ (dotted line), and $-0.07 \times 10^{18}\,\mathrm{cm}^{-3}$ (dashed-dotted line) around $N = 1.8 \times 10^{18}\,\mathrm{cm}^{-3}$. The electron and hole temperature is room temperature in these calculations.

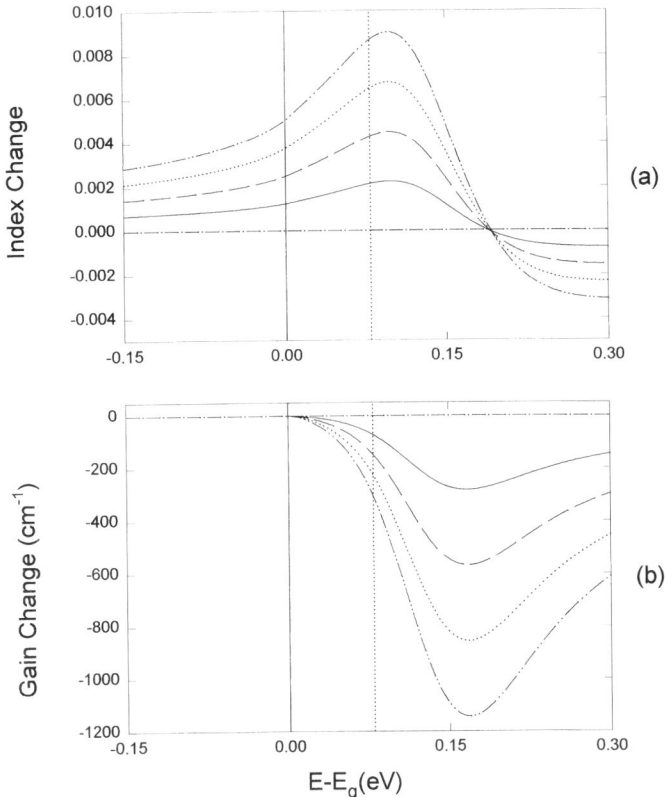

FIG. 16. Computed change in (a) refractive index and (b) gain for carrier density changes (electrons and holes) of +10 K (solid line), +20 K (dashed line), +30 K (dotted line), and +40 K (dashed-dotted line). The carrier density in this calculation is $N = 1.8 \times 10^{18}\,\text{cm}^{-3}$. The dotted vertical line shows the position of E_{trans} at room temperature.

Refractive index nonlinearities resulting from carrier temperature changes can be predicted in a similar manner. In this case, we calculate

$$\Delta g_0 = g_0(E, T_1) - g_0(E, T_2) \tag{24}$$

Figure 16 shows the computed change in (a) refractive index and (b) gain for carrier temperature changes (electrons and holes) of +10 K (solid line), +20 K (dashed line), +30 K (dotted line), and +40 K (dashed-dotted line). From these calculations of gain and index changes for carrier density and carrier temperature changes, we can obtain estimates for the linewidth enhancement factor and the temperature linewidth enhancement factor. For carrier density changes, the linewidth enhancement factor α, is defined as

$$\alpha = -\frac{\delta\chi_r}{\delta N} \bigg/ \frac{\delta\chi_i}{\delta N} = -\frac{4\pi}{\lambda} \frac{\delta n}{\delta N} \bigg/ \frac{\delta g_0}{\delta N} \qquad (25)$$

where χ_r and χ_i are the real and imaginary parts of the complex susceptibility, and N is the carrier density (Henry, 1982). Note that α is wavelength-dependent and that very near the band edge of the material $\delta g_0/\delta N$ is small, causing α to be relatively large. Because changes in the carrier temperature affect the gain and refractive index of the material in a manner similar to carrier density changes, we also can define an effective linewidth enhancement factor, α_T that is related to temperature:

$$\alpha_T = -\frac{\delta\chi_r}{\delta T} \bigg/ \frac{\delta\chi_i}{\delta T} = -\frac{4\pi}{\lambda} \frac{\delta n}{\delta T} \bigg/ \frac{\delta g_0}{\delta T} \qquad (26)$$

Again, note that α_T is wavelength-dependent and that its value also diverges near the band edge of the material.

In addition to the refractive index changes resulting from carrier density and carrier temperature changes, TPA and other effects (such as the optical Stark effect) that produce absorption changes are expected to contribute to the nonlinear refractive index (Sheik-Bahae et al., 1991; Sheik-Bahae and Van Stryland, 1994). There are few experimental or theoretical results that demonstrate or predict the effects of spectral hole burning on the nonlinear refractive index. One reason for this is that the shape and size of the spectral holes that are burned are not well understood and therefore are difficult to model. However, if the k-selection rules hold and if the spectral hole is symmetric around a center wavelength ω_0, the corresponding refractive index change at ω_0 will be zero. Illustrations of this basic property of the KK transformation can be found in many texts (Pankove, 1975; Butcher and Cotter, 1991).

Figure 17 shows the measured probe phase shift as a function of pump-probe delay for a bulk diode biased in the (a) absorption, (b) transparency, and (c) gain regimes. Notice that, as was the case in the measured gain dynamics, the fast dynamics occurring near zero time delay are not as sharp in the cross-polarized data as in the co-polarized data. This blurring of the response is attributed to the same combination of effects: pump-probe pulse walk-off, anisotropy in the nonlinearity, and differences in the coherent coupling terms. However, both the co-polarized and cross-polarized sets of data are qualitatively the same.

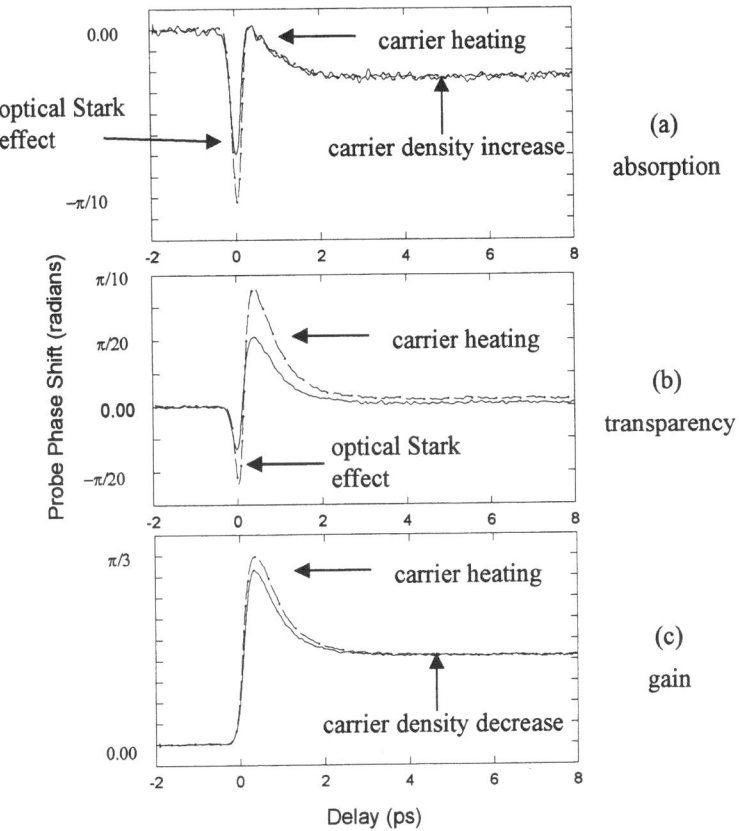

FIG. 17. Measured probe phase shift as a function of pump-probe delay for co-polarized (dashed line) and cross-polarized (solid line) pump-probe beams for a bulk diode biased in the (a) absorption, (b) transparency, and (c) gain regimes.

As was the case in the transmission experiments, the long-lived change or step change in refractive index is associated with pump-induced carrier density changes. Decreasing the carrier density increases the refractive index, and vice versa. This result is consistent with the calculations shown in the preceding section and Fig. 15. In addition to the long-lived change in refractive index, a transient increase in refractive index is observed in all three regimes of operation, recovering with a time constant of ∼600 fs, consistent with the carrier heating signal observed in the gain data. This result agrees with calculations in Fig. 16 and with previous studies of AlGaAs devices that showed that for photon energies near the bandgap energy, heating of the carrier distribution increased the refractive index (Hultgren and Ippen, 1991; Henry et al., 1981; Manning et al., 1983; Bennett et al., 1990). Finally, in all three regimes of operation we observe an instantaneous decrease in the refractive index. This component has been attributed to a virtual process such as the optical Stark effect (LaGasse et al., 1990; Anderson et al.., 1990; Hultgren and Ippen, 1991; Hultgren et al., 1992).

It is interesting to note that in an experiment by Hultgren et al., (1992), the magnitude of the virtual component attributed to the optical Stark effect was dependent on both wavelength and carrier density. In particular, they observed a resonant enhancement of the instantaneous nonlinearity as the pump-probe pulse wavelength was tuned from below the bandedge toward, and into, the band. This resonance behavior had been observed previously in passive waveguides (LaGasse et al., 1990). Also, the magnitude of the instantaneous nonlinearity was greatly reduced when higher bias currents (carrier densities) were applied to the semiconductor optical amplifier. This effect can also be observed in the data shown in Fig. 17. Increasing the bias current increases the carrier density and fills states near the band edge. This band filling pushes the absorption edge to higher energies, away from the probe wavelength. Thus, for a given probe wavelength, the magnitude of the instantaneous nonlinearity is reduced at increased carrier densities. This behavior agrees with a theory of ultrafast refractive index dynamics that incorporates the optical Stark effect (Sheik-Bahae and Van Stryland, 1994).

V. Data Analysis and Interpretation

The gain and refractive index in a semiconductor optical amplifier can be modeled phenomenologically by a set of rate equations (Icsevgi and Lamb, 1969; Frigo, 1983; Tucker and Pope, 1983; Bowers et al., 1986; Hansen et al., 1989; Saitoh and Mukai, 1990; Lai et al., 1990; Hall et al., 1990a). For example, an ordinary set of rate equations that track the changes in carrier density N, and the photon flux S, in the active region is given by

$$\frac{dS}{dt} = -\frac{S}{\tau_p} + \Gamma[g_d(N - N_t)]S + \frac{\gamma \Gamma N}{\tau_s} \qquad (27)$$

and

$$\frac{dN}{dt} = -g_d(N - N_t)S - \frac{N}{\tau_s} + \frac{I}{qV} \qquad (28)$$

where Eq. (28) is the same as Eq. (12), and Γ is the mode confinement factor, γ is the fraction of spontaneous emission coupled into the lasing mode, and τ_p is the photon lifetime in the cavity. To model the nonlinear behavior of the active material, we modify the rate equations by adding two more variables (and equations), one that describes carrier heating and one that describes the ultrafast components.

$$\frac{dS}{dz} = \Gamma[g_d(N - N_t - N_{ch} + N_u) - \alpha_{fc}N]S - \beta S^2 \quad (29)$$

$$\frac{dN}{dt} = -g_d(N - N_t - N_{ch} + N_u)S - \frac{N}{\tau_s} + \frac{I}{qV} \quad (30)$$

$$\frac{dN_{ch}}{dt} = \gamma_{ch}\alpha_{fc}NS - \frac{N_{ch}}{\tau_{ch}} \quad (31)$$

$$\frac{dN_u}{dt} = \gamma_u g_d S - \frac{N_u}{\tau_u} \quad (32)$$

Here, S is the photon flux, N_{ch} is the effective carrier density change at the emission wavelength caused by carrier heating, N_u is the effective carrier density change caused by the ultrafast dynamic, α_{fc} is the free carrier absorption coefficient, γ_u determines the relative rate at which the ultrafast absorption component is produced, γ_{ch} represents how efficiently free carrier absorption contributes to carrier heating, τ_{ch} is the relaxation time of the heated carriers, and τ_u is the relaxation time for the ultrafast dynamic. Notice that we have added a term, βS^2, where β is the TPA coefficient, to the equation for the photon flux, to account for nonlinear absorption. Also, we have adapted the equation for the photon density to model an amplifier, rather than a laser. This model assumes that the carrier density is heated by free carrier absorption and that carriers generated by TPA make a negligible contribution to the gain in the active region and to the heating of the carrier distribution. The validity of these assumptions depends on several parameters, such as the mode confinement factor for the active waveguide, Γ, the temporal width of the optical pulse traversing the amplifier, the bias current applied to the semiconductor diode, and the wavelength of the input light relative to the average energy wavelength of the carrier distribution (Fig. 7).

Some of the terms in the rate equations, $g_d = 2.5 \times 10^{-16}\,\text{cm}^2$, $N_t = 10^{18}\,\text{cm}^{-3}$, and $\Gamma = 0.05$ (Agrawal and Dutta, 1986) for the MQW device and $\Gamma = 0.30$ (Koren et al., 1987) for the bulk device, are estimated from the literature. Other terms (i.e., γ_u, γ_{ch}, α_{fc}, τ_{ch}, τ_u, and β) are phenomenological and must be determined experimentally by measuring the pump-probe response of the amplifier using pulses similar (in pulsewidth and wavelength) to those being used to measure the saturation behavior. The probe transmission, or probe output intensity, as a function of pump-probe delay, can be determined by the differential gain in the amplifier at position z and time t, given by

$$G(t, z) = \Gamma\{g_d[N(t, z) - N_t - N_{ch}(t, z) + N_u(t, z)] - \alpha_{fc}N(t, z)\} - \beta S(t, z) \tag{33}$$

where

$$s_0(t - \tau) = s_i(t - \tau)e^{[\int_0^{z_0} G(t,z)dz]} \tag{34}$$

and where $s_0(t - \tau)$ is the probe output intensity, $s_i(t - \tau)$ is the probe input intensity, z_0 is the length of the diode, and τ is the delay between pump and probe pulses. The measured quantity, the gain seen by the probe, is given by

$$G_{\text{probe}}(\tau) = \frac{\int s_0(t - \tau)dt}{\int s_i(t - \tau)dt} \tag{35}$$

The top plot in Fig. 18 shows the experimental pump-probe results (solid line) along with the fit (dashed line) for an MQW diode operating in the gain regime that has been investigated using optical pulses with widths of $\Delta t = 150$ fs. It is clear from the fit that the rate equation model is a good working model for the gain in the active semiconductor material. In fact, the fit is so good that it is difficult to distinguish from the data. We have plotted the fit in the bottom plot, to highlight the excellent agreement between the fit, and the data.

It is important to stress that N_{ch} and N_u are not actual carrier populations. To the extent that carrier heating reduces the probe transmission, the behavior can be modeled as a carrier density reduction and quantified using N_{ch}. Similarly, to the extent that absorption bleaching increases the probe transmission, it can be modeled as a carrier density increase and quantified using N_u. Other rate equation models describing the nonlinear effects in terms of energy changes have been published (Gomatam and DeFonzo, 1990; Mark and Mork, 1992; Wang and Schweizer, 1996). However, those results assumed a time constant for spectral hole burning that was less than or equal to 70 fs, and there were discrepancies between the data and the modeled fits. More recently, semiclassical density-matrix equations have been used to model the response of active semiconductor waveguides.

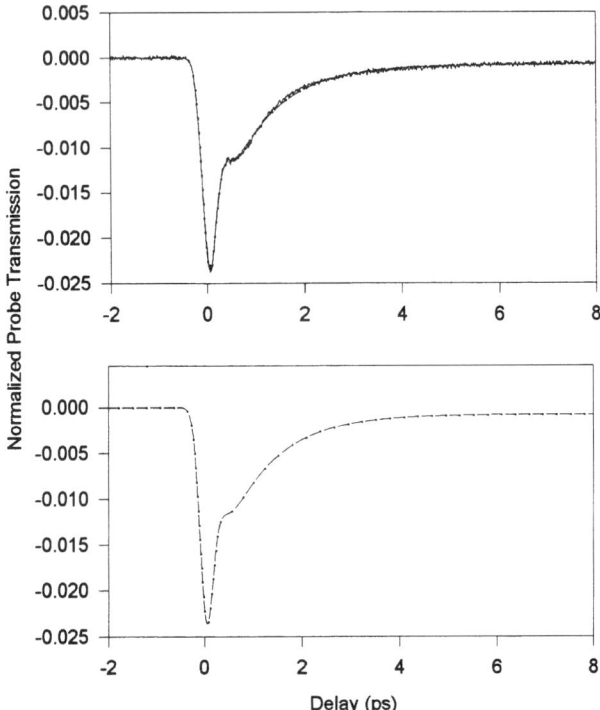

FIG. 18. Change in probe transmission as a function of delay for an MQW diode biased in the gain regime. In the top trace, the data are shown as a solid line and the fit is shown as a dashed line. The fit alone is shown in the bottom trace, to highlight the excellent agreement between experiment and theory.

However, these equations are still phenomenological in the sense that some physical quantities in the model are treated as fitting parameters. All of the models must be phenomenological to some extent, because the original density-matrix and rate equations were unable to predict the existence of the carrier heating nonlinearity. An important point resulting from the validity of the rate equation model is that one may use an impulse response function analysis to analyze and fit the pump-probe data. This analysis is completely consistent with the rate equation analysis because the solutions to those rate equations are a series of exponentials. While the rate equation models are the simplest to implement numerically for data analysis, the semiclassical density-matrix equations provide a useful framework for theoretical evaluation of the dynamic gain and index coupling effects that are described later in this section.

Using the impulse response function analysis, we investigate the contributions of different physical processes to these pump-probe signals by assuming a total impulse response function that consists of a sum of individual exponential responses. The relative amplitudes and time constants of the individual responses are associated with a physical mechanism and are adjusted until the total assumed response, convolved with an intensity autocorrelation of the pump-probe pulses, fits the data. This approach assumes that polarization dephasing is rapid on the timescale of the pulses. For the purposes of this discussion, we will initially separate the response due to pump intensity alone from that due to artifacts produced by the coherent interference between pump and probe. To do this, we write the change in gain produced by the pump photon flux alone as

$$\Delta g(t) = \int_{-\infty}^{+\infty} h(t - t')S(t')dt' \tag{36}$$

where $h(t - t')$ is the response of $\Delta g(t)$ to an impulse of light, and $S(t')$ is the pump photon flux. Note that the integration, written for analytical convenience from $-\infty$, to $+\infty$, need only be over an interval that is long in comparison with the dynamic of interest. In this model, we assume that recovery of the material response is complete between pump-probe pulse pairs. If the probe pulse is a replica of the pump pulse, we can then also write

$$\Delta W(\tau) = \int S(\tau - t) \int h(t - t')S(t')dt' dt \tag{37}$$

to describe the experimentally measured change in probe energy, $\Delta W(\tau)$, caused by the pump, as a function of probe delay. This signal scales with the product of the pump and probe powers. Equation (37) can be rewritten as

$$\Delta W(\tau) = \int h(\tau - t)G^2(t)dt \tag{38}$$

where $G^2(t)$ is the experimentally determined intensity autocorrelation function

$$G^2(t) = \int S(t')S(t + t')dt' \tag{39}$$

This expression is convenient, because $G^2(t)$ can be measured accurately using a background-free intensity autocorrelator and one usually does not have accurate information about the actual pump and probe pulse shapes, $S(t)$.

The gain impulse response function is only the imaginary part of the total response function. There is also a real part

$$h(t) = h_r(t) + ih_i(t) \tag{40}$$

corresponding to the refractive index nonlinearities. To some extent, the functional forms of the real and imaginary portions of the impulse response function mirror each other. For example, from the data we have already presented, it is clear that both the gain and refractive index responses have components resulting from carrier density changes, carrier heating, and virtual processes. However, spectral hole burning, which may impact the probe transmission, is not expected to impact the probe phase (Sheik-Bahae and Van Stryland, 1994). Therefore, no spectral hole burning term will be included in the response function for the refractive index.

To analyze the pump-induced gain and refractive index changes in a semiconductor amplifier, we use impulse response functions of the form

$$h_{i,r}(t) = u(t)(a_0 + a_1 e^{-t/\tau_1} + a_2 e^{-t/\tau_2}) + a_3 \delta(t) \tag{41}$$

where a_0 is the amplitude of the long-lived change in gain or refractive index resulting from carrier density changes, a_1 is the amplitude of the ultrafast component that recovers with a time constant of $\tau_1 = 200\,\text{fs}$, a_2 is the amplitude of the carrier heating component that recovers with a time constant of $\tau_2 = 600\,\text{fs}$, and a_3 is the amplitude of the instantaneous component. The carrier density changes modeled by a_0 recover with a time constant of approximately $\tau_0 = 1\,\text{ns}$. However, on a typical pump-probe timescale of 10 ps, the exponential e^{-t/τ_0} appears constant, and we say that the pump-induced carrier density change causes a "step" $[u(t)]$ change in probe transmission or refractive index.

Another effect that influences the measurement of the nonlinearities is the so-called spectral artifact. In physical terms, a dynamic coupling of the gain and index arises from the fact that the wavelength-dependent gain profile of the diode amplifier can act as a filter and convert phase or frequency changes induced by the pump into amplitude changes measured in the transmission experiments (Mork et al., 1996). This conversion of phase or frequency variations to amplitude variations in a filter is a well-known effect. In semiconductor amplifiers, it has recently been proposed as a method for ultrafast all-optical switching (Nakamura and Tajima, 1997). Because

$\Delta f(t) \sim d\phi/dt$, it is easy to understand that this dynamic coupling will produce a term in the transmission measurement (ΔT) that is proportional to the derivative of the nonlinear index change

$$\Delta T \propto \frac{\delta h_r}{\delta t} \qquad (42)$$

(Mørk and Mecozzi, 1994). The magnitude of this dynamic coupling term depends on several parameters. For example, it is expected to be larger for shorter pump-probe pulses, because the more rapidly induced index changes will produce larger dynamic frequency shifts. It is also expected to be wavelength-dependent, smaller at the diode's gain peak wavelength, where the gain slope is zero, and larger away from that wavelength, as the gain slope increases. A priori, it is difficult to quantify the size of this effect.

It might also be expected that amplitude changes could bring about phase changes and create a dynamic coupling term in the refractive index measurements. However, theoretical predictions (Mecozzi and Mork, 1996; Mork and Mecozzi, 1996) suggest that this dynamic coupling is a less significant portion of the total nonlinear phase response, and these predictions have been verified by the experimental data we present here.

This calculation of the photoinduced nonlinear absorption described above is still incomplete, because it ignores an important measurement artifact — the so-called coherent artifact. The coherent artifact, which arises from coherent interference between the pump and probe fields while they are overlapped in time within the diode, can be related to the electric field correlation of the pulses (Ippen and Shank, 1977; Vardeny and Tauc, 1981). From the analysis by Vardeny and Tauc, in which nonlinear changes in probe transmission are related to changes in the third-order polarizability tensor, we can write

$$\Delta I_{pr\|} \propto (\gamma_\| + \beta_\|) \qquad (43)$$

where $\Delta I_{pr\|}$ is the total pump-induced intensity modulation on a parallel polarized probe,

$$\gamma_\| = \int dt dt' |E(t-\tau)|^2 h_{xxxx}(t-t')|E(t')|^2 \qquad (44)$$

is the nonlinear impulse response of the material convolved with the pulse intensity autocorrelation (compare with Eqs. 37 and 38), and

$$\beta_\| = \int dt dt' E^*(t-\tau)E(t)h_{xxxx}(t-t')E^*(t')E(t'-\tau) \qquad (45)$$

is the artifact coherent coupling term (Ippen and Shank, 1977). Here, we use the electric field [$E(t)$] representation for the optical pulses and we consider only the transmission changes for simplicity. However, the refractive index nonlinearities can be treated in the same fashion. The shape of the coherent coupling term is determined only by the electric field autocorrelation function and therefore contributes to the signal only where the pump and probe are mutually coherent. Here the subscripts of the impulse response function $h_{xxxx}(t)$, suggest its relation to the appropriate tensor component of the nonlinear susceptibility. Note that for any impulse response function, $h_{xxxx}(t)$, $\gamma_\|(0) = \beta_\|(0)$. Therefore, for parallel polarized pump and probe pulses, half of the total measured response at zero time delay is the actual material response and half of the signal is attributable to the coherent coupling artifact. In other words, all the components in the total response contribute to the magnitude of the artifact, but there is no information about the time behavior of the system in this term.

For orthogonally polarized pulses, the actual material response is given by

$$\gamma_\perp = \int dt\, dt' |E(t-\tau)|^2 h_{xxyy}(t-t') |E(t')|^2 \tag{46}$$

and the coherent coupling response is given by

$$\beta_\perp = \int dt\, dt' E^*(t-\tau) E(t) h_{xyyx}(t-t') E^*(t') E(t'-\tau) \tag{47}$$

Notice that in the case of orthogonally polarized pump-probe beams, $\gamma_\perp$ and $\beta_\perp$, depend on χ_{xxyy} and χ_{xyyx}, and it is not necessarily true that $\gamma_\perp(0) = \beta_\perp(0)$. In fact, χ_{xxyy} and χ_{xyyx} are not well known for active semiconductor media. Therefore the cross-polarized data are more complicated to interpret because the magnitude of the coherent coupling artifact cannot be accurately determined. Still, some approximations can be made. For example, in cross-polarized pump-probe experiments, the interaction of the pump and probe beams causes the total field polarization to change as a function of pump-probe delay. The total field polarization is important because it preferentially excites dipoles with certain momentum vectors. If the momentum dephasing time is short compared with the optical pulsewidth, then the material has no polarization memory and the coherent artifact vanishes (Vardeny and Tauc, 1981). In semiconductor diodes, the dephasing time is assumed to be less than 10 fs (Knox et al., 1988). For this reason, the coherent coupling artifact is not expected to contribute to the cross-polarized pump-probe results. This assumption may break down for the instantaneous portion of the response — for example, for the portion of the response attributable to TPA or the optical Stark effect. Ignoring the coherent coupling artifact may lead to errors in determining the absolute magnitude of the TPA coefficient β or of the nonlinear refractive index coefficient n_2. However, it will not affect the physical interpretation of the data.

At this point, we have postulated impulse response functions based on the observed nonlinear responses for gain and refractive index in active media. The next step is to use these impulse response functions to fit the measured responses and to try to associate physical mechanisms with the various components in the fit. Unfortunately, the impulse response functions, as we have proposed them, contain enough adjustable parameters to fit almost any curve. In order to get a handle on this problem, and to begin to make sense of the data presented so far, we choose specific sets of data to analyze first, in an attempt to simplify, or at least contain, the adjustable parameters in the impulse response functions.

For example, consider the nonlinear refractive index response measured in a cross-polarized pump-probe experiment. To fit these data, we assumed

$$h_r(t) = u(t)(a_0 + a_1 e^{-t/\tau_1} + a_2 e^{-t/\tau_2}) + a_3 \delta(t) \qquad (48)$$

Here the coefficients and time constants represent the same effects as described in Eq. (41). This impulse response function does not contain a coherent coupling artifact term because the pump and probe pulses are cross-polarized. Neither does it contain a significant dynamic coupling term. As we shall see from the data and the fits, the magnitude of the dynamic coupling term in the refractive index response is negligibly small. Initially, we will fit the measured response at the transparency point, because then $a_0 = 0$, further simplifying $h(t)$. A curve fit is generated by guessing the values of the a_j's and τ_j's in $h(t)$, and convolving that $h(t)$ with the experimentally determined intensity autocorrelation function. Then the curve fits are plotted as overlapping figures on the data and compared. A recursive routine is used to obtain a best fit. The trace in Fig. 19 shows the measured pump-induced probe phase shift for an MQW diode biased at the transparency point (solid line), along with the calculated fit (dashed line). As was the case with the rate equation model, the fitted curve is difficult to distinguish because of the excellent agreement between the data and the fit. In this case, the time constants were $\tau_1 = 200$ fs and $\tau_2 = 1$ ps, and the amplitudes were $a_0 = 0$, $-a_1 = +a_2 = 7.6 \times 10^{-3}$ and $a_3 = -0.04$. Note that excellent agreement is achieved without considering the spectral artifact. Also note that $a_2 = -a_1$. The fit is quite sensitive to the precise

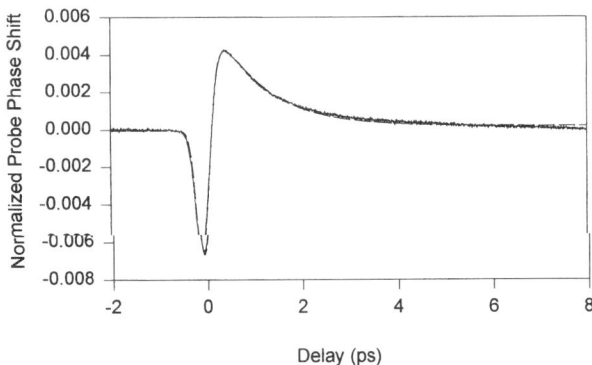

FIG. 19. Measured probe phase shift as a function of pump-probe delay for a MQW semiconductor optical amplifier biased at the transparency point. The trace shows the data (solid line) along with the fit (dashed line).

knowledge of zero time delay. In fact, if there is a discrepancy in the assumed zero time delay of more than 10% of the pulsewidth, good fits cannot be achieved. In addition, good fits are not achieved if the time constants corresponding to the best fits are changed by much more than 10%, regardless of how the a_j's are changed. Such effects are noticeable only if the curve fits (predicted response) are overlaid on the data. Plotting the predicted responses and measured responses side-by-side can indicate only qualitative agreement. Figures 20 and 21 show the measured probe phase shift for the same diode biased in the absorption and gain regimes, respectively. Again, the data and fit are overlapped in the traces. It is interesting to note that in all three regimes of operation, the best fits are achieved when $a_2 = -a_1$. This is not to say that the a_1's and a_2's are the same in every regime, which they are not. However, their ratio is constant. This relationship has suggested an alternative interpretation for the impulse response function. For example, based on the observed behavior, we can rewrite $h(t)$ as

$$h_r(t) = u(t)[a_0 + a_2 e^{-t/\tau_2}(1 - e^{-(t/\tau_{eff})})] + a_3 \delta(t) \tag{49}$$

where $1/\tau_{\text{eff}} = 1/\tau_1 - 1/\tau_2$. Physically, this would associate τ_1 with the time it takes the initial nonequilibrium carrier distribution to evolve into a new hot Fermi distribution. The hot distribution then cools back to the lattice temperature with the time constant of τ_2. To reflect the physical interpretation of the response function, we call $\tau_1 = \tau_{\text{therm}}$, the thermalization time and $\tau_2 = \tau_{\text{ch}}$ the carrier heating recovery time. This time constant is also

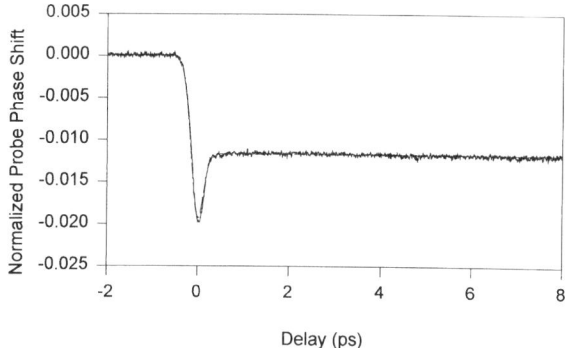

FIG. 20. Measured probe phase shift as a function of pump-probe delay for a MQW semiconductor optical amplifier biased in the absorption regime. The trace shows the data (solid line) along with the fit (dashed line).

sometimes referred to as the carrier cooling time, and it is the same time constant used in the rate equation model. We point out that this risetime of carrier heating has been observed and identified previously (Hall *et al.*, 1992b; Hultgren *et al.*, 1992; Hall *et al.*, 1994). In fact, the risetime of carrier heating is partially responsible for the apparent absorption bleaching that was observed in the transmission data. Figure 22 shows the fits (solid lines) to the index data in all three regimes of operation, along with the individual components from the fits.

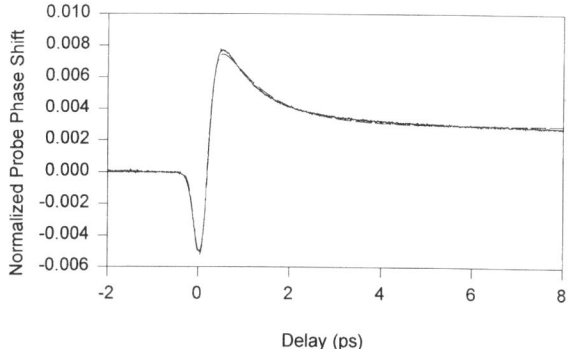

FIG. 21. Measured probe phase shift as a function of pump-probe delay for a MQW semiconductor optical amplifier biased in the gain regime. The trace shows the data (solid line) along with the fit (dashed line).

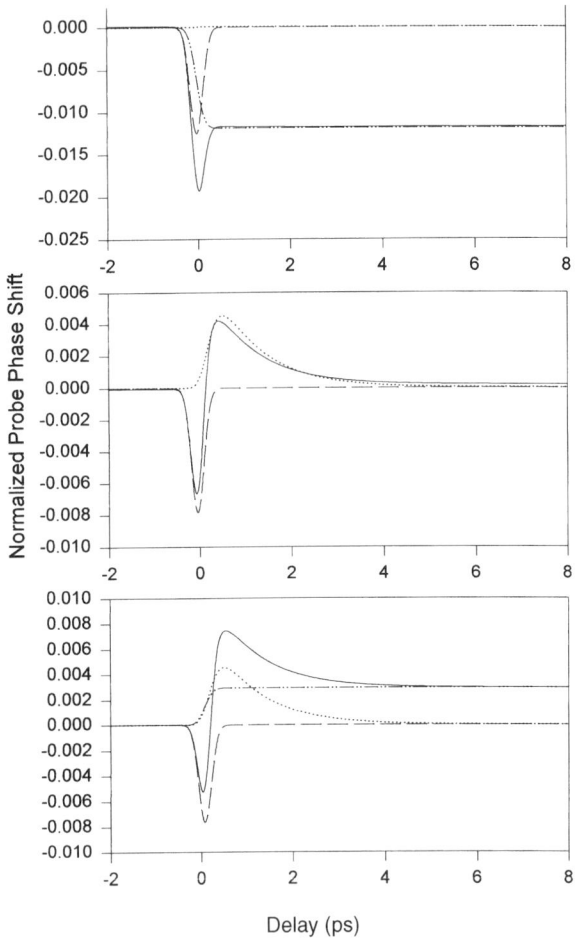

FIG. 22. Total fits (solid lines) for the nonlinear index measurements in the absorption (top), transparency (middle), and gain (bottom) regimes, along with the individual components (dashed lines, optical Stark effect; dotted lines, delayed carrier heating; and dashed-dotted lines, carrier density changes).

The impulse response function for the transmission data contains many more terms. However, we can constrain the fit in a number of ways. For example, the index and transmission are obtained simultaneously, using two ham radio receivers, one for AM reception and one for FM. Therefore, the position of zero time delay is the same in both sets of data. Also, the delay in the thermalization of the heated carrier distribution and its subsequent recovery back to the lattice temperature should have the same time constants in the transmission data as in the phase data. Therefore, when fitting the transmission data, we assume

$$h_i(t) = u(t)[a_4 + a_{ch}e^{-t/\tau_{ch}}(1 - e^{-(t/\tau_{eff})}) + a_{SHB}e^{-t/\tau_{SHB}}] + a_5\delta(t) \quad (50)$$

where a_4 is the magnitude of the step change in transmission, a_{SHB} and τ_{SHB} are the magnitude and time constant for the spectral hole burning component of the response, and a_5 is the magnitude of the TPA coefficient. To compute the fit, this $h_i(t)$ is convolved with the autocorrelation function and then the dynamic coupling term, proportional to the derivative of the index response (Δn), is added in as

$$\Delta T \propto \int h_i(\tau - t)G^2(t)dt + a_6 \frac{\delta(\Delta n)}{\delta t} \quad (51)$$

Figure 23 shows the measured change in probe transmission (solid line) and

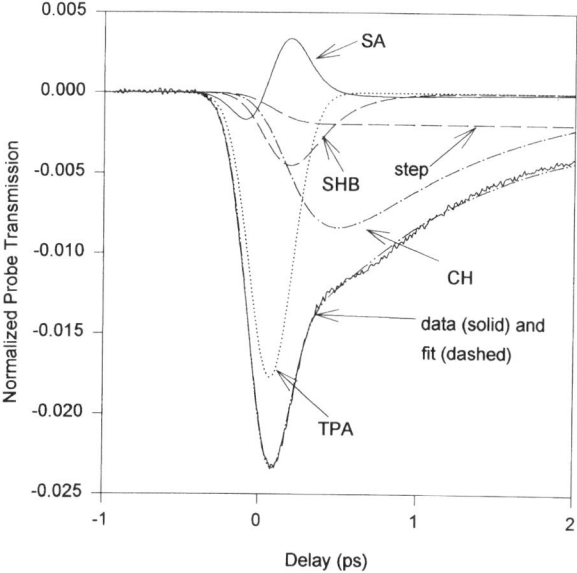

FIG. 23. Measured change in probe transmission (solid line) as a function of pump-probe delay for an MQW diode biased in the gain regime, along with the total fit (dash-dot-dot line). The individual fitting components are labeled on the figure. The term attributed to the dynamic coupling of the gain and index is denoted "SA" for "spectral artifact."

the curve fit (dashed line) as functions of pump-probe delay for an MQW diode biased in the gain regime. The agreement between the data and the fit is so good that it is difficult to distinguish the two. Also shown are the individual components in the total fit. The long-lived "step" component as well as the spectral hole burning (SHB) component are attributable to stimulated emission-induced carrier density changes. Also shown are the carrier heating (CH) component, the two photon absorption (TPA) component, and the dynamic gain-index coupling component labeled "SA" for "spectral artifact." In the process of fitting the data, we have observed that the thermalization time constant τ_1, and the spectral hole burning time constant τ_{SHB}, are always approximately the same. This result is not surprising, because both the thermalization of the carrier distribution and the filling of the spectral hole are effects dominated by carrier-carrier scattering. Note however, that these time constants are similar to the pump-probe optical pulsewidth and are therefore close to the time resolution of the experiment.

In Figs. 24 and 25, we show the measured change in probe transmission as a function of pump-probe delay for the same MQW SOA biased in the transparency and absorption regimes, respectively. Note that at the trans-

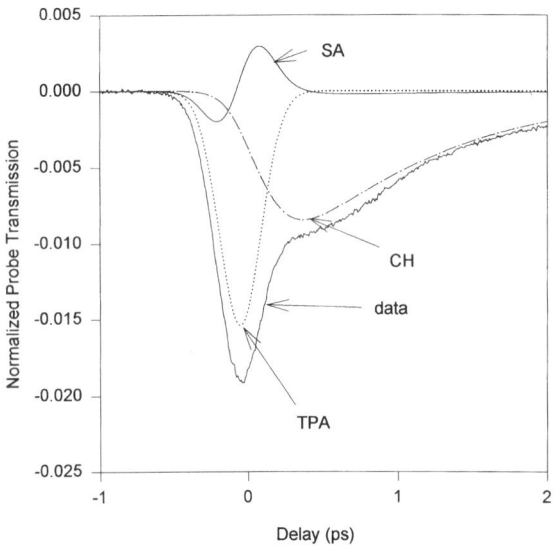

FIG. 24. Measured change in probe transmission (solid line) as a function of pump-probe delay for an MQW diode biased at the transparency point. The individual fitting components are labeled on the figure.

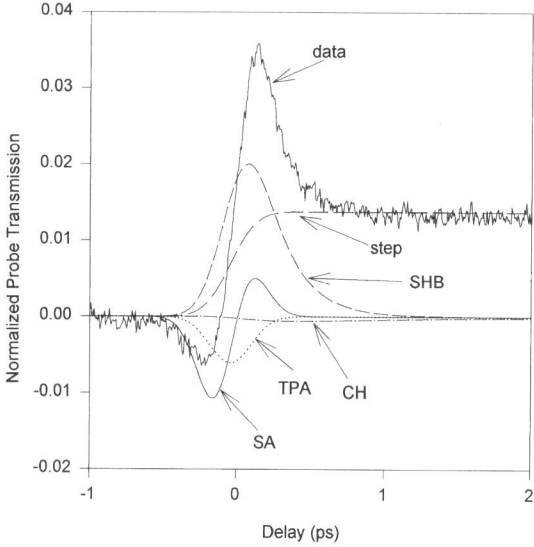

FIG. 25. Measured change in probe transmission (solid line) as a function of pump-probe delay for an MQW diode biased in the absorption regime. The individual fitting components are labeled on the figure. The term attributed to the dynamic coupling of the gain and index is denoted "SA" for "spectral artifact."

parency point, there is no step or spectral hole burning component of the response, by definition. In the absorption regime, the signs of the step and spectral hole burning components have reversed, relative to the gain regime. Also note that the amplitude of the carrier heating term in the absorption regime has been reduced, but the effect is still a reduction in probe transmission, recovering with the characteristic 1-ps time constant. This large reduction in the amplitude of the carrier heating effect at low carrier densities suggests that free carrier absorption is the dominant source of heating in this experiment.

All of the data shown in Figs. 19–21, and 23–25 have been fit using the impulse response function method. The physical mechanisms responsible for the various components in the response are consistent with the scaling of the amplitudes — the a_j's — in the fits. For example, the carrier heating term has the same sign, causing a reduction in probe transmission in the gain, transparency, and absorption regimes. In addition, the carrier heating contribution to the response increases with increasing bias currents, or carrier densities, suggesting that free carrier absorption is an important heating mechanism. The spectral hole burning term and the long-lived change in transmission owing to carrier density changes are positive in the absorption regime, zero at the transparency point, and negative in the gain regime. In addition, we see that TPA causes a transient decrease in probe transmission in all regimes of operation, and that the magnitude of this term is relatively insensitive to carrier density. Finally, we see the dynamic coupling of the gain and index (spectral artifact) in all three regimes of operation. The magnitude of this term is projected to be a maximum when the pump-probe wavelengths are away from the gain peak of the device, going to zero at the peak gain wavelength. The trend we observe, by tuning the bias current to the device, agrees with this prediction.

Our understanding of the nonlinear response of active semiconductor media has changed over the past 12 years. Before the seminal work of Stix *et al.* (1986), theoretical studies of the gain nonlinearities considered only population dynamics and spectral hole burning effects. However, early experiments showed conclusive evidence of an unpredicted nonlinearity — i.e., carrier heating. It was not until 6 years later, using significantly shorter (and more nearly to transform-limited) pulses, that Hall *et al.* (1992b) observed the delay in the turn-on of the carrier heating. That work also showed the first experimental evidence of spectral hole burning in active media. Recently, advances in experimental techniques (Hall *et al.*, 1992a) have allowed simultaneous measurement of gain and index nonlinearities. These new measurements allow more precise fitting routines and less ambiguity in the fitting coefficients. These new techniques have provided the first experimental verification of the dynamic coupling of the index and gain nonlinearities described by Mørk and Mecozzi (1994). Still, we can expect more changes in our understanding in the *next* 12 years. More accurate determination of the carrier thermalization time, and of the spectral artifact, will be obtained when shorter optical pulses are utilized in pump-probe experiments. However, the length of the optical waveguide being tested must then also be reduced, as the optical pulsewidths get shorter, to avoid significant pulse spreading as a result of dispersion (Hall *et al.*, 1992c). Still, the physical mechanisms responsible for interesting dynamics recovering with time constants of less than 200 fs might be further clarified using shorter pump-probe pulses.

VI. Shaping and Saturation of Short Pulses in Active Waveguides

Saturation is a phenomenon that is common to all laser materials. In active semiconductor materials, saturation occurs because stimulated emission decreases the population inversion within the active region of the diode. Carriers that are removed by means of stimulated emission are not instantaneously replaced, and therefore the gain is temporarily depleted. If the gain behaves as a homogeneously broadened system, then in the limit where the photon density that stimulates emission is high, the carrier concentration can be driven to the transparency point, where stimulated emission and stimulated absorption of incident photons are equally probable. At this point, the small signal gain of the material is equal to 1, and the power into the device is equal to the power out of the device (except, of course, for coupling and scattering losses). A simple rate equation model for the gain saturation (Siegman, 1986) is given by

$$g = \frac{g_0}{1 + I/I_{sat}} \quad (52)$$

where g_0 is the small-signal, single-pass gain, I is the intensity of the light in the diode, and I_{sat} is the saturation intensity given by

$$I_{sat} = \frac{\hbar\omega}{\sigma\tau_s} \quad (53)$$

where σ is the stimulated transition cross section. For semiconductor materials, it is more typical to replace the stimulated transition cross section σ, with its equivalent term in the semiconductor rate equations, the differential gain g_d. Experimental results have shown that theoretical models treating the gain of a semiconductor optical amplifier as a homogeneously broadened system (Frigo, 1983; Thedrez et al., 1988; de Rougemont and Frey, 1988) are accurate for continuous wave (cw) signals and broad (>10 ps) pulses. Wiesenfeld et al. (1988) showed that pulses of 3 to 9 ps could propagate through a semiconductor optical amplifier and undergo no measurable pulse distortion, even while reducing the gain by 3 dB. They also showed that, for these pulsewidths, the gain of the device saturated at a pulse energy of

$$E_{sat} = \frac{\hbar\omega A}{2g_d \Gamma} \quad (54)$$

where Γ is the mode confinement factor and A is the cross-sectional area of the device. The saturation energy E_{sat}, is the pulse energy at which the probability of carrier recombination by stimulated emission is $(1 - 1/e)/2$, and is related to both material and device parameters (Siegman, 1986). The saturation energy can be related to the 3-dB output saturation energy E_{3dB}

of the device, defined as the output energy at which the gain has been reduced by 3 dB with respect to its small signal value, by

$$E_{3dB} = \left[\frac{1.38}{(1 + \alpha/g_0)}\right] E_{sat} = \left[\frac{1.38}{(1 + \alpha/g_0)}\right]\left(\frac{\hbar\omega A}{g_d \Gamma}\right) \quad (55)$$

where α is the loss coefficient for the active region (Eisenstein et al., 1990). Likewise, the saturation power, $P_{sat} = E_{sat}/\tau_s$, can be related to the 3-dB output saturation power P_{3dB} by

$$P_{3dB} = \left[\frac{0.691}{(1 + \alpha/g_0)}\right] P_{sat} \quad (56)$$

Several sets of experimental results have confirmed that gain saturation for cw light, and for optical pulses longer than a few picoseconds with repetition periods much greater than the upper state lifetime τ_s, can be explained using a simple rate equation model assuming a homogeneously broadened two-level system (Eisenstein et al., 1988; Wiesenfeld et al., 1988; Saitoh et al., 1989; Saitoh and Mukai, 1990; Eisenstein et al., 1990).

The saturation characteristics are somewhat different for trains of optical pulses with high repetition rates. The repetition rate, $R = 1/T$, is considered high if the pulse period T, is short compared with the upper state lifetime τ_s. In this case, the gain of the device does not completely recover between optical pulses. The homogeneously broadened, two-level system still can be used to model the device behavior, with the exception that certain parameters must be generalized to be time-dependent. For example, the carrier density, and therefore the small signal gain, become time-dependent quantities. It has been shown by Hansen et al., (1989) that, in this case, the device can be described by an effective saturation energy,

$$E_{sat,eff} = \tanh\left(\frac{T}{2\tau_s}\right) E_{sat} \quad (57)$$

which is repetition-rate-dependent. Note that $E_{sat,eff}$ reduces to E_{sat}, in the limit where $T \gg \tau_s$.

The saturation characteristics become quite different when the optical pulses that are incident on the amplifier are short, with durations that are comparable to the nonlinear gain (carrier heating) recovery time. In this case, the gain saturation of the device is not accurately described by a simple two-level system. Rather, the rate equations must be modified to include the effects of carrier heating on the carrier density. As an example of the importance of the nonlinear gain on the saturation behavior of an amplifier, consider Fig. 26, which shows the amplifier gain for two devices — a bulk active region amplifier and an MQW amplifier — as a function of output pulse energy, measured using two optical pulsewidths, 15 ps (diamonds) and 150 fs (circles). In this experiment, optical pulses with varying pulse energies were launched into the amplifier, and the gain, E_{out}/E_{in}, was measured as a function of output energy. The optical pulses were generated by a modelocked laser system with a pulse repetition rate of 100 MHz. The 10-ns pulse-to-pulse spacing is much longer than the upper state lifetime for these devices, so the repetition rate dependence of the saturation energy does not explain these results. Rather, the saturation behavior of the device is pulsewidth-dependent.

This saturation behavior is well modeled by the rate equations described

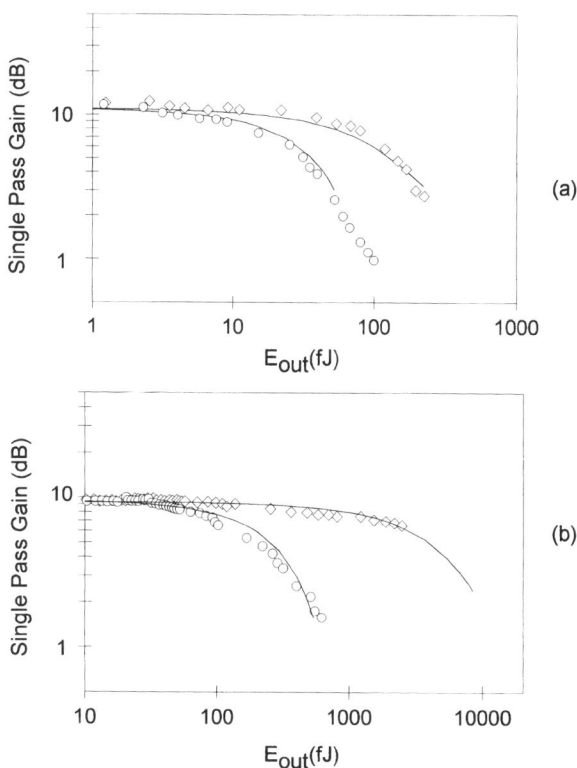

FIG. 26. Amplifier gain as a function of output energy for pulse widths of $\Delta t = 15$ ps (diamonds) and $\Delta t = 150$ fs (circles). (a) The bulk amplifier bias current was three times the threshold, and the pulse wavelength was $\lambda = 1.515\,\mu$m. (b) the MQW amplifier bias current was two times the threshold, and the pulse wavelength was $\lambda = 1.515\,\mu$m. The solid lines are the theoretical predictions of the saturation behavior obtained from the modified rate equation model.

in Section V. For example, remember that the rate equation model was able to fit the experimental pump-probe data, as shown in Fig. 18. Those data (and the associated fit) were measured (calculated) using optical pulses with widths of $\Delta t = 150$ fs. Figure 27 shows the calculated pump-probe result for a 20-ps pulse propagating in the diode. The total delay scale is much larger than in previous plots—200 ps in this plot versus 10 ps in the previous plots. The data are plotted in terms of the normalized change in probe transmission and are roughly independent of pump energy as long as the device is not highly saturated. Notice that the ultrafast dynamics caused by the carrier heating, which were a significant portion of the 150-fs pulses's response, are virtually nonexistent in the 20-ps pulse's response. Similar pulsewidth-dependent responses have been observed in GaAs diodes (Delfyett et al., 1991). It is this difference in response for the two pulsewidths that is responsible for the pulsewidth-dependent saturation behavior.

One can calculate the gain for the device, as given in Eq. (35), using the values of γ_u, γ_{ch}, α_{fc}, τ_{ch}, τ_u, and β that have been obtained by fitting the pump-probe data. The results of the calculation appear as solid lines in Fig. 26. Here, the initial carrier density N is adjusted to fit the observed linear gain. No other material parameters are changed once the pump-probe data have been fitted. Notice that the roll-off of the gain curves as predicted by the theory is in good agreement with the experiment. From these experiments, we conclude that carrier heating will play an important role in the amplification of sub-picosecond pulses. The so-called ultrafast term, γ_u, did not contribute significantly in determining the output saturation energy in these experiments, but it may be an important contribution at other wavelengths (D'Ottavi et al., 1994; Willatzen et al., 1991; Willatzen et al., 1992; Uskov et al., 1992) and for shorter pulsewidths.

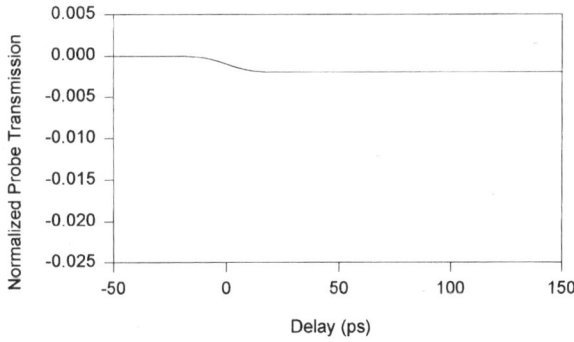

FIG. 27. Predicted change in probe transmission versus pump-probe delay (on a 200-ps timescale) for 20-ps pulses. This curve was calculated using the impulse response function from Fig. 18.

It is important to note that even though the saturated output energy for ultrashort pulses is less than for longer pulses, partial gain recovery following such pulses is very fast (~ 1 ps). Thus a rapid sequence of ultrashort pulses can be used to extract the same net energy as that of a long pulse. Figure 28 shows the calculated saturation curves for a bulk amplifier and for pulsewidths ranging from 200 fs to 20 ps. Note that for pulses longer than approximately 10 ps, the carrier heating nonlinearity has negligible influence on the gain. For these longer pulses, the saturation behavior of the diode becomes pulsewidth-independent. For short optical pulses that contain most of their energy in a broad pedestal, the saturation behavior will also be relatively independent of pulsewidth (Saitoh et al., 1989; Saitoh and Mukai, 1990). However, clean, pedestal-free pulses, even those as long as 5 ps, have reduced output saturation energies owing to the carrier heating nonlinearity.

In addition to nonlinear transmission changes, optical pulses also undergo nonlinear phase changes and associated spectral changes when propagating through an active semiconductor waveguide. Most pump-probe measurements of the refractive index nonlinearities were performed using sub-picosecond pulses in the small-signal regime. Typical measured nonlinear phase shifts were less than $\pi/2$. These phase shifts were too small to effect any noticeable change in the spectrum of the pump-probe pulses (Hultgren and Ippen, 1991; Hall et al., 1993). Such small phase shifts, however, may impart chirp on optical pulses propagating in a semiconduc-

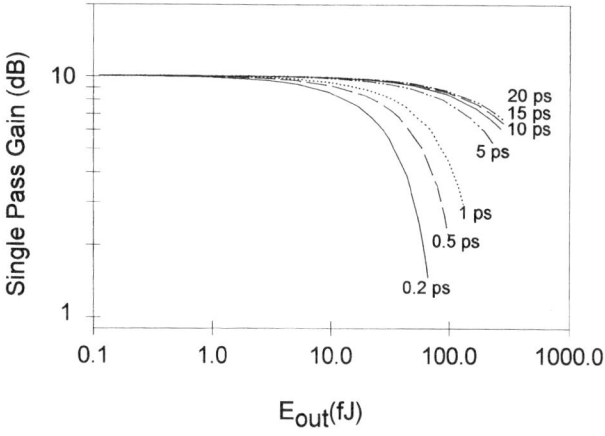

FIG. 28. Predicted gain for a semiconductor amplifier as a function of output energy for a variety of input wavelengths. The solid line shows the result for 0.2-ps pulses, the dashed line for 0.5-ps pulses, the dotted line for 1-ps pulses, and so on, as labeled on the plot.

tor amplifier (Koch and Linke, 1986; Agrawal and Olsson, 1989a). However, spectral broadening as a result of larger induced nonlinear phase shifts was observed for picosecond pulses with tens of picojoules of pulse energy (Olsson and Agrawal, 1989; Grant and Sibbett, 1991). If one assumes that the observed broadening of the spectrum can be attributed to an instantaneous refractive index nonlinearity following the temporal envelope of the pulse, one can use the standard method for relating spectral changes to induced nonlinear phase shifts (Stolen and Lin, 1978). Using this method, Grant and Sibbett (1991) calculated a nonlinear refractive index coefficient, n_2,

$$n_2 = \frac{\lambda \Delta \Phi}{2\pi I L} = 2 \times 10^{-11} \, \text{cm}^2/\text{W} \tag{58}$$

where λ is the center wavelength of the pulses, L is the length of the diode, and I is the pulse intensity.

These large nonlinear phase shifts and spectral changes were observed for pulses that previously had been shown to propagate through a similar amplifier with no temporal distortion of the pulse profile. The spectral broadening may not translate into large distortions in the time domain because the group velocity dispersion in these diodes is low (Hall et al., 1992c) and because the diode amplifier lengths are short, typically less than 1 mm.

In GaAs, spectral distortion associated with pulse propagation through a semiconductor amplifier was observed in single-pulse as well as dual-pulse (pump-probe) experiments. Delfyett and coworkers (Delfyett et al., 1991) showed that measured changes in pulse frequency could be used to characterize nonlinearities in refractive index. The instantaneous pulse frequency, ω_{inst} is given by

$$\omega_{\text{inst}} = \omega_0 - \frac{\omega_0 L}{c} \left(\frac{\partial n(t)}{\partial t} \right) \tag{59}$$

where ω_0 is the optical carrier frequency, L is the interaction length, and $n(t)$ is the time-dependent refractive index (Shen, 1984). As explained in the preceding section, when pulses that propagate through the diode amplifier undergo gain, they reduce the carrier density by stimulating emission. Reducing the carrier density increases the refractive index (Thompson, 1985; Henry et al., 1981; Manning et al., 1983). If a carrier density change resulting from stimulated emission is the only effect, an instantaneous frequency red shift, proportional to the intensity envelope of the optical pulse, will be imparted to the pulse (Agrawal and Olsson, 1989b; Finlayson *et al.*, 1990). However, if other mechanisms contribute to the gain and refractive index changes in the diode, the resulting shifts in the instantaneous frequency are more complicated. For example, in agreement with previous pump-probe experiments, Delfyett *et al.* (1991) showed that the instantaneous frequency shifts induced on a propagating pulse were pulsewidth-dependent. Pulses that were short enough to be affected by the carrier heating dynamic showed, in addition to the red shift owing to carrier density changes, a blue shift of the carrier frequency, owing to the partial recovery of the gain through carrier thermalization and the carrier heating recovery.

Based on the experimental evidence of refractive index nonlinearities resulting from carrier heating, new models of pulse propagation in semiconductor amplifiers were developed that incorporated these effects (Dienes *et al.*, 1992; Mørk and Mecozzi, 1996; Hong *et al.*, 1996). In these theories, the nonlinear terms are derived semiphenomenologically, as they were in the original rate equation models matching experiments and theory (Lai *et al.*, 1990).

VII. Four-Wave Mixing

Up until this point, we have concentrated on the time domain characterization and interpretation of the nonlinear responses of active semiconductor media. However, frequency domain measurements are a complementary way to characterize the gain and refractive index nonlinearities in semiconductor diode lasers and amplifiers. In frequency domain studies, optical beams with different carrier frequencies are simultaneously coupled into the waveguide of interest, and the generation of new mixing frequencies is analyzed. This process is called four-wave mixing (FWM). The original theoretical treatments of wave mixing effects in semiconductor lasers can be traced to Bogatov *et al.* (1975), who showed that the efficiency of asymmetric interaction of spectral modes in a diode decreased as spectral distances increased. Subsequent theoretical treatments describing nearly degenerate FWM in amplifiers and lasers considered only two physical mechanisms: modulation of the carrier density at the beat frequency of the two input waves, and spectral hole burning (Agrawal, 1987; Agrawal, 1988). These theoretical treatments were later modified to include the effects of carrier heating (Kikuchi *et al.*, 1994; Uskov *et al.*, 1994; Mecozzi *et al.*, 1995b).

A simple schematic diagram of an FWM spectroscopy experiment in a semiconductor optical amplifier is shown in Fig. 29. Two input beams, with carrier frequencies of ω and $\omega + \Delta$, are launched into the device. Typically,

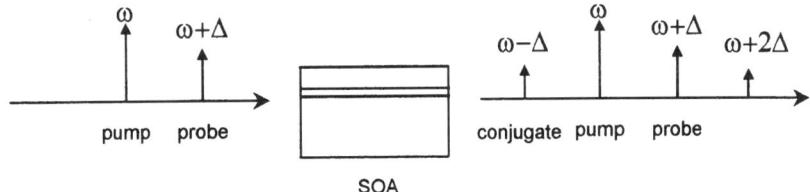

FIG. 29. Schematic diagram of a four-wave mixing (FWM) experiment. Two input tones, a pump at ω and a probe at $\omega + \Delta$, are launched into an active semiconductor waveguide. FWM in the active waveguide generates a conjugate signal at $\omega - \Delta$. Typically, additional sidebands, such as the one at $\omega + 2\Delta$, may be observed in FWM experiments.

one field, the pump field, is much stronger than the other field, the probe field. The beat frequency of these two fields drives carrier distributions in the material and modulates the gain (and refractive index). This induced gain modulation couples energy from the pump and probe fields to a third field, the so-called conjugate field, with a carrier frequency of $\omega - \Delta$. The phase of the conjugate field is the conjugate of the probe field. The efficiency with which the beat frequency modulates the gain and index depends on the various resonances in the medium response. Measuring the strength of the generated conjugate signal as a function of pump-probe detuning reveals various frequency roll-offs related to the characteristic time constants of the material nonlinearities.

The susceptibility determined by FWM experiments, $\chi^{(3)}(\Delta)$, can be related to the susceptibility determined by pump-probe experiments, $\chi^{(3)}(t)$, by the Fourier transform

$$\chi^{(3)}(\Delta) = \int_{-\infty}^{\infty} \chi^{(3)}(t) e^{(-j\Delta t)} dt \tag{60}$$

For example, analysis of pump-probe data yields

$$\chi^{(3)}(t) = a_1 e^{(-t/\tau_1)} + a_2 e^{(-t/\tau_2)} + a_3 e^{(-t/\tau_3)} + \cdots \tag{61}$$

where the relative strengths of the various nonlinear effects are represented by complex amplitudes a_j and the characteristic time constants are represented by τ_j. Figure 30 shows a calculation of $|\chi^{(3)}(\Delta)|$ versus Δ given $|a_1| = |a_2| = |a_3| = 1$, and $\tau_1 = 1\,\text{ns}$, $\tau_2 = 1\,\text{ps}$, and $\tau_3 = 10\,\text{fs}$. The amplitudes have been chosen arbitrarily, but the time constants are approximations of the carrier lifetime, the carrier heating recovery time, and a

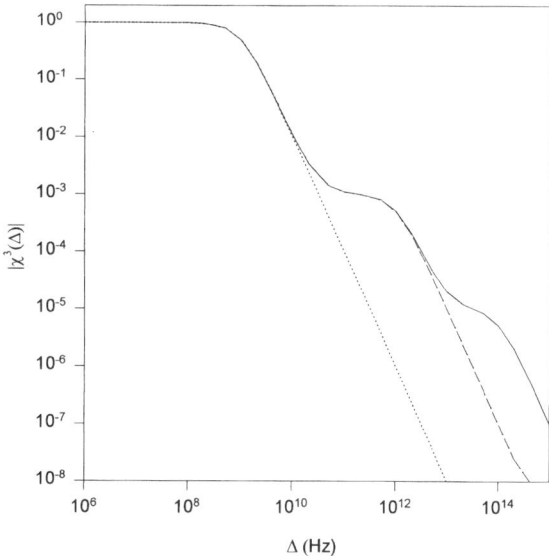

FIG. 30. Example calculation of the frequency dependence of the third-order susceptibility in the presence of carrier population dynamics only (dotted line); carrier population and heating dynamics (dashed line); and carrier population, heating, and scattering effects (solid line).

carrier scattering time. Note that the time constants are revealed as roll-offs in the frequency response. It is important to note that the shape of this curve will change depending on the observed time constants and on whether the contributing nonlinear effects are reactive, resistive, or some combination of both (Mecozzi et al., 1995b).

Experimentally, FWM in a semiconductor amplifier was first reported by Nakajima and Frey (1986). This experiment was performed in a laser diode rather than an amplifier, and the frequency detuning was chosen to coincide with a cavity resonance so that the nonlinearity was enhanced by resonant amplification (Bergmann et al., 1978). Cavity-enhanced FWM is efficient only when the pump and probe wavelengths coincide closely with cavity modes (Provost and Frey, 1989), and so this technique has little spectroscopic value. However, this technique has been applied to high-bit-rate wavelength conversion (Iannone and Prucnal, 1994; Iannone et al., 1994) and has demonstrated wavelength shifts exceeding 1 THz (Murata et al., 1991).

The first experimental verification of the frequency response of the third-order susceptibility at frequency separations greater than 10 GHz in a diode amplifier was achieved by Tiemeijer (1991), who showed a FWM result consistent with a strong carrier heating nonlinearity. However, the frequency detuning in this experiment (600 GHz) was too low for any effects of spectral hole burning to be observed. Later work by Kikuchi et al. (1992), showed that, under some experimental conditions, carrier heating has little effect on the gain of the diode and that spectral hole burning (or some ultrafast mechanism) is the strongest nonlinear effect. More recent experiments have shown that effects resulting from both carrier heating and spectral hole burning are usually observed (Zhou et al., 1993; Kikuchi et al., 1994; D'Ottavi et al., 1994; Uskov et al., 1994). It is interesting to note that a nonlinear response has been frequently attributed to spectral hole burning in FWM spectroscopy experiments, but has been much more difficult to confirm in time domain studies. One possible explanation is the wavelength dependence of the relative strengths of the carrier heating and spectral hole burning nonlinearities (Willatzen et al., 1991; D'Ottavi et al., 1994). Theoretically, it has been suggested that carrier heating dominates the nonlinear response of the diode when the pump-probe wavelengths and the diode bias currents are near their transparency values. Near the band edge, however, spectral hole burning is expected to dominate. This prediction is supported qualitatively by the FWM results and by tunable pump-probe experiments performed near the band edge of a strained InGaAs amplifier (Sun et al., 1993). However, it is important to note that the predictions of the theory are very sensitive to the time constants assigned to various nonlinear effects and to assumed material parameters such as the free carrier absorption coefficient, the carrier density, and the carrier temperature.

While FWM spectroscopy is a complementary measurement to the pump-probe measurements of gain and index nonlinearities, it is not as accurate in determining the amplitudes and time constants of the nonlinear effects. Presumably, the ambiguity in the determination of the a_j's and τ_j's originates in the fitting procedure used to extract these fundamental parameters. As was pointed out originally by Tiemeijer (1991), the experimentally measured relative sideband power, given by P_0/P_1, where P_0 is the power in the conjugate mode and P_1 is the power of the pump, is related to the nonlinear parameters by

$$\frac{P_0}{P_1} = P_1 P_2 \sum_j K_i \frac{1}{1 + (\tau_j^2 \Delta^2)} \tag{62}$$

where the subscript j denotes the physical effect responsible for the FWM, P_2 is the power in the probe, and K and τ denote the strength and characteristic time constant associated with the effect. However, with this fitting technique, FWM experiments have yielded inconsistent results. For example, some FWM results have predicted an α_T that is negative (Zhou et al., 1993) while others have predicted an α_T that is positive (Uskov et al., 1994). Direct measurements by pump-probe techniques have all shown that the sign of α_T, like the sign of Henry's linewidth enhancement factor α, is positive (Hultgren and Ippen, 1991; Hall et al., 1993). There is also an inconsistency in the determined interwell transport time in a polarization-insensitive device. FWM measurements suggest an interwell transport time of 16 ps (Paiella et al., 1996), whereas pump-probe measurements suggest a 7-ps time constant in an identical device (Lenz et al., 1996). Additional measurements need to be made to resolve the differences. Certainly, the FWM measurements lend critical insight into the frequency domain performance of semiconductor optical amplifiers in the presence of optical signals. Some FWM applications for semiconductor optical amplifiers are discussed in the next section.

VIII. Applications

In optical communications systems, gain and refractive index nonlinearities in semiconductor optical amplifiers (SOAs) may limit or enhance system performance. For example, these gain nonlinearities can influence the modulation response (Tucker and Kaminow, 1984; Tucker, 1985; Bowers, 1987) and mode stability (Hendow and Sargent, 1985) of lasers. Refractive index nonlinearities affect the chirp (Koch and Linke, 1986; Cartledge, 1990) and the spectrum (Agrawal and Olsson, 1989b; Grant and Sibbett, 1991) of short optical pulses propagating through the semiconductor amplifier. However, these nonlinearities may also be exploited in some applications. As mentioned previously, FWM and conjugate signal generation may be exploited for midspan phase conjugation, significantly increasing the propagation distances in long-haul transmission systems (Yariv et al., 1979; Jopson and Tench, 1993; Iannone et al., 1994; Tatham et al., 1994; Lacey et al., 1995). Also, FWM may be used to characterize cross-talk and intermodulation distortion in wavelength division multiplexed systems (Lassen et al., 1988; Durhuus et al., 1992). The pulsewidth-dependent saturation behavior can be used as a pulse discriminator in code division multiplexed systems (Salehi et al., 1989). Gain saturation effects have also been used in fiber loop memory applications to provide pulse energy stability (Doerr et al., 1994) and to set the data rate and provide timing stability (Hall et al., 1995). Currently, the most common applications for active semiconductor nonlinearities are wavelength converters and optical switches.

1. WAVELENGTH CONVERTERS

In wavelength division multiplexed (WDM) systems, data can be encoded on different wavelength channels and propagated through the system. WDM networks can scale to higher rates and larger numbers of users by reusing wavelengths in separate parts of the network (Alexander et al., 1993). Whether or not wavelength converters increase the capacity of a network depends on the topology and geographical extent of the network (Kaminow et al., 1996). For wide area networks and mesh topologies, models predict modest benefits when wavelength converters are employed (Barry and Humblet, 1996; Ramaswami and Sivarajan, 1996). Even though their potential benefits are unclear from an architectural standpoint, development of high-speed wavelength converters has been an active area of research.

The function of a wavelength converter is to take an optical data signal at one wavelength and translate it to another wavelength, while maintaining the integrity of the original signal. Ideally, this function is performed in a way that is transparent to, or insensitive to, the bit-rate and modulation format of the original signal. Opto-electronic techniques in which the optical data signal is detected, filtered, and amplified electronically, and then used to modulate light at a different wavelength, are limited in speed by the opto-electronic conversion. All-optical techniques based on cross-gain modulation, cross-phase modulation, and FWM in SOAs are more promising.

Of these three techniques, cross-gain modulation (XGM) is the simplest. In this technique, an intensity modulated signal, referred to as the pump, propagates through an SOA and reduces the gain. The induced gain fluctuations are impressed on a second input to the SOA, a cw beam at a different wavelength called the probe (Koga et al., 1988; Glance et al., 1992; Joergensen et al., 1993). In this case, the wavelength-converted data are the complement of the original data signal. The complementary data may also be encoded on multiple cw probe beams simultaneously (Wiesenfeld and Glance, 1992). While wavelength conversion to both shorter and longer wavelengths has been demonstrated (Wiesenfeld et al., 1993; Joergensen et al., 1993), the extinction ratio of signals converted to shorter wavelengths is always better because the gain compresses asymmetrically as a result of band-filling effects (Wiesenfeld, 1996). Because the technique relies on gain modulation, as the optical data rate increases, steps must be taken to decrease the upper-state lifetime of the carriers in the SOA so that the carrier-density modulation can follow the pump signal. As discussed previously, the upper-state lifetime can be decreased in the presence of an intense optical holding beam (Manning and Davies, 1994; Patrick and Manning, 1994). In many XGM demonstrations, the probe beam itself is used as the optical holding beam (Mikkelsen et al., 1993; Wiesenfeld et al., 1993; Wiesenfeld et al., 1994a).

The necessary presence of an intense optical holding beam decreases the steady state gain of the SOA and reduces the extinction ratio of the wavelength-converted signal. Therefore, as the optical data rate increases, wavelength conversion by XGM is hindered by a reduced extinction ratio and by intersymbol interference owing to the finite carrier lifetime. Still, conversion at data rates as high as 20 Gb/s have been demonstrated (Wiesenfeld et al., 1994a). An additional drawback of this technique is that at high input powers and large gain reductions, significant phase changes accompany the gain changes. These phase changes may impart a chirp to the optical data stream and limit the signal transmission distance. Also note that this technique is applicable only to amplitude modulated signals. Still, wavelength conversion by XGM is a useful technique because it requires only moderate input powers and it can be a polarization-insensitive technique if the gain of the SOA is polarization-insensitive. Also, if the pump and probe propagate in opposite directions through the SOA, no filter or polarizer is needed to separate the pump and probe beams at the SOA output. Finally, this technique has been demonstrated using semiconductor lasers rather than SOAs (Ottolenghi et al., 1993; Braagaard et al., 1994). The use of lasers requires higher input powers and yields less flexibility in the range of converted wavelengths.

Cross-phase modulation (XPM) also can be used to achieve wavelength conversion. In this technique, the pump compresses the gain and changes the refractive index of the semiconductor amplifier. A probe beam propagating through the SOA acquires a variable phase shift, depending on whether or not the pump is present. If the SOA is placed in one arm of an interferometer, the induced phase change or phase modulation can be converted to an intensity modulation (Mikkelsen et al., 1994; Durhuus et al., 1994). There are several advantages of achieving wavelength conversion by XPM rather than by XGM. One advantage is that the interferometer can be configured for either "inverting" and "noninverting" operation, depending on the initial phase bias. Inverting operation, like XGM, generates a complementary copy of the input signal at the converted wavelength, whereas noninverting operation maintains the original data stream exactly. Another advantage is that very high extinction ratios can be realized in the interferometer (Wiesenfeld, 1996) and counterpropagating pump and probe beams can be used to eliminate the need for a filter or polarizer at the output of the converter. Also, the wavelength dependence of the phase change is weaker than that of the gain change (Fig. 15), so conversion to shorter and longer wavelengths is more uniform. In addition, the sign of the chirp imparted to the wavelength converted signal depends on whether the interferometer is biased as an inverting or noninverting wavelength converter. For noninverting operation, the chirp imparted to the converted signal causes pulse compression in standard optical fiber so that no dispersion penalties are observed in transmission experiments (Ratovelomanana et al., 1995; Idler et al., 1995). The drawbacks associated with the XPM technique are the interferometric design of the device, the high sensitivity of device performance to changes in input parameters such as power level, polarization, and wavelength, and the fact that only amplitude modulated signals can be converted. Still, wavelength conversion at rates up to 40 Gb/s has been demonstrated in integrated interferometric converters employing refractive index nonlinearities in SOAs (Danielsen et al., 1996).

FWM is the only all-optical wavelength conversion technique that is independent of the data modulation format (Vahala et al., 1996). Besides amplitude modulated signals, it converts analog signals and phase modulated signals, but the phase modulation is inverted because the wavelength converted stream is the phase conjugate of the input. Recall that the phase conjugation of the wavelength converted beam allows the "undoing" of transmission induced spectral distortions in midspan spectral inversion schemes (Tatham et al., 1994). However, relative to XGM and XPM, FWM has received less consideration as a feasible wavelength conversion technique for systems applications. One reason is that it is complicated to make the technique polarization-insensitive (Jopson and Tench, 1993). Also, some sort of filtering is required at the output of the FWM device to separate the pump, probe, and conjugate beams. Another drawback is that the conversion efficiency is highly wavelength-dependent (Fig. 30), as well as being asymmetric (Zhou et al., 1993). However, recent experiments have shown low-noise wavelength conversion efficiencies of 0 dB for wavelength shifts exceeding 5 nm (Girardin et al., 1997). In addition to these results, FWM wavelength conversion of data streams has been demonstrated at a rate of 10 Gb/s (Ludwig and Raybon, 1994; Lee et al., 1997).

2. ALL-OPTICAL SWITCHING

All-optical switches can perform logic operations at data rates greatly exceeding those achieved by electronic switches. These switches will find applications in all-optical networks, in high-speed data encryption and analog-to-digital (A/D) converters, and in other specialized processors. Currently, electronic gates operating at bit rates exceeding 50 Gb/s do not seem practical. Instead, all-optical logic gates, in which both "logic" inputs to the gate are optical pulse streams, can be implemented. One of the great attractions of all-optical switches is the potential for scaling their operation to rates higher than a few hundred gigabits per second. However, optical switches will be feasible only if they can be reliably and repeatedly manufactured. As a result, considerable recent research has centered around the use of semiconductor optical amplifiers as the all-optical switching elements. The most attractive features of SOAs are that short device lengths (~ 1 mm) are possible and that there is potential for monolithic integration with other components.

All-optical switches require materials whose transmission or refractive index is intensity-dependent. The simplest switches utilize parametric processes based on $\chi^{(3)}$ nonlinearities such as four-wave mixing (FWM) and cross-gain modulation (XGM) in a single semiconductor waveguide. As described above, switches based on cross-gain saturation can operate as NOT gates or inverters because the transmission of one pulse through the device can be extinguished in the presence of another pulse. FWM, on the other hand, provides AND functionality when the conjugate signal is the optical output from the switch. The conjugate is generated only when both optical input pulses are simultaneously present in the waveguide. As is the case for wavelength converters, increased functionality and some switching performance benefits can be realized by using refractive index nonlinearities. In such cases, physical implementation of the switch is more complicated because the semiconductor waveguide must be placed in an interferometer to convert phase changes to amplitude or polarization changes. In addition, the transmission nonlinearities associated with the refractive index nonlinearities (as described by α, α_T, and the Kramers-Kronig relation) can degrade the contrast ratio of the switch (Patel *et al.*, 1997). However, interferometric switches can be configured to achieve a greater range of logical operations and higher contrast ratios than the single waveguide switches can achieve.

In some sense, the wavelength conversion experiments described in the preceding section are optical switching experiments. However, those experiments have relied on the slower transmission and refractive index effects resulting from carrier population changes. Short-pulse all-optical switching experiments try to circumvent these carrier population changes, because they are slow, and seek instead to capitalize on the high-speed nonlinear responses caused by carrier heating, SHB, TPA, and the optical Kerr effect. Figure 31 shows an optical switch consisting of a two-arm interferometer with a nonlinear material in one arm. The interferometer operates as follows: signal pulses incident on the interferometer are split into two equal-power pulses, which traverse the two arms and are recombined at the output coupler. The transmission of the interferometer depends on the path length difference between the two arms, according to

$$\frac{P_{\text{out}}}{P_{\text{in}}} = \cos^2\left(\frac{\Phi_1 + \Delta\Phi}{2}\right) \tag{63}$$

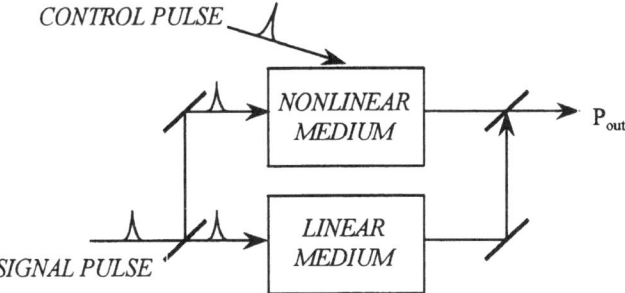

FIG. 31. Schematic diagram of a two-arm interferometer with a nonlinear material in one arm.

where $\Delta\Phi$ is the nonlinear phase shift described in Eq. (58) and $\Phi_1 = {}^2\pi\Delta L/\lambda$ is the linear phase bias. All-optical switching is achieved when a control pulse with the intensity to induce a nonlinear phase change of π is incident on the nonlinear medium. In this scenario, the signal and control beams are the logical inputs to the gate. The interferometer acts as an AND gate if Φ_1 is π, and as a NOT gate if Φ_1 is 0. If the nonlinear medium is an SOA, the operating speed of the switch is limited to $1/\tau_s$, the population recovery rate, by intersymbol interference effects. More specifically, if the period of the bit stream is less than τ_s, the nonlinear medium will not recover completely between control pulses (Fig. 17), and residual phase shifts will be imparted to subsequent signal pulses, even in the absence of a control pulse.

One way to circumvent the speed limitations of these long-lived effects is to use a balanced interferometer and to impose a differential nonlinear phase shift between the arms of the interferometer (Sokoloff et al., 1993). In a balanced device, there are SOAs in both arms, offset in relative pulse arrival time by Δt, as shown in Fig. 32. The switching behavior is given by

$$\frac{P_{out}}{P_{in}} = \cos^2\left[\frac{\Phi_1 + \Delta\Phi_1(t) + \Delta\Phi_2(t + \Delta t)}{2}\right] \quad (64)$$

where $\Delta\Phi_1(t)$ is the nonlinear phase shift induced in one arm, $\Delta\Phi_2(t + \Delta t)$ is the nonlinear phase shift induced in the other arm, and Φ_1 is the linear bias of the interferometer, as described previously. Control pulses launched into the interferometer are split into two components, upper and lower, that traverse the separate arms and arrive at the two SOAs at different times. The upper control pulse component should arrive at SOA1 before the signal pulse component in that arm, but after the signal pulse component in the

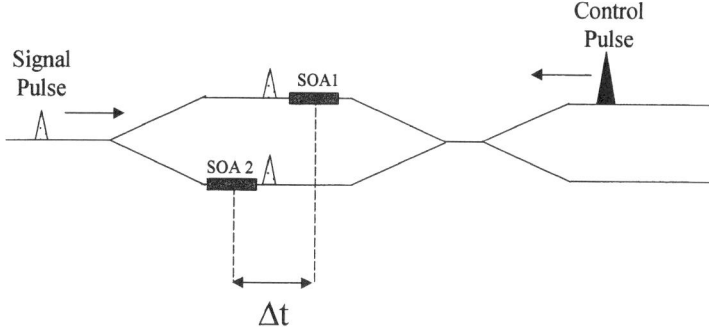

FIG. 32. Schematic diagram of a balanced nonlinear interferometer using SOAs as the nonlinear medium.

lower arm has traversed SOA2, as shown in Fig. 32. In this way, the control pulse in the upper arm of the interferometer induces a phase shift, $\Delta\Phi_1(t)$, and imbalances, or switches, the state of the interferometer. The interferometer remains imbalanced until the other (lower) component of the control pulse arrives at the second SOA (SOA2) and induces a phase shift $\Delta\Phi_2(t + \Delta t)$. If the control pulse intensity is split equally between the two arms, the two SOAs are identical, and Δt is on the order of 10 ps, then $\Delta\Phi_1(t) = \Delta\Phi_2(t + \Delta t)$. In this case, the second component of the control pulse switches the state of the interferometer back to the initial state, and the interferometer is balanced again, even though the long-lived refractive index nonlinearities have not recovered completely (Adams *et al.*, 1995). A balanced interferometer of this sort is characterized by the so-called "switching window." The switching window is defined as the time period during which the interferometer is imbalanced, Δt. The switching window must be shorter than the time constant of the response being balanced, and shorter than the bit period of the data streams that are being switched. The switching window may be limited by the recovery times associated with refractive index nonlinearities caused by carrier heating.

One drawback of the balanced interferometer arrangement described above is that the interferometer has two separate paths and therefore requires two SOAs. Often, interferometers of this type require external stabilization (Kang *et al.*, 1995). Monolithically integrated Mach-Zehnder interferometers are more stable, but their performance is critically dependent on the similarity of the nonlinear responses of the two SOAs (Jahn *et al.*, 1995). More robust interferometric switch designs are based on single-arm interferometers, such as the fiber Sagnac interferometer or nonlinear optical loop mirror (Doran and Wood, 1988). The original single-arm interferometric switches utilized fiber nonlinearities. Recently, it was shown that by utilizing an SOA as the nonlinear element in the loop, the switch could be made more compact (Eiselt, 1992). In addition, by placing the SOA away from the center of the loop, balanced operation could be achieved (Sokoloff et al., 1993). The drawback of these semiconductor loop switches is that at least some portion of the control pulse is present at the output of the switch and must be extinguished by a filter or a polarizer. Therefore, the output from this switch may not directly feed the control or signal port of a subsequent switch. This switch is not cascadable.

Cascadable switches are essential for the construction of optical circuits and processors containing more than a few gates. The dual-arm Mach-Zehnder is cascadable, when the control pulses counterpropagate through the SOAs, but has the limitations outlined above. A single-arm interferometric switch that is also cascadable has been demonstrated (Patel et al., 1996a). This switch design, shown in Fig. 33, is called the ultrafast nonlinear interferometer or UNI and is based on a time-division interferometer technique that was originally used to measure refractive index nonlinearities in semiconductor devices (Shirasaki et al., 1987; LaGasse et al., 1988; 1989). A signal pulse at the input of the interferometer is split by a polarization-sensitive delay into orthogonal components, separated in time by approximately the optical pulsewidth. The two signal components travel through the SOA, are recombined in a second polarization delay stage, and are interfered in a polarizer set at 45 degrees with respect to the orthogonal signal polarizations. Counterpropagating control pulses, coupled into the SOA by means of a 50/50 splitter, are temporally coincident with the trailing signal pulse. The control pulse imparts sub-picosecond, as well as long-lived, phase changes to the trailing signal pulse. The leading pulse is unaffected by the induced sub-picosecond nonlinearities. In the absence of a control pulse,

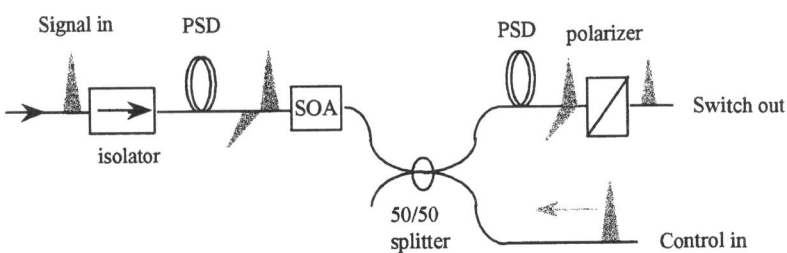

FIG. 33. Schematic of the ultrafast nonlinear interferometer (UNI). PSD, polarization sensitive delay; SOA, semiconductor optical amplifier. The implementation shown utilizes fiber couplers and PSDs.

any lingering refractive index changes affect both signal components equally. Therefore, the interferometer is balanced, or insensitive to long-lived effects. Also, because the control pulse counterpropagates through the SOA, the switch is cascadable. Cascadability has been verified in a recirculating shift register experiment (Hall *et al.*, 1997). The speed of the device is determined by the width of the optical pulses used in the experiment, and by the recovery time of the sub-picosecond nonlinearities. Recently, the record switching speed for bitwise logic has been extended from 40 Gb/s (Patel *et al.*, 1996a; Hall *et al.*, 1997) to 100 Gb/s (Hall and Rauschenbach, 1998), both demonstrated with the UNI. The drawback of this switch design is that the interferometer accepts only a limited range of input polarizations, and the switching window has limited tunability.

Extending the operating speed of these all-optical switches past 100 Gb/s should be straightforward. First, it will be important for the signal wavelengths to be positioned near the band edge of the SOA, where α and α_T, are large. This way, the unwanted effects of gain changes associated with the desired refractive index changes will be minimized. In cascadable switch designs, where the control pulses counterpropagate through the SOA, the length of the SOA must be less than or equal to the separation of the orthogonal signal component pulses, so that the control pulse interacts with only one component. Therefore, as the bit rate increases and the optical pulsewidths decrease, the length of the SOA may need to be reduced. If the interaction length is reduced, the strength of the nonlinearity must be increased by using higher input powers, higher bias currents, or amplifiers with larger overlap integrals, Γ. Extending the operating bit rates to 500 Gb/s will require new switch designs that can mitigate the effects of the carrier heating recovery time.

IX. Summary

In summary, we have discussed the gain and refractive index nonlinearities in active media. These nonlinearities are of interest for several reasons. They influence the modulation response and mode stability of diode lasers; they limit the speed of, and produce crosstalk between, multiplexed signals in optical amplifiers; and they are useful in the design of nonlinear waveguide devices such as modulators and switches. We have concentrated our discussion on nonlinear effects observed in InGaAsP waveguide amplifiers, but similar nonlinearities have been observed in a wide variety of material systems and device structures. We have shown that many physical mechanisms contribute to the total nonlinear responses of these active media. We have characterized the behavior of a variety of mechanisms, including carrier density changes, carrier heating, spectral hole burning, two-photon absorption, the optical Stark effect, and a dynamic coupling between the refractive index nonlinearities and the gain nonlinearities. Each of these effects has been shown to have a predictable behavior, attributed to a simple physical mechanism in the active media. However, several important questions remain unanswered.

It is clear that recent experiments have solidified much of what we know about the nonlinear gain and refractive index in SOAs. For example, we know that optical fields propagating through an active semiconductor waveguide can heat the carrier distribution by means of free carrier absorption, two-photon absorption, and stimulated transitions. Depending on the energy of the optical field and the carrier density in the device, stimulated transitions may also cool the carrier distribution. There is a thermalization time associated with the time it takes to establish the carrier distribution with an effective temperature that is either higher or lower than that of the equilibrium distribution. This hot or cold distribution then recovers back to the equilibrium temperature with a time constant of approximately 1 ps. This carrier heating nonlinearity was not predicted by early theoretical work on nonlinearities in semiconductor diodes and amplifiers, and is often modeled phenomenologically today using rate equation models or semiclassical density-matrix calculations. The carrier heating (cooling) nonlinearity causes a transient decrease (increase) in the probe transmission, which recovers with a time constant of 0.5 to 2 ps, depending on the device structure and the material system. Carrier heating (cooling) causes a transient increase (decrease) in the refractive index that recovers with the same time constant as the gain nonlinearity.

We have found that spectral hole burning, an energy-selective gain or absorption depletion that follows the magnitude and sign of the induced long-lived gain change, recovers with a time constant of approximately 100–200 fs. Spectral hole burning is an effect that is observed in many gain media, not only in semiconductor diode lasers and amplifiers. Still, it is not clear why the measured time constant is so long compared with the time constant of 10 to 50 fs predicted for the intraband scattering time. Simple data fitting routines suggest that spectral hole burning does not contribute to the nonlinear response of the refractive index.

There are instantaneous contributions to the nonlinear gain and refractive index of semiconductor optical amplifiers, but they depend on diode parameters in different ways. For example, there is an instantaneous decrease in probe transmission that is relatively insensitive to carrier density that we attribute to two-photon absorption. There is also an instantaneous decrease in the refractive index of these devices that is highly sensitive to the carrier density. Because of the sign of the nonlinearity, and because of its dependence on carrier density, we attribute this dynamic to the optical Stark effect. Quantitative comparison of these instantaneous effects is complicated by the coherent interactions of the pump and probe pulses.

Furthermore, there is a dynamic coupling of the gain and index that affects the measurement of the gain nonlinearities. Nonlinear, time-dependent pump-induced phase shifts lead to probe frequency shifts. These frequency shifts, coupled with the wavelength-dependent nature of the gain, yield refractive index-induced gain changes. These gain changes have been observed in the experiments presented in this chapter.

We believe that there are many interesting experiments still to be done and many advantages to be gained from a better understanding of the mechanisms responsible for nonlinear gain and refractive index in semiconductor optical amplifiers. Besides the better design of components such as laser sources and optical amplifiers, advances in all-optical switches, wavelength converters, and modulators may be possible. The nonlinear responses of the active media we have described in this chapter are rich. We expect that future experiments will confirm what has already been described, and will illuminate ultrafast effects that are currently obscured by the optical pump-probe pulses themselves.

ACKNOWLEDGEMENT

This work was supported in part by AFOSR and DARPA.

REFERENCES

Adams, M. J., Davies, D. A. O., Tatham, M. C., and Fisher, M. A. (1995). *Opt and Quantum Electron.* **27**, 1.
Agrawal, G. P. (1987). *Appl. Phys. Lett.*, **51**, 302.
Agrawal, G. P. (1988). *J. Opt. Soc. Am. B* **5**, 147.
Agrawal, G. P. and Dutta, N. K. (1986). *Long Wavelength Semiconductor Lasers*, (1986), Van Nostrand Reinhold; New York.
Agrawal, G. P., and Olsson, N. A. (1989a). *Opt. Lett.*, **14**, 500.
Agrawal, G. P. and Olsson, N. A. (1989b). *IEEE J. Quantum Electron.* **25**, 2297.
Alexander, S. B., Bondurant, R. S., Byrne, D., Chan, V. S. W., Finn, S. G., Gallager, R., Glance, B. S., Haus, H. A., Humblet, P., Jain, R., Kaminow, I. P., Karol, M., Kennedy, R. S., Kirby, A., Le, H. Q., Saleh, A. A. M., Schofield, B. A., Shapiro, J. H., Shankaranarayanan, N. K., Thomas, R. E., Williamson, R. C., Wilson, R. W. (1993). *J. Lightwave Technol.* **1**, 714.
Anderson, K. K., LaGasse, M. J., Wang, C. A., Fujimoto, J. G., and Haus, H. A. (1990). *Appl. Phys. Lett.* **56**, 1834.
Ashcroft, N. W. and Mermin, N. D. (1976). *Solid State Physics.* Saunders College; Philadelphia.
Barry, R. A. and Humblet, P. A. (1996). *IEEE J. Selec. Areas in Comm.* **14**, 858.

Basov, N. G., Kroklin, O. N., and Popov, Y. M. (1961). *Sov. Phys. JETP* **13**, 1320.
Bastard, G. (1988). *Wave Mechanics Applied to Semiconductor Heterostructures*, Halsted Press; New York.
Bennett, B. R., Soref, R. A., Del Alamo, J. A. (1990). *IEEE J. Quantum Electron.* **26**, 113.
Bergmann, E. E., Bigio, I. J., Feldman, B. J., and Fischer, R. A. (1978). *Opt. Lett.* **3**, 82.
Bernard, M. G. A. and Duraffourg, G. (1961). *Phys. Stat. Solid.* **1**, 667.
Bogatov, A. P., Eliseev, P. G., and Sverdlov, B. N. (1975). *IEEE J. Quantum Electron.* **QE-11**, 510.
Bowers, J. E. (1987). *Solid State Electron.* **30**, 1.
Bowers, J. E., Hemenway, B. R., Gnarck, A. H., and Wilt, D. P. (1986). *IEEE J. Quantum. Electron.* **QE-22**, 833.
Braagaard, C., Mikkelsen, B., Durhuus, T. and Stubkjaer, K. E. (1994). *J. Lightwave Technol.*, **12**, 943.
Butcher, P. N., and Cotter, D. (1991). *The Elements of Nonlinear Optics*. Cambridge University Press; Cambridge.
Cartledge, J. C. (1990). *IEEE Photon. Technol. Lett.*, **2**, 835.
Casey, H. C. and Panish, M. B. (1978). *Heterostructure Lasers, Part A: Fundamental Principles.* Academic Press, New York.
Danielsen, S. L., Joergensen,C., BVaa, M., Mikkelsen, B., Stubkjaer, K. E., Doussiere, P., Pommerau, F., Goldstein, L., Ngo, R., and Goiz, M. (1996). *Electron Lett.* **32**, 1688.
Delfyett, P. J., Silberberg, Y., and Alphonse, G. A. (1991). *Appl. Phys. Lett.* **59**, 10.
de Rougemont, F. and Frey, R. (1988). *Phys. Rev. B.* **37**, 1237.
Dienes, A., Heritage, J. P., Hong, M. Y., and Chang, Y. H. (1992). *Opt. Lett.* **17**, 1602.
Doerr, C. R., Wong, W. S., Haus, H. A., and Ippen, E. P. (1994). *Opt. Lett.* **19**, 1747.
Doran, N. J. and Wood, D. (1988). *Opt. Lett.* **13**, 56.
D'Ottavi, A., Iannone, E., Mecozzi, A., Scotti, S., Spano, P., Spano, P., Dall'Ara, R., Guekos, G., and Eckner, J. (1994). *Appl. Phys. Lett.* **65**, 2633.
Durhuus, T., Mikkelsen, B. and Stubkjaer, K. E. (1992). *J. Lightwave Technol.* **10**, 1056.
Durhuus, T., Joergensen, C., Mikkelsen, B., Pedersen, R. J. S., and Stubkjaer, K. E. (1994). *IEEE Photon Technol. Lett.* **6**, 53.
Dutta, N. K. (1980). *J. Appl. Phys.* **51**, 6095.
Dutta, N. K. (1981). *J. Appl. Phys.* **52**, 55.
Eiselt, M. (1992). *Electron. Lett.* **28**, 1505.
Eisenstein, G., Raybon, G., and Schulz, L. W. (1988). *J. Lightwave Technol.* **6**, 12.
Eisenstein, G., Tucker, R. S., Wiesenfeld, J. M., Hansen, P. B., Raybon, G., Johnson, B. C., Bridges, T. J., Storz, F. G., and Burrus, C. A. (1989). *Appl. Phys. Lett.* **54**, 454.
Eisenstein, G., Koren, U., Raybon, G., Koch, T. L., Wiesenfeld, J. M., Wegener, M., Tucker, R. S., and Miller, B. I. (1990). *Appl. Phys. Lett.* **56**, 1201.
Eisenstein, G., Wiesenfeld, J. M., Wegener, M., Sucha, G., Chemla, D. S., Weiss, S., Raybon, G., and Koren, U. (1991). *Appl. Phys. Lett.* **58**, 158.
Finlayson, N., Wright, E. M., and Stegeman, G. I. (1990). *IEEE J. Quantum. Electron.* **26**, 770.
Frigo, N. (1983). *IEEE J. Quantum. Electron.* **QE-19**, 511.
Girardin, F., Eckner, J., Guekos, G., Dall'Ara, R., Mecozzi, A., D'Ottavi, A., Martelli, F., Scotti, S., and Spano, P. (1997). *IEEE Photon. Technol. Lett.* **9**, 746.
Glance, B. S., Wiesenfeld, J. M., Koren, U., Gnauck, A. H., Presby, H. M., and Jourdan, A. (1992). *Electron. Lett.* **28**, 1714.
Gomatam, B. N. and DeFonzo, A. P. (1990). *IEEE J. Quantum Electron.* **26**, 1689.
Grant, R. S. and Sibbett, W. (1991). *Appl. Phys. Lett.* **58**, 1119.
Hall, K. L., Ippen, E. P., Mark, J., and Eisenstein, G. (1989). *OSA Proc. Picosecond Electronics Optoelectronics* **4**, 73.

Hall, K. L., Lai, Y., Ippen, E. P., Eisenstein, G., and Koren, U. (1990a). *Appl. Phys. Lett.* **57**, 2888.
Hall, K. L., Mark, J., Ippen, E. P., and Eisenstein, G. (1990b). *Appl. Phys. Lett.* **56**, 1740.
Hall, K. L., Ippen, E. P., and Eisenstein, G. (1990c). *Appl. Phys. Lett.* **57**, 129.
Hall, K. L., Lenz, G., Ippen, E. P., and Raybon, G. (1992a). *Opt. Lett.* **17**, 874.
Hall, K. L., Lenz, G., Ippen, E. P., Koren, U., and Raybon, G. (1992b). *Appl. Phys. Lett.* **61**, 2512.
Hall, K. L., Lenz, G., and Ippen, E. P. (1992c). *J. Lightwave Technol.* **10**, 616.
Hall, K. L., Darwish, A. M., Ippen, E. P., Koren, U., and Raybon, G. (1993). *Appl. Phys. Lett.* **62**, 1320.
Hall, K. L., Lenz, G., Darwish, A. M., and Ippen, E. P. (1994). *Opt. Comm.* **111**, 589.
Hall, K. L., Moores, J. D., Rauschenbach, K. A., Wong, W. S., Ippen, E. P., and Haus, H. A. (1995). *IEEE Photon. Technol. Lett.* **7**, 1093.
Hall, K. L. and Rauschenbach, K. A. (1998). *Technical Digest for the Optical Fiber Communications Conference*, Postdeadline PD5.
Hall, R. N., Fenner, G. E., Kingsley, J. D., Soltys, T. J., and Carlson, R. O. (1962). *Phys. Rev. Lett.* **9**, 366.
Hansen, P. B, Wiesenfeld, J. M., Eisenstein, G., Tucker, R. S., and Raybon, G. (1989). *IEEE J. Quantum Electron.* **QE-25**, 2611.
Hendow, S. T. and Sargent, M. (1985). *J. Opt. Soc. Am. B* **2**, 84.
Henry, C. H. (1982). *IEEE J. Quantum Electron.* **18**, 259.
Henry, C. H., Levine, B. F., Logan, R. A., and Bethea, C. G. (1983). *IEEE J. Quantum Electron.* **QE-19**, 905.
Henry, C. H., Logan, R. A., and Bertness, K. A. (1981). *J. Appl. Phys.* **52**, 4457.
Holonyak, N. and Bevacqua, S. F. (1962). *Appl. Phys. Lett.* **1**, 82.
Hong, M. Y., Chang, Y. H., Dienes, A., Heritage, J. P., Delfyett, P. J., Dijaili, S., and Patterson, F. G. (1996). *IEEE J. Select. Topics in Quantum Electron.* **2**, 523.
Hultgren, C. T. and Ippen, E. P. (1991). *Appl. Phys. Lett.* **59**, 635.
Hultgren, C. T., Dougherty, D. J., and Ippen, E. P. (1992). *Appl. Phys. Lett.* **61**, 2767.
Hutchings, D. C., Sheik-Bahae, M., Hagan, D. J., and Van Stryland, E. W. (1992). *Opt. and Quantum Electron.* **24**, 1.
Iannone, P. P. and Prucnal, P. R. (1994). *IEEE Photon. Technol. Lett.* **6**. 988.
Iannone, P. P., Gnauck, A. H., and Prucnal, P. R. (1994b). *IEEE Photon. Technol. Lett.* **6**. 1046.
Idler, W., Schilling, M., Daub, K., Baums, D., Koerner, U., Lach, E., Laube, G., and Wunstel, K. (1995). *Electron. Lett.* **31**, 454.
Ippen, E. P. and Shank, C. V. (1977). In *Ultrashort Light Pulses, Picosecond Techniques and Applications*. Springer-Verlag, New York, p. 83.
Icsevgi, A. and Lamb, J. W. E. (1969). *Physical Review* **185**, 517.
Ikegami, T. and Suematsu, Y. (1970). *Electron Comm. Japan*, **53**, 69.
Jahn, E., Agrawal, N., Arbert, M., Ehrke, H. J., Franke, D., Ludwig, R., Pieper, W., Weber, H. G., and Weinert, C. M. (1995). *Electron. Lett.* **31**, 1857.
Joergensen, C., Durhuus, T., Braagaard, C., Mikkelsen, B., and Stubkjaer, K. E. (1993). *IEEE Photon. Technol. Lett.* **5**. 657.
Jopson, R. M. and Tench, R. E. (1993). *Electron. Lett.* **29**, 2216.
Kaminow, I. P., Doerr, C. R., Dragone, C., Koch, T., Koren, U., Saleh, A. A. M., Kirby, A. J., Ozveren, C. M., Schofield, B., Thomas, R. E., Barry, R. A., Castagnozzi, D. M., Chan, V. W. S., Hemenway, B. R., Marquis, D., Parikh, S. A., Stevens, M. L., Swanson, E. A., Finn, S. G. and Gallager, R. G. (1996). *IEEE J. Selec. Areas Comm.* **14**, 780.
Kan, S. C., Vassilovski, D., Wu, T. C., and Lau, K. Y. (1992). *Appl. Phys. Lett.* **61**, 752.

Kang, K. I., Glesk, I., Chang, T. G., Prucnal, P. R., and Boncek, R. K. (1995). *Electron. Lett.* **31**, 749.
Kesler, M. P. (1988). Ph. D. Thesis, Massachusetts Institute of Technology, Cambridge.
Kesler, M. P. and Ippen, E. P. (1987). *Appl. Phys. Lett.* **51**, 1765.
Kikuchi, K., Amano, M., Zah, C. E., and Lee, T. P. (1994). *Appl. Phys. Lett.* **64**, 548.
Kikuchi, K., Kakui, M., Zah, C. E., and Lee, T. P. (1992). *IEEE J. Quantum Electron.* **QE-28**, 151.
Knox, W. H., Chemla, D. S., Livescu, G., Cunningham, J. E., and Henry, J. E. (1988). *Phys. Rev. Lett.* **61**, 1290.
Knox, W. H., Hirlimann, C., Miller, D. A. B., Shah, J., Chemla, D. S., and Shank, C. V. (1986). *Phys. Rev. Lett.* **56**, 1191.
Koch, T. L. and Linke, R. A. (1986). *Appl. Phys. Lett.* **48**, 613.
Koga, M., Tokura, N., and Nawata, K. (1988). *Appl. Opt.* **27**, 3964.
Koren, U., Miller, B. I., Su, Y. K., Koch, T. L., and Bowers, J. E. (1987). *Appl. Phys. Lett.* **51**, 1744.
Koren, U., Oron, M., Young, M. G., Miller, B. I., De Miguel, J. L., Raybon, G., and Chien, M. (1990). *Electron. Lett.* **26**, 465.
Lacey, J. P. R., Madden, S. J., Summerfield, M. A., Tucker, R. S., and Faris, A. I. (1995). *Electron. Lett.* **31**, 743.
LaGasse, M. J., Anderson, K. K., Haus, H. A., and Fujimoto, J. G. (1989). *Appl. Phys. Lett.* **54**, 2068.
LaGasse, M. J., Anderson, K. K., Wang, C. A., Haus, H. A., and Fujimoto, J. G. (1990). *Appl. Phys. Lett.* **56**, 417.
LaGasse, M. J., Liu-Wong, D., Fujimoto, J. G., and Haus, H. A. (1988). *Opt. Lett.* **6**, 311.
Lai, Y., Hall, K. L., Ippen, E. P., and Eisenstein, G. (1990) *IEEE Photon. Technol. Lett.* **2**, 711.
Lassen, H. E., Hansen P. B., and Stubkjaer, K. E. (1988). *J. Lightwave Technol.* **6**, 1559.
Lee, R. B., Geraghty, D. F., Verdiell, M., Ziari, M., Mathur, A., and Vahala, K. J. (1997). *IEEE Photon. Technol. Lett.* **9**, 752.
Lenz, G., Ippen, E. P., Wiesenfeld, J. M., Newkirk, M. A., and Koren, U. (1996). *Appl. Phys. Lett.* **68**, 2933.
Ludwig, R. and Raybon, G. (1994). *Electron. Lett.* **30**, 338.
Manning, R. J., and Davies, D. A. O. (1994). *Opt. Lett.* **19**, 889.
Manning, R. J., Davies, D. A. O., Cotter, D., and Lucek, J. K. (1994a). *Electron. Lett.* **30**, 787.
Manning, R. J., Davies, D. A. O., and Lucek, J. K. (1994b). *Electron. Lett.* **30**,1233.
Manning, J., Olshansky, R., and Su, C. B. (1983). *IEEE J. Quantum Electron.* **19**, 1525.
Mark, J., Liu, L. Y., Hall, K. L., Haus, H. A., and Ippen, E. P. (1989). *Opt. Lett.* **14**, 48.
Mark, J. and Mørk, J. (1992). *Appl. Phys. Lett.* **61**, 2281.
Mecozzi, A., and Mork, J. (1996). *J. Opt. Soc. Am. B* **13**, 2437.
Mecozzi, A., Scotti, S., D'Ottavi, A., Iannone, E., and Spano, P. (1995b). *IEEE J. Quantum Electron.* **31**, 689.
Mikkelsen, B., Durhuus, T., Joergensen, C., Pedersen, R. J. S., Danielsen, S. L., Stubkjaer, K. E., Gustavsson, M., van Berlo, W., and Janson, M. (1994). *European Conf. on Optical Communications* **4**, 67.
Mikkelsen, B., Vaa, M., Pedersen, J., Durhuus, T., Joergensen, C., Braagaard, C., Storkfelt, N., Stubkjaer, K. E., Doussiere, P., Garabedian, G., Graver, C., Derouin, E., Fillion, T., and Klenk, M. (1993). *European Conf. on Optical Communications*, Postdeadline Thp12. 6.
Mørk, J., Mark, J., and Seltzer, C. P. (1994). *Appl. Phys. Lett.* **64**, 2206.
Mørk, J., and Mecozzi, A. (1994). *Appl. Phys. Lett.* **65**, 1736.
Mørk, J., Mecozzi, A., and Hultgren, C. T. (1996). *Appl. Phys. Lett.* **68**, 449.
Mørk, J., and Mecozzi, A. (1996). *J. Opt. Soc. Am. B* **13**, 1803.

Mozer, A., Romanek, K. M., Hildebrand, O., Schmid, W., and Pilkhun, M. H. (1983). *IEEE J. Quantum Electron.* **QE-19**, 913.
Murata, S., Tomita, A., Shimizu, J., and Suzuki, A. (1991). *IEEE Photon. Technol. Lett.* **3**, 1021.
Nagarajan, R., Ishikawa, M., Fukushima, T., Geels, R. S., and Bowers, J. E. (1992). *IEEE J. Quantum Electron.* **28**, 1990.
Nagarajan, R. and Bowers, J. E. (1993). *IEEE J. Quantum Electron.* **29**, 1601.
Nakajima, H. and Frey, R. (1986). *IEEE J. Quantum Electron.* **QE-22**, 1349.
Nakamura, S. and Tajima, K. (1997). *Appl. Phys. Lett.* **70**, 3498.
Nathan, M. I., Dumke, W. P., Burns, G., Dill, F. H., and Lasher, G. J. (1962). *Appl. Phys. Lett.* **1**, 62.
Olshansky, R., Su, C. B., Manning, J., and Powaznik, W. (1984). *IEEE J. Quantum Electron.* **QE-20**, 838.
Olsson, N. A. and Agrawal, G. P. (1989). *Appl. Phys. Lett.* **55**, 13.
Ottolenghi, P. P., Jourdan, A., and Jacquet, J. (1993). *Tech. Dig. ECOC '93*, **2**, 141.
Paiella, R., Hunziker, G., Vahala, K. J., and Koren, U. (1996). *Appl. Phys. Lett.* **69**, 4142.
Pankove, J. I. (1975). *Optical Processes in Semiconductors.* Dover, New York.
Patel, N. S., Hall, K. L., and Rauschenbach, K. A. (1996). *Opt. Lett.* **21**, 1466.
Patel, N. S., Hall, K. L., and Rauschenbach, K. A. (1997). *IEEE Photon. Technol. Lett.* **9**, 1277.
Patel, N. S., Hall, K. L., and Rauschenbach, K. A. (1998). *Appl. Opt.* **37**, 2831.
Patrick, D. M., and Manning, R. J. (1994). *Electron. Lett.* **30**, 252.
Provost, J. G. and Frey, R. (1989). *Appl. Phys. Lett.* **55**, 519.
Quist, T. M., Rediker, R. H., Keyes, R. J., Krag, W. E., Lax, B., McWhorter, A. L., and Zeiger, H. J. (1962). *Appl. Phys. Lett.* **1**, 91.
Ramaswami, R., and Sivarajan, K. N. (1996). *IEEE J. Select. Areas in Comm.* **14**, 840.
Ratovelomanana, F., Vodjdani, N., Enard, A., Glaste, G., Rondi, D., Blondeau, R., Joergensen, C., Durhuus, T., Mikkelsen, B., Stubkjaer, K. E., Jourdan, A., and Soulage, G. (1995). *IEEE Photon. Technol. Lett.* **7**, 992.
Rideout, W, Sharfin, W. F., Koteles, E. S., Vassell, M., and Elman, B. (1991). *IEEE Photon. Technol. Lett.* **3**, 7847.
Saitoh, T., Itoh, H., Noguchi, Y., Sudo, S. and Mukai, T. (1989). *IEEE Photon. Technol. Lett.*, **1**, 297.
Saitoh, T. and Mukai, T. (1990). *IEEE J. Quantum Electron.* **26**, 2086.
Salehi, J. A., Weiner, A. M., and Heritage, J. P. (1989). *J. Lightwave Technol.* **8**, 478.
Sermage, B., Eichler, H. J., Heritage, J., Nelson, R. J., and Dutta, N. K. (1983). *Appl. Phys. Lett.* **42**, 259.
Sheik-Bahae, M., Hutchings, D. C., Hagan, D. J., Van Stryland, E. W. (1991). *IEEE J. Quantum Electron.* **27**, 1296.
Sheik-Bahae, M. and Van Stryland, E. W. (1994). *Phys. Rev. B.* **50**, 14171.
Shen, Y. R., (1984). *Nonlinear Optics.* John Wiley & Sons, New York.
Shirasaki, M., Haus, H. A., Wong, D. L. (1987). In *Conference on Lasers and Electro-Optics* **14**, paper THO1.
Siegman, A. E. (1986). *Lasers.* University Science Books; Mill Valley, CA.
Sokoloff, J. P., Prucnal, P. R., Glesk, I., and Kane, M. (1993). *IEEE Photon. Technol. Lett.* **5**, 787.
Stix, M. S., Kesler, M. P., and Ippen, E. P. (1986). *Appl. Phys. Lett.* **48**, 1722.
Stolen, R. H. and Lin, C. (1978). *Phys. Rev. A.* **17**, 1448.
Su, C. B., Schlafer, J., Manning, J., and Olshansky, R. (1982). *Electron. Lett.* **18**, 595.
Sun, C. K., Choi, H. K., Wang, C. A., and Fujimoto, J. G. (1993). *Appl. Phys. Lett.* **62**, 747.
Tang, J. M., Spencer, P. S., and Shore, K. A. (1996). *Electron. Lett.* **32**, 1293.

Tatham, M. C., Gu, X., Westbrook, L. D., Sherlock, G., and Spirit, D. M. (1994). *Electron. Lett.* **30**, 1335.
Tatum, J. A., MacFarlane, D. L., Bowen, R. C., Klimeck, G., and Frensley, W. R. (1996). *IEEE J. Quantum Electron.* **32**, 664.
Tessler, N. and Eisenstein, G. (1993). *IEEE J. Quantum Electron.* **29**, 1586.
Thedrez, B., Jones, A., and Frey, R. (1988). *IEEE J. Quantum Electron.* **24**, 1499.
Thompson, G. H. B. (1983). *Electron. Lett.* **19**, 154.
Thompson, G. H. B. (1985). *Physics of Semiconductor Laser Devices.* John Wiley & Sons, Chichester.
Tiemeijer, L. F. (1991). *Appl. Phys. Lett.* **59**, 499.
Tucker, R. S., and Pope, D. J. (1983). *IEEE J. Quantum Electron.* **QE-19**, 1179.
Tucker, R. S., and Kaminow, I. P. (1984). *J. Lightwave Technol.* **LT-2**, 385.
Tucker, R. S. (1985). *J. Lightwave Technol.* **LT-3**, 1180.
Uji, T., Iwamoto, K., and Lang, R. (1983). *IEEE Trans. Electron Devices* **ED-30**, 316.
Uskov, A., Mork, J., and Mark, J. (1992). *IEEE Photon. Technol. Lett.* **4**, 443.
Uskov, A., Mork, J., Mark, J., Tatham, M. C., and Sherlock, G. (1994). *Appl. Phys. Lett.* **65**, 944.
Vahala, K. J., Zhou, J., Geraghty, D., Lee, R., Newkirk, M., and Miller, B. I. (1996). *Int. J. High Speed Electron.* **7**, 153.
Vardeny, Z. and Tauc, J. (1981). *Optics Comm.* **39**, 396.
Wang, J. and Schweizer, H. (1996). *IEEE Photon. Technol. Lett.* **8**,1441.
Weiss, S., Wisenfeld, J. M., Chemla, D. S., Raybon, G., Sucha, G., Wegener, M., Eisenstein, G., Burrus, C. A., Dentai, A. G., Koren, U., Miller, B. I., Temkin, H., Logan, R. A., and Tanbun-Ek, T. (1992). *Appl. Phys. Lett.* **60**, 9.
Wiesenfeld, J. M. (1996). *Int. J. High Speed Electron.* **7**, 179.
Wiesenfeld, J. M., Eisenstein, G., Tucker, R. S., Raybon, G., and Hansen, P. B. (1988). *Appl. Phys. Lett.* **53**, 1239.
Wiesenfeld, J. M. and Glance, B. S. (1992). *IEEE Photon. Technol. Lett.* **4**, 1168.
Wiesenfeld, J. M., Glance, B., Perino, J. S., Gnauck, A. H. (1993). *IEEE Photon. Technol. Lett.* **5**, 1300.
Wiesenfeld, J. M., Perino, J. S., Gnauck, A. H. and Glance, B. (1994a). *Electron. Lett.* **30**, 720.
Wiesenfeld, J. M., Weiss, S., Botkin, D., and Chemla, D. S. (1994b). *Opt. Quantum Electron.* **26**, S731.
Willatzen, M., Uskov, A., Mork, J., Olesen, H., Tromborg, B., and Jauho, A. P. (1991). *IEEE Photon. Technol. Lett.* **3**, 606.
Willatzen, M., Takahashi, T., and Arakawa, Y. (1992). *IEEE Photon. Technol. Lett.* **4**, 682.
Willatzen, M., Mark, J., Mork, J., and Seltzer, C. P. (1994). *Appl. Phys. Lett.* **64**, 143.
Wilt, D., Karlicek, R., Strege, K., Dautremont-Smith, W., Dutta, N., Flynn, E., Johnson, W., and Nelson, R. (1984). *J. Appl. Phys.* **56**, 710.
Wintner, E., and Ippen, E. P. (1984). *Appl. Phys. Lett.* **44**, 999.
Yariv, A., Fekete, D. and Pepper, D. M. (1979). *Opt. Lett.* **4**, 52.
Zhou, J., Park, N., Dawson, J. W., Vahala, K. J., Newkirk, M. A., and Miller, B. I. (1993). *Appl. Phys. Lett.* **63**, 1179.

CHAPTER 3

Optical Responses of Quantum Wires/Dots and Microcavities

Eiichi Hanamura

OPTICAL SCIENCES CENTER
THE UNIVERSITY OF ARIZONA
TUCSON, ARIZONA

I. THEORETICAL ASPECTS 161
   1. Superradiance of Excitons from Quantum Dots and Wires 164
   2. Excitonic Optical Nonlinearity 169
   3. Excitons in Quantum Wells 178
   4. Coherent Light Emission from Quantum Wires 182
II. EXPERIMENTAL RESULTS 189
   1. Superradiance of Excitons 189
   2. Figure of Merit for Nonlinear Optical Responses 191
   3. Biexcitons as a Nonlinear Medium 195
   4. Weak Localization and Nonlinear Optical Responses 201
   LIST OF ACRONYMS 208
   REFERENCES 208

I. Theoretical Aspects

The exciton is a coherent elementary excitation that can propagate over a crystal. Here we consider an exciton in several systems, such as quantum dot, wire, and well. We distinguish between Frenkel and Wannier excitons (Knox, 1963). An electronic state in a crystal can be described by Bloch functions or by Wannier functions, which can be transformed from one to the other. The crystal ground state Ψ_g can be described, for example, by a product of Wannier functions $w_i^v(\mathbf{r}_i)$:

$$\Psi_g = \prod_{i=1}^{N} w_i^v(\mathbf{r}_i) \tag{1}$$

Here, $w_i^v(\mathbf{r})$ is localized around the ith lattice point and is made of

List of Acronyms can be located preceeding the references to this chapter.

superpositions of Bloch functions in the valence band v. The Frenkel exciton with wave vector $\mathbf{K}$ is made of a coherent superposition of a single excitation from the valence band state $w_i^v(\mathbf{r})$ into the conduction band, with Wannier function $w_j^c(\mathbf{r})$:

$$\Psi_\mathbf{K} = \frac{1}{\sqrt{N}} \sum_j e^{i\mathbf{K}\cdot\mathbf{R}_j} w_j^c(\mathbf{r}_j) \prod_{i \neq j} w_i^v(\mathbf{r}_i) \tag{2}$$

Then the transition dipole moment of this exciton has macroscopic enhancement $\sqrt{N}$:

$$\mathbf{P}_{\mathbf{K}_0} = \left\langle \Psi_\mathbf{K} \left| \sum_i e^{i\mathbf{K}_0 \cdot \mathbf{R}_i}(-e\mathbf{r}_i) \right| \Psi_g \right\rangle = \sqrt{N}\,\mu\delta(\mathbf{K}, \mathbf{K}_0) \tag{3}$$

where N is the number of unit cells in the crystal, $\mathbf{K}_0$ corresponds to the wave vector of the radiation field, and μ is the transition dipole moment per unit cell. This means that the oscillator strength is strongly accumulated on the single exciton level $\Psi_\mathbf{K}$.

On the other hand, the Wannier exciton is made of a pair consisting of a conduction band electron at w_j^c and a valence band hole at $w_{j'}^v$, and is described by the electron-hole relative motion coordinate $\mathbf{r} = u(\mathbf{j} - \mathbf{j}')$, where u is the size of the unit cell, and by the center-of-mass motion coordinate $\mathbf{R} = u(m_e\mathbf{j} + m_h\mathbf{j}')/(m_e + m_h)$, as follows:

$$\Psi_{n\mathbf{K}} = \frac{1}{\sqrt{N}} \sum_j \sum_{j'} e^{i\mathbf{K}\cdot\mathbf{R}} \phi_n(\mathbf{r}) w_j^c(\mathbf{r}_j) \prod_{i \neq j'} w_i^v(\mathbf{r}_i) \tag{4}$$

In both Eq. (2) and Eq. (4), the product of Wannier functions should be replaced by their Slater determinants in a strict sense. The transition dipole moment of the lowest electron-hole relative motion, $n = 1s$, also has macroscopic enhancement, as follows:

$$\mathbf{P}_\mathbf{K} = \sqrt{N}\,\mu\sqrt{\frac{u^3}{\pi a_B^3}}\,\delta(\mathbf{K}, \mathbf{K}_0) \tag{5}$$

where u^3 is the unit cell volume and a_B is the exciton Bohr radius.

There are quantitative as well as qualitative differences between Frenkel and Wannier excitons. As the quantitative difference, the exciton Bohr radius a_B of the Wannier exciton, which measures the average separation between the electron and hole forming the exciton, is much larger than u, the size of the unit cell, whereas the excitation is well localized within the same unit cell or the same molecule in the case of the Frenkel exciton. A good example of the latter is the lowest-energy excitation in J-aggregates of dyes. The qualitative difference, which results in the quantitative one, comes from a different mechanism of the excitation propagation. In semiconductors, both the electron excited into a conduction band and the hole created in a valence band can propagate by themselves in the band picture, and the electron and hole relative motion has bound states as a result of the coulombic attraction between the electron and hole. The center-of-mass motion is described by a plane wave reflecting the translation symmetry and originating from the band nature of the electron and hole. On the other hand, the electronic excitation is well localized within a single molecule or a unit cell in the Frenkel exciton—e.g., in a dye molecule in a J-aggregate. Here this excitation interacts with neighboring molecules by coulombic force so that one of these neighboring molecules is excited while the original excited molecule goes down to the ground state. As a consequence, only excitation propagates over the crystal, reflecting the translation symmetry, and this propagation is described by the transition dipole–transition dipole interaction.

The charge transfer exciton has characteristics intermediate between those of the Wannier and Frenkel excitons: the lowest electronic excitation in an organic semiconductor anthracene-PMDA (pyromellitic acid dianhydride) crystal is an example of this. An electron in the HOMO (highest occupied molecular orbital) state of anthracene is transferred into the LUMO (lowest unoccupied molecular orbital) state of PMDA. This excitation complex can propagate by Frenkel-type or Wannier-type interactions, both reflecting the translational symmetry of the crystal. Note that the propagation matrix elements have very strong anisotropy for the J-aggregates of dyes and for charge transfer systems such as anthracene-PMDA. Therefore, these excitons sometimes can be treated as one-dimensional systems.

In these treatments, we have assumed that the coherency of the exciton extends over the crystal. This is the case in quantum dots or microcrystallites, but in the case of finite dimensions, the coherent volume $N_c u^3$ is limited by the scattering of the exciton with phonons or defects. In the latter case, the macroscopic enhancement in Eqs. (3) and (5) is limited to a mesoscopic size N_c instead of to N. In any case, however, these mesoscopic or macroscopic transition dipole moments result in superradiant rapid decay of the exciton and enhancement of the third-order susceptibility $\chi^{(3)}(\omega;\omega,-\omega,\omega)$ under near-resonant excitation of the exciton. These phenomena will be discussed, respectively, in Subsections 1 (Hanamura, 1988b) and 2 (Hanamura, 1988a) below. In Subsections 3 and 4, respectively, we will discuss the effects of finite coherent volume of excitons in a quantum well and coherent light emission from a cavity consisting of quantum wires.

1. SUPERRADIANCE OF EXCITONS FROM QUANTUM DOTS AND WIRES

In this subsection, we derive the theoretical prediction of superradiant decay of the Wannier exciton in semiconductor dots embedded in an insulating or glass matrix, and the time-resolved superradiant decay spectrum from a highly excited system of Frenkel excitons.

a. Superradiance of a Wannier Exciton in a Quantum Dot

The advantage of quantum dots is that the coherent volume of a Wannier exciton at low temperature is essentially determined by the size of the quantum dot. Then the transition dipole moment of the Wannier exciton with quantum number n for the center-of-mass motion at $\mathbf{K} = 0$ is evaluated within the effective-mass approximation as

$$\mathbf{P}_n = \frac{2\sqrt{2}}{\pi} \left(\frac{R}{a_B}\right)^{3/2} \frac{1}{n} \boldsymbol{\mu}_{cv} \qquad (n = 1, 2, \ldots) \tag{6}$$

where R is the radius of the quantum dot (assuming a sphere), $\boldsymbol{\mu}_{cv}$ is the electric transition dipole per unit cell, and the size R is assumed to be much smaller than a wavelength λ of the relevant radiation field, but much larger than the exciton Bohr radius a_B. Here n is a principal quantum number of the center-of-mass motion, and the lowest energy state both for the relative and the center-of-mass motion has the largest oscillator strength. Owing to this large transition dipole moment, a single Wannier exciton can decay spontaneously with decay rate

$$\frac{1}{T_1} = 64\pi \left(\frac{R}{a_B}\right)^3 \frac{4\mu_{cv}^2}{3\hbar\lambda^3} \tag{7}$$

This rate has mesoscopic enhancement $64\pi(R/a_B)^3$, in comparison with the spontaneous emission rate $4\mu_{cv}^2/3\hbar\lambda^3$ of band-to-band transitions. Although this is the radiative decay process of a single excitation, we can call it superradiance because this Wannier exciton is a coherent superposition of all molecules composing the quantum dot, with the maximum cooperation number. The experimental confirmation of this prediction will be introduced in Section II (see Hanamura 1988b).

b. Superradiance from Many Frenkel Excitons in Linear Chains

Conventional quantum wires are one-dimensional versions (Wegscheider *et al.*, 1993) of quantum wells, in which sizes in two directions are smaller than the exciton Bohr radius a_B, but the size in the third direction is much larger than a_B. However, in this subsection we treat the system of quasi-one-dimensional Frenkel excitons (Tokihiro *et al.*, 1993, 1995; Manabe *et al.*, 1993), which is the case, for example, in J-aggregates of dyes. The results obtained mathematically for a one-dimensional chain of atoms (molecules) may be applicable to higher-dimensional systems of Frenkel excitons but are not applicable to the system of Wannier excitons, which interact with each other in more complicated ways.

When the excited atoms (molecules) share a common polarization direction and phase, this system behaves as a macroscopic dipole moment that is proportional to the number N of involved atoms. As a result, the excitation energy of the atomic system is emitted as a short and intense pulse of radiation field. The peak intensity of the emitted pulse is proportional to N^2, whereas the pulse width is inversely proportional to N. Here we have assumed that the atoms interact, not directly, but only through the radiation field. This is called Dicke's superradiance (Dicke, 1954). Here arises a question of whether or not superradiance is possible when the atomic excitation can propagate as a Frenkel exciton. If it is possible, how different is the superradiance of Frenkel excitons from that of atoms? We will answer these questions in this subsection.

The system of Frenkel excitons is described, in terms of spin $\frac{1}{2}$ operators $\hat{s}_j^\pm = \hat{s}_j^x \pm i\hat{s}_j^y$ and $\hat{s}_j^z$, by the following Hamiltonian:

$$H = \hbar\omega_o \sum_j \hat{s}_j^z - \hbar J \sum_j (\hat{s}_j^+ \hat{s}_{j+1}^- + \hat{s}_j^- \hat{s}_{j+1}^+) - 2\hbar J_z \sum_j \left(\hat{s}_j^z + \frac{1}{2}\right)\left(\hat{s}_{j+1}^z + \frac{1}{2}\right) \quad (8)$$

where the magnetic quantum numbers $m = -\frac{1}{2}$ and $m = \frac{1}{2}$ correspond, respectively, to the ground and excited states of the atom j; the energy separation is $\hbar\omega_o$; $\hbar J$ describes the excitation propagation between the neighboring atoms by dipolar interaction; and $\hbar J_z$ is the interaction between excitons. Here we assume a one-dimensional system of atoms with periodic boundary conditions, because we know the exact solutions for any degree of excitation for this system. When the excited state is accompanied by a static dipole moment μ_z, the interaction between two excited atoms (molecules) J_z in Eq. (8) is expressed by the dipolar interaction proportional to μ_z^2, and it works only when both eigenvalues $\hat{s}_j^z$ and $\hat{s}_{j+1}^z$ are $\frac{1}{2}$ — i.e., when both the jth and $(j+1)$th atoms are in the excited states. The case of finite J_z is discussed in Manabe *et al.* (1993) and Tokihiro *et al.* (1995). The second term of Eq. (8) describes the dipole-dipole interaction, in which the excitation — e.g., at the $(j+1)$th atom — is annihilated by $\hat{s}_{j+1}^-$ and consequently the jth atom is excited by $\hat{s}_j^+$. The nearest-neighbor interaction J is expressed by

$$2\hbar J = -\frac{d^2}{4\pi\varepsilon_0 a^3}\left[1 - \frac{3(\hat{\varepsilon}_d \cdot \mathbf{a})^2}{a^2}\right] \tag{9}$$

Here $\mathbf{a}$ is the vector connecting the nearest neighbor, $a \equiv |\mathbf{a}|$ is the nearest-neighbor distance, and $\hat{\varepsilon}_d$ is the unit polarization vector of the transition dipole $\mathbf{d}$. Without the static dipolar interaction J_z between the excited excitons, our system can be presented by an xxo model (Baxter, 1982) (see also Eq. 12).

The master equation of superradiance is obtained by eliminating the coordinates of the radiation field adiabatically through projecting the equation of motion of total density matrix on the radiation vacuum (Gross and Haroche, 1982). Then the density matrix $\rho(t)$ of the atomic system obeys the master equation

$$\frac{d\rho}{dt} = -\frac{\Gamma}{2}\sum_{ij}[\hat{s}_i^+\hat{s}_j^-, \rho]_+ + \Gamma\sum_{ij}\hat{s}_j^-\rho\hat{s}_i^+ - i\hbar^{-1}[H_{xxo}, \rho]_- \tag{10}$$

where $[A, B]_\pm \equiv AB \pm BA$. Here,

$$\Gamma = \frac{8\pi^2 d^2}{3\hbar\varepsilon_0 \lambda^3} \tag{11}$$

is the radiative decay rate of a single atom with transition dipole moment d, and

$$H_{xxo} = \hbar\omega_o\sum_i \hat{s}_i^z - \hbar J\sum_i(\hat{s}_i^+\hat{s}_{i+1}^- + \hat{s}_{i+1}^-\hat{s}_i^+) \tag{12}$$

Dicke's superradiance (Dicke, 1954) of the atomic system could be described by Eq. (10) with vanishing J in Eq. (12). However, the fact that $|J| \gg \Gamma$ is derived by comparing Eqs. (9) and (11), because $\lambda \gg a$ in the crystal, so that the propagation effect J is inevitable for the superradiance of excitons from the crystal.

By solving the master equation, we have found that superradiance of Frenkel excitons is possible from the crystal with characteristics that are very different from those of the atomic system. First, the size dependence of the superradiant pulse is evaluated, as shown in Fig. 1. The peak intensity increases superlinearly with N, but not in proportion to N^2, as does atomic superradiance. In fact, the peak signal is proportional to $N^{1.5}$ (Suzuura et

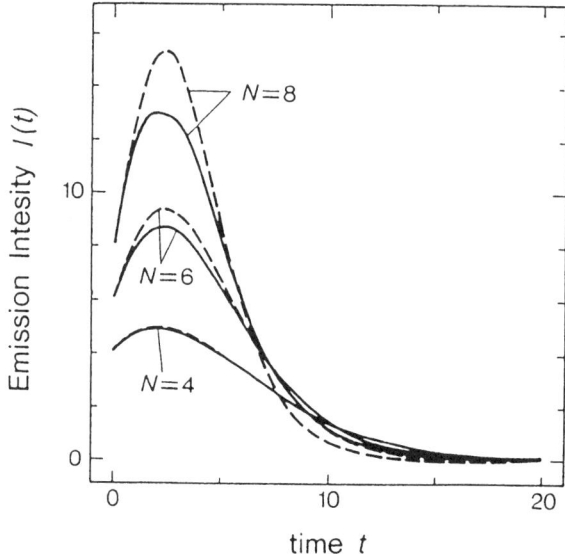

FIG. 1. The superradiant pulse profile of a fully population-inverted system of Frenkel excitons (solid line) in a linear chain and that of Dicke's system (dotted line) — i.e., a dilute gas of atoms. The number N denotes the system size ($N = 4$, 6, or 8), and the decay rate Γ of the single atom is chosen to be 0.1. The exciton effect is measured by the deviation of the solid lines from the dotted lines, which increases as N increases (Tokihiro et al., 1993).

al., 1994). This means that superradiance is possible from the crystal but is different from that of the atomic system.

We have also studied the effect of an induced static dipole moment on the excited state, which results in finite J_z. Depending on the relative signs and magnitudes of J and J_z, the superradiant behavior is different, as shown in Fig. 2. First, case (a), $J = J_z$, which is equivalent to the Heisenberg spin model, shows the same profile of superradiance as that in Dicke's model. Case (b), $J_z = 0$, is the xxo model, while case (c), $J = -J_z$, corresponds to the xxz model (Baxter, 1982). In case (d), $J = 0$ and $J_z > 0$, which is equivalent to the Ising model, the superradiant emission intensity oscillates in time.

As the first difference between our xxo model and Dicke's model ($J = J_z = 0$), we have pointed out that the signal peak is proportional to $N^{1.5}$, in contrast to N^2. This is evidence that excitons can show superradiance, because the signal peak should be linearly proportional to N when the emission of each excitation is incoherent and independent. As a result of the $N^{1.5}$ dependence of the emission peak, the emission intensity $I(t)$ in the

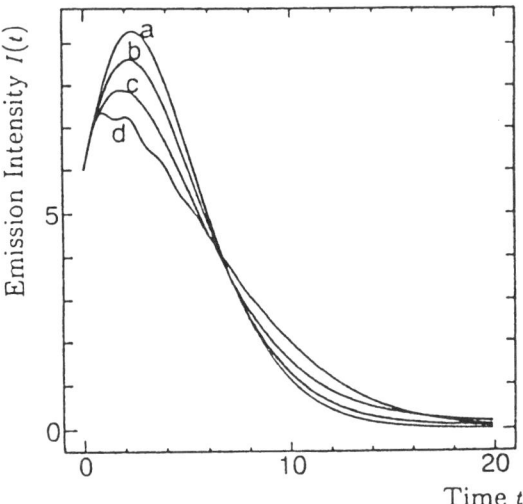

FIG. 2. Superradiant pulse profile $I(t)$: (a) Heisenberg model ($J = J_z = 1.41$), (b) xxo model ($J = 2.00, J_z = 0$), (c) xxz model ($J = 1.41, J_z = -1.41$), and (d) Ising model ($J = 0, J_z = 2.00$) (Manabe et al., 1993).

xxo model shows a longer tail than does Dicke's superradiance, as shown in Fig. 3. The origin of the long tail and the weaker size-dependence can be explained as follows. In the case of Dicke's superradiance, the total Hamiltonian H, including the electron-radiation interaction H' (i.e., $H = H_o + H'$, with $H_o \equiv \Sigma_{i=1}^{N} \hbar\omega_o s_i^z$ and $H' = -\mu \Sigma_{i=1}^{N} \mathbf{s}_j \cdot \mathbf{E}$), commutes with $\mathbf{S}^2 = (\Sigma_{j=1}^{N} \mathbf{s}_j)^2$, so that the cooperation number $S = N/2$ is kept at a maximum over the whole emission process, with all atoms prepared in the excited state as the initial state of the atomic system. On the other hand, the cooperation number is not conserved in the excitonic system because of the excitation propagation so that the atomic system goes through states with smaller cooperation numbers and the slow emission components of these states result in a long tail.

The second characteristic of the excitonic superradiance is observed in the time-analyzed emission spectra. The emission peak frequency shifts monotonically in the superradiant pulse of the excitonic system, whereas it is fixed in Dicke's superradiance. This is shown in Fig. 4. When the excitation-transfer matrix-element J is positive, the emission peak frequency is blue-shifted in the first half of the emission pulse, not shifted at the peak, and red-shifted in the second half of the emission pulse. The magnitude of frequency change is $4J$ for the case $J_z = 0$. This chirping depends strongly on the signs and relative magnitudes of J and J_z.

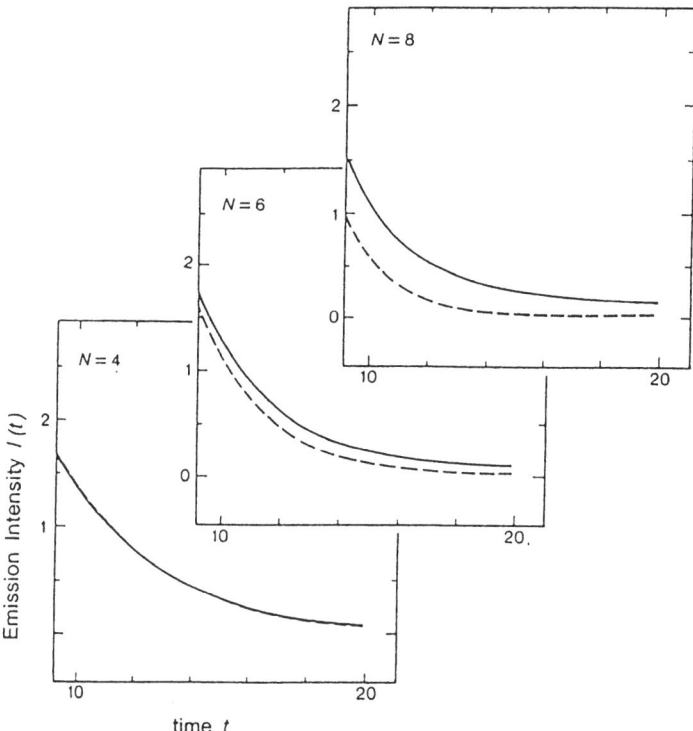

FIG. 3. The slow components of the superradiant decay intensity $I(t)$ of Frenkel excitons in a system with sizes $N = 4$, 6, and 8. The same parameters as in Fig. 1 are used. The slow component increases as the system size N increases (Tokihiro et al., 1993).

Although the results in this subsection were obtained for a linear chain because of mathematically exact treatment, many aspects of these results may be applicable to systems of any number of dimensions.

2. EXCITONIC OPTICAL NONLINEARITY

As discussed in the preceding section, the exciton has a mesoscopic transition dipole moment in quantum dots. When two excitons are excited in a single quantum dot, they interact strongly with each other so that large optical nonlinearity is expected under nearly resonant pumping of this exciton level (Hanamura, 1988a). In this subsection we evaluate the third-order optical susceptibility $\chi^{(3)}(\omega;\omega,-\omega,\omega)$, the real and imaginary parts of

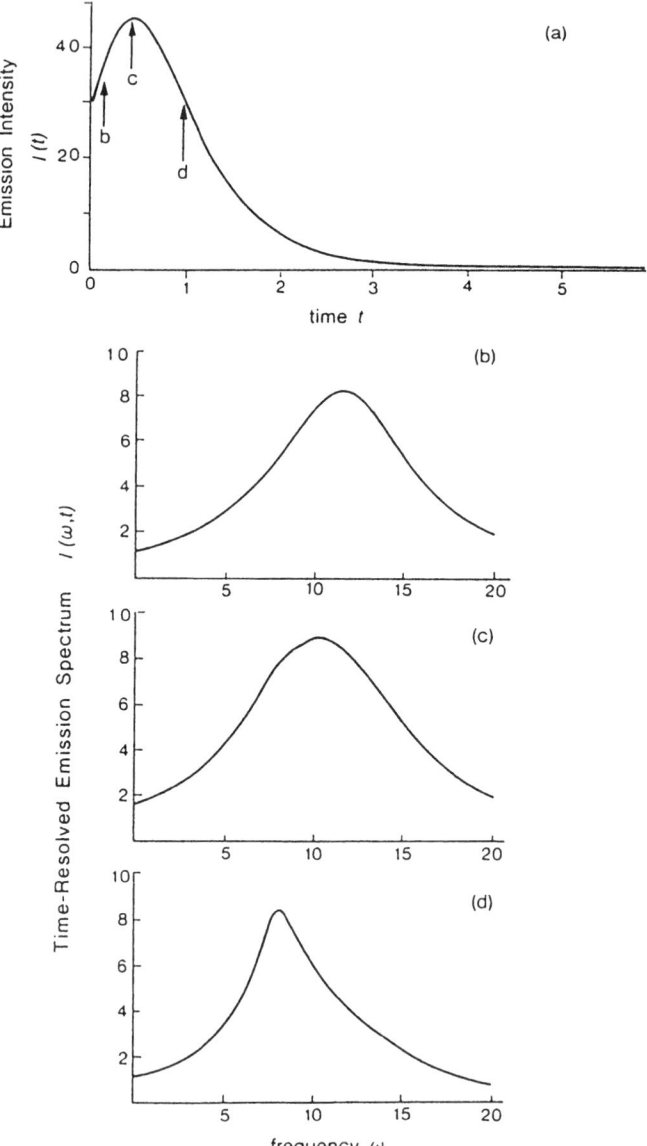

FIG. 4. (a) The superradiant pulse profile from a linear chain with $N = 6$, and the time-resolved emission spectra at (b) $t = 0.15$, (c) $t = 0.44$ (at peak), and (d) $t = 1.0$. Material constants: $\omega_0 = 10.0$, $\tau = 1.0$, and $\Gamma = 0.5$ (Tokihiro et al., 1993).

which describe the optical Kerr effect and absorption saturation, respectively.

We evaluate the third-order polarization:

$$\langle \mathbf{P}^{(3)}(\omega) \rangle = \text{Tr}[\rho^{(3)}(t)\mathbf{P}] \tag{13}$$

where the density matrix $\rho(t)$ is evaluated to the third order in the exciton-photon interaction H':

$$H' = -\mathbf{P} \cdot \mathbf{E}_\omega(t) \tag{14}$$

Here the external field $\mathbf{E}_\omega(t) = \mathbf{E}(\omega)\exp(-i\omega t) + c.c.$, and $\mathbf{P}$ is the electric dipole moment operator derived in the preceding subsection. We are interested only in nearly resonant pumping of the exciton with eigenfrequency ω_o and the largest dipole moment. Therefore we use the single exciton state and the rotating-wave approximation. Then eight of the 48 terms in these approximations contribute to $\chi^{(3)}$, and these eight terms are diagrammed in Fig. 5. We consider two incident beams with frequencies ω_1

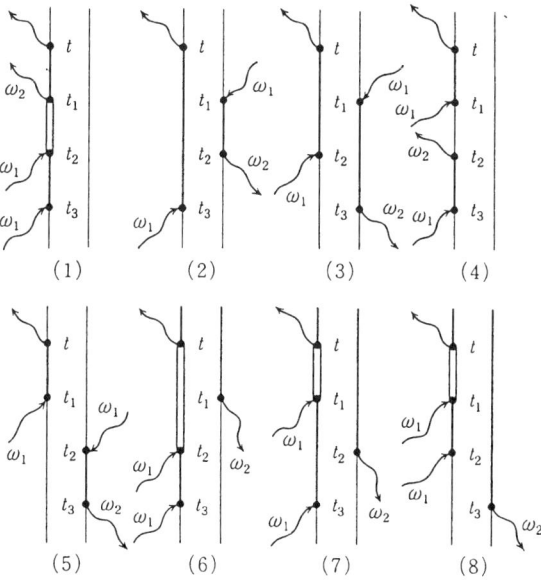

FIG. 5. Eight diagrams describing the time development of the density matrix, which contributes to the third-order susceptibility $\chi^{(3)}(2\omega_1 - \omega_2; \omega_1, -\omega_2, \omega_1)$. Thin, solid, and double lines describe the ground state, a single excited state, and a double excited state in the rotating-wave approximation, respectively.

and ω_2, and we denote the electronic ground state by g, the excited states pumped by a single photon by n, and those pumped by two photons by m. Then we evaluate the third-order polarization $P^{(3)}$ with frequency $2\omega_1 - \omega_2$ and the third-order susceptibility $\chi^{(3)}(2\omega_1 - \omega_2; \omega_1, -\omega_2, \omega_1)$, which are related through

$$\mathbf{P}^{(3)}(2\omega_1 - \omega_2) = \chi^{(3)}(2\omega_1 - \omega_2; \omega_1, -\omega_2, \omega_1) : \mathbf{E}^2(\omega_1)\mathbf{E}^*(\omega_2) \quad (15)$$

Of course, this gives $\chi^{(3)}(\omega; \omega, -\omega, \omega)$ when $\omega_1 = \omega_2 = \omega$.

To first order in the external field, we have the following equation of motion for the density matrix:

$$\frac{d\rho_{ng}^{(1)}}{dt} = \frac{1}{i\hbar}[\hbar\omega_{ng}\rho_{ng}^{(1)} + H'_{ng}\rho_{gg}^{(0)}] - \Gamma_{ng}\rho_{ng}^{(1)} \quad (16)$$

Here we may choose H'_{ng} having the time dependence $\exp(-i\omega_1 t)$ under the rotating wave approximation, so that, for the stationary response,

$$\rho_{ng}^{(1)}(\omega_1) = \frac{H'_{ng}(\omega_1)\rho_{gg}^{(0)}}{\hbar(\omega_1 - \omega_{ng} + i\Gamma_{ng})} \quad (17)$$

This describes the first-step perturbation that develops the ket-vector to the state $|n\rangle$, starting from the initial ground state $\rho_{gg}^{(0)} = |g\rangle\langle g|$, and corresponds to the first step in diagrams (1), (2), (4), (6), and (7) in Fig. 5.

The density matrix $\rho(t) \equiv |\psi(t)\rangle\langle\psi(t)|$ describes the time development of ket-state $|\psi(t)\rangle$ and bra-state $\langle\chi(t)|$, starting from the electronic ground state $\rho_{gg}^{(0)} = |g\rangle\langle g|$. The double-sided Feynman diagrams in Fig. 5 present the time development of $|\psi(t)\rangle$ and $\langle\psi(t)|$, respectively, by the left-hand and right-hand lines in the upward direction. Single and double solid lines indicate the single and double electronic excitations. In Fig. 5(1), for example, photon ω_1 is absorbed at times t_3 and t_2 and the crystal is excited into single and double excited states, respectively, and photon ω_2 is emitted by deexciting the crystal from the double into single exciton states at time t_1. Finally the third-order polarization with frequency $2\omega_1 - \omega_2$ is induced at time t. These interactions between radiation field and electronic systems are brought about on the ket-state — i.e., on the left-hand side — as Fig. 5(1) shows. On the right-hand side — e.g., in Fig. 5(2) — the photon-electron interaction is understood as follows. The photon ω_2 is absorbed at t_2 and the crystal is excited from the ground state into a single exciton state, while the photon ω_1 is emitted and the single exciton state comes down into the ground state at t_1. However, time development of the state on the right-hand side obeys the time-reversed Schrödinger equation. Therefore we may understand that the state propagates from the top to the bottom. In contrast to the description of Shen (1984), the photon ω_1 is absorbed and a single exciton state is created at t_1. The time propagates downward, the ω_2 photon is emitted, and the single exciton state goes down to the ground state at t_2. The other first-order perturbation by $E(\omega_2)^* \exp(i\omega_2 t)$ propagates the bra-vector state:

$$\rho_{gn}^{(1)}(-\omega_2) = \frac{\rho_{gg}^{(0)} H'_{ng}(-\omega_2)}{\hbar(\omega_2 - \omega_{ng} - i\Gamma_{ng})} \quad (18)$$

which describes the first steps in diagrams (3), (5), and (8) in Fig. 5.

After the second-order perturbation under the rotating-wave approximation, we have three kinds of density matrices. First,

$$\frac{d\rho_{mg}^{(2)}}{dt} = \frac{1}{i\hbar}[\hbar\omega_{mg}\rho_{mg}^{(2)} + H'_{mn}(\omega_1)\rho_{ng}^{(1)}(\omega_1)] - \Gamma_{mg}\rho_{mg}^{(2)} \quad (19)$$

where Γ_{mg} denotes the transverse relaxation rate that dephases the second-order polarization $\rho_{mg}^{(2)}$. Here, $\rho_{ng}^{(1)}(\omega_1)$ and $H'_{mn}(\omega_1)$ have the time dependence $\exp(-i\omega_1 t)$, and $\rho_{mg}^{(2)} \sim \exp(-2i\omega_1 t)$. As a result, we have

$$\rho_{mg}^{(2)}(2\omega_1) = \frac{H'_{mn}(\omega_1)H'_{ng}(\omega_1)\rho_{gg}^{(0)}}{\hbar^2(2\omega_1 - \omega_{mg} + i\Gamma_{mg})(\omega_1 - \omega_{ng} + i\Gamma_{ng})} \quad (20)$$

We have two other kinds of the second-order density matrices $\rho_{nn}^{(2)}$ and $\rho_{gg}^{(2)}$, which obey the following equations of motion:

$$\frac{d\rho_{nn}^{(2)}}{dt} = \frac{1}{i\hbar}[H'_{ng}(\omega_1)\rho_{gn}^{(1)}(-\omega_2) - \rho_{ng}^{(1)}(\omega_1)H'_{gn}(\omega_2)] - \Gamma_{n\to g}\rho_{nn}^{(2)} \quad (21)$$

and

$$\frac{d\rho_{gg}^{(2)}}{dt} = \frac{1}{i\hbar}[H'_{gn}(-\omega_2)\rho_{ng}^{(1)}(\omega_1) - \rho_{gn}^{(1)}(-\omega_2)H'_{ng}(\omega_1)] + \Gamma_{n\to g}\rho_{nn}^{(2)} \quad (22)$$

where $\Gamma_{n\to g}$ is the longitudinal decay rate describing the transition from the excited state $|n\rangle$ to the ground state $|g\rangle$. For the stationary pumping of ω_1 and ω_2, both $\rho_{nn}^{(2)}$ and $\rho_{gg}^{(2)}$ have the time dependence $\exp[-i(\omega_1 - \omega_2)t]$, so that

$$\rho_{nn}^{(2)}(\omega_1 - \omega_2) = -\rho_{gg}^{(2)}(\omega_1 - \omega_2) = \frac{H'_{ng}(\omega_1)\rho_{gn}^{(1)}(-\omega_2) - \rho_{ng}^{(1)}(\omega_1)H'_{gn}(-\omega_2)}{i\hbar[\Gamma_{n\to g} - i(\omega_1 - \omega_2)]}$$

(23)

The family tree of the density matrix is drawn in Fig. 6. From this tree, we have two kinds of third-order density matrices with the time dependence $\exp[-i(2\omega_1 - \omega_2)t]$. These matrices obey the following equations of motion:

$$\frac{d\rho_{ng}^{(3)}}{dt} = -i(\omega_{ng} - i\Gamma_{ng})\rho_{ng}^{(3)} + \frac{1}{i\hbar}[H'_{nm}(-\omega_2)\rho_{mg}^{(2)}(2\omega_1)$$
$$+ H'_{ng}(\omega_1)\rho_{gg}^{(2)}(\omega_1 - \omega_2) - \rho_{nn}^{(2)}(\omega_1 - \omega_2)H'_{ng}(\omega_1)] \quad (24)$$

and

$$\frac{d\rho_{mn}^{(3)}}{dt} = -i(\omega_{mn} - i\Gamma_{mn})\rho_{mn}^{(3)} + \frac{1}{i\hbar}[H'_{mn}(\omega_1)\rho_{nn}^{(2)}(\omega_1 - \omega_2) - \rho_{mg}^{(2)}(2\omega_1)H'_{gn}(-\omega_2)]$$

(25)

For the stationary response, we can solve Eqs. (24) and (25) as

$$\rho_{ng}^{(3)}(2\omega_1 - \omega_2) = \frac{H'_{nm}(-\omega_2)\rho_{mg}^{(2)}(2\omega_1) - 2\rho_{nn}^{(2)}(\omega_1 - \omega_2)H'_{ng}(\omega_1)}{\hbar(2\omega_1 - \omega_2 - \omega_{ng} + i\Gamma_{ng})} \quad (26)$$

and

$$\rho_{mn}^{(3)}(2\omega_1 - \omega_2) = \frac{H'_{mn}(\omega_1)\rho_{nn}^{(2)}(\omega_1 - \omega_2) - \rho_{mg}^{(2)}(2\omega_1)H'_{gn}(-\omega_2)}{\hbar(2\omega_1 - \omega_2 - \omega_{mn} + i\Gamma_{mn})} \quad (27)$$

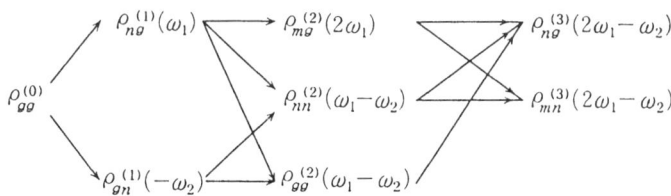

FIG. 6. Tree diagrams describing the third-order perturbations that contribute to $\chi^{(3)}(2\omega_1 - \omega_2; \omega_1, -\omega_2, \omega_1)$. Each root corresponds to one of the eight diagrams in Fig. 5.

Finally, we have the third-order polarization combining these results into Eq. (13) (Hanamura, 1988b):

$$\langle P^{(3)}(2\omega_1 - \omega_2)\rangle = P_{gn}\rho_{ng}^{(3)}(2\omega_1 - \omega_2) + P_{nm}\rho_{mn}^{(3)}(2\omega_1 - \omega_2) + c.c. \quad (28)$$

Let us discuss the third-order optical susceptibility of semiconductor microcrystallites—e.g., CuCl embedded in an insulating crystal NaCl—based on Eq. (28). The size of CuCl microcrystallites is well controlled, and the average radius R ranges from 1.3 to 10 nm. The exciton Bohr radius $a_B = 0.67$ nm, so that the electron-hole relative motion is in the lowest 1s state and the center-of-mass motion is well quantized at low temperature. The bandgap of NaCl is 7 eV and is much larger than the exciton energy, 3.2 eV, in the bulk crystal. As a result, the center-of-mass motion is quantized within the microcrystallite by assuming an infinite potential barrier, and the quantized energy E_n is given by

$$E_n = E_g - R_y + \frac{\hbar^2}{2M}\left(\frac{\pi n}{R}\right)^2 \quad (n = 1, 2, \ldots) \quad (29)$$

where the integer n is the principal quantum number for the center-of-mass motion, E_g is the bandgap of the CuCl microcrystallite, and R_y is the exciton binding energy, 200 meV, in the microcrystallite. The center-of-mass quantization energy is on the order of 10 meV.

The transition dipole moment is expressed in terms of the quantum number n in Eq. (6). Therefore, this dipole moment has the mesoscopic enhancement $(8R^3/\pi^2 a_B^3)^{1/2}$, and the oscillator strength is concentrated in the lowest state (1s, $n = 1$), the energy of which is denoted as $E_1 \equiv \hbar\omega_o$. The transition dipole moment is expressed as

$$P_{ng} \equiv P_1, \quad P_{mn} = \sqrt{2}\,P_1, \quad P_{nm} = \sqrt{2}\,P_1^* \quad (30)$$

where P_1 is a relevant component of $\mathbf{P}_1$ and the factor $\sqrt{2}$ comes from the bosonic nature of the exciton. This is understood as follows. The transition dipole moment P_{mn} contains a raising operation $\langle 2|a_{1s}^\dagger|1\rangle = \sqrt{2}$, and P_{nm} contains a lowering operation $\langle 1|a_{1s}|2\rangle = \sqrt{2}$, where a_{1s} and $a_{1s}^\dagger$ are an annihilation operator and a creation operator of the 1s exciton, respectively. Here we neglect spins of electrons and holes for simplicity.

The deviation of the exciton from an ideal boson is represented by the interaction energy $\hbar\omega_{int}$ between two excitons, and the eigenenergies are written as

$$\omega_{ng} \equiv \omega_o, \quad \omega_{mg} = 2(\omega_o + \omega_{int}) \quad (31)$$

and

$$\omega_{mn} = \omega_{mg} - \omega_{ng} = \omega_o + 2\omega_{\text{int}} \tag{32}$$

The relaxation constants are also described by the longitudinal decay rate 2γ of a single exciton and its pure dephasing rate γ' as

$$\Gamma_{n \to g} = 2\gamma, \quad \Gamma_{ng} \equiv \Gamma = \gamma + \gamma', \quad \Gamma_{mg} = 2(\gamma + \gamma'), \quad \Gamma_{mn} = \Gamma + 2\gamma \tag{33}$$

These constants are obtained from the following consideration. The longitudinal decay rate $\Gamma_{n \to g}$ describes the decay rate of the excited state population $\psi_n^\dagger \psi_n$ and is defined by 2γ. The transverse relaxation rate Γ_{ng} comes from disturbance working on the wavefunction ψ_n in the excited state—i.e., $\rho_{ng} \equiv |\psi_n\rangle\langle\psi_g|$—so that Γ_{ng} consists of half of the longitudinal decay rate $\Gamma_{n \to g}$ and a pure dephasing rate γ'—i.e., $\Gamma_{ng} = \gamma + \gamma'$. The relaxation rate Γ_{mg} comes from the sum of two transverse relaxation rates of two independent excitons. The density matrix ρ_{mn} describes the state in which two excitons are in the ket while a single exciton is in the bra. Therefore, the corresponding relaxation is described by the sum of population decay 2γ and single transverse relaxation Γ. The longitudinal decay rate 2γ consists of the sum of the nonradiative and superradiant decay rates; the latter were discussed in the preceding subsection. For a single incident beam $\omega = \omega_1 = \omega_2$, the third-order susceptibility is simplified into the following form (Hanamura, 1988b):

$$\chi^{(3)}(\omega; \omega, -\omega, \omega) = \frac{N_c}{\hbar^3} \frac{|P_1|^4}{(\omega - \omega_o)^2 + \Gamma^2}$$

$$\times \left[\frac{1}{\omega - \omega_o + i\Gamma} - \frac{1}{\omega - \omega_o - 2\omega_{\text{int}} + i(\Gamma + 2\gamma)} \right] \tag{34}$$

$$\times \left[1 + \frac{2\gamma'}{\gamma} + \frac{2i\Gamma - \omega_{\text{int}}}{\omega - \omega_o - \omega_{\text{int}} + i\Gamma} \right]$$

Here N_d is the number density of microcrystallites and is expressed as $N_d = r/v \equiv 3r/(4\pi R^3)$, with a volume ratio r of microcrystallites over the matrix.

We will be able to evaluate the size dependence of $\chi^{(3)}$ by considering three limiting cases.

a. $\omega_{\text{int}} > |\omega - \omega_o| > \Gamma$

3 OPTICAL RESPONSES OF QUANTUM WIRES/DOTS AND MICROWAVES

$$\chi^{(3)} = \frac{2N_d|P_1|^4}{\hbar^3(\omega - \omega_o)^3}\left(1 + \frac{\gamma'}{\gamma}\right) \propto \left(\frac{R}{a_B}\right)^3 \tag{35}$$

Keeping the volume ratio r constant and increasing the size R of microcrystallites, $\chi^{(3)}$ increases in proportion to the volume of a single microcrystallite. This occurs because the effect of mesoscopic dipole moment $|P_1|^4 \propto R^6$ overcomes the size dependence R^{-3} of the number density N_d. However, note that the exciton-exciton interaction $\hbar\omega_{int}$, which is inducing the optical nonlinearity, increases when the size R decreases, as shown in the first Born approximation as

$$\hbar\omega_{int} = \frac{13}{4} R_y \left(\frac{a_B}{R}\right)^3 \tag{36}$$

Therefore, the mesoscopic enhancement of Eq. (35) is limited by the condition $\omega_{int} > \Gamma$—i.e., $(R/a_B)^3 < 13Ry/4\hbar\Gamma$. Because $\hbar\Gamma = 0.02$ meV at low temperature and $\hbar\omega_{int} = 3$ meV for a CuCl sphere with $R = 3$ nm, we can choose a frequency region that satisfies this condition.

b. $\omega_{int} > \Gamma > |\omega - \omega_o|$

$$\chi^{(3)} = -i\frac{2N_d|P_1|^4}{\hbar^3\Gamma^2\gamma} \tag{37}$$

Here, $\chi^{(3)}$ is purely imaginary and describes the role of absorption saturation. When the longitudinal decay 2γ is determined by superradiance, $\chi^{(3)}$ in Eq. (34) loses the size dependence but $\chi^{(3)}$ keeps the mesoscopic enhancement, as will be discussed in the next subsection for a quantum well system. As long as Γ and γ are determined by an R-independent process, $\chi^{(3)}$ has R-dependent enhancement. For a dilute system of CuCl microcrystallites with the average radius $R = 8$ nm and volume ratio $r = 0.1$, Im $\chi^{(3)}$ is estimated to be -10^{-3} esu, using the observed values $\hbar\Gamma = 0.5$ meV and $2\hbar\gamma = 0.03$ meV at a temperature of 77 K.

c. $|\omega - \omega_o| > \Gamma > \omega_{int}$

$$\chi^{(3)} = \frac{2iN_d|P_1|^4(2\gamma' + \gamma)}{\hbar^3(\omega - \omega_o)^4} \propto \left(\frac{R}{a_B}\right)^3 \tag{38}$$

This $\chi^{(3)}$ has the same size dependence as in case a, but its absolute value is reduced by $(2\gamma' + \gamma)/|\omega - \omega_o|$, and its phase differs by $\pi/2$, in comparison with case a.

3. Excitons in Quantum Wells

In this section we discuss the superradiance and excitonic enhancement of $\chi^{(3)}$ for two-dimensional (2D) excitons in a quantum well (Hanamura, 1989b). Here we must take into account the effect of the finite coherent length of 2D excitons (owing to their scattering by defects or phonons) on the superradiant decay rate and $\chi^{(3)}$.

First let us obtain the expression for a coherent length L^* of an exciton in terms of its spectrum width Γ. The center-of-mass motion of a 2D exciton in a quantum well with area L^2 is described by a plane wave in an ideal limit

$$\Psi_{K_o}(R) = \frac{1}{\sqrt{L^2}} \exp(i K_o \cdot R) \tag{39}$$

or in its Fourier component

$$\Psi_{K_o}(K) = \sqrt{L^2}\, \delta_{KK_o} = \frac{(2\pi)^2}{\sqrt{L^2}} \delta(K - K_o) \tag{40}$$

When the 2D exciton has the spectrum width Γ, the exciton state is described by a superposition of wave number states K with $E \equiv \hbar^2 K^2/2M \leq \hbar\Gamma$, where M is an effective mass of excitonic center-of-mass motion. Taking into account the constant density-of-states characteristic of 2D center-of-mass motion, we derive the following wavefunction in the wavevector space:

$$\Psi(K) = \begin{cases} \sqrt{4(L^*)^2/\pi}: & K \leq \pi/L^* \\ 0 & \text{otherwise} \end{cases} \tag{41}$$

Here we define the coherent length of the 2D exciton by L^*. This may be accepted from the real-space expression—i.e., the inverse Fourier transform—of Eq. (41):

$$\Psi(R) = \frac{1}{\sqrt{\pi R}} J_1\left(\frac{\pi R}{L^*}\right) \tag{42}$$

This means that the center-of-mass motion is well localized within $R \leq L^*$.

Then we can derive the relation between the spectrum width Γ and the coherent length L^* by requiring that this exciton be described by superposing the density-of-states $D^*(E)$ over the energy width $\hbar\Gamma$:

$$\int_0^{\hbar\Gamma} D^*(E)\, dE = 1 \tag{43}$$

where the 2D area L^2 is divided into $L^2/\pi(L^*)^2$ regions and the state density $D^*(E)$ within $\pi(L^*)^2$ is derived per unit energy as

$$D^*(E) = \pi(L^*)^2 \frac{M}{2\pi\hbar^2} \tag{44}$$

Inserting this expression into Eq. (43), we derive the expression for the coherent length L^* as

$$L^* = \sqrt{\frac{2\hbar}{M\Gamma}} \tag{45}$$

The partially coherent exciton with coherent length L^* is derived to have the transition dipole moment

$$\langle \Psi | \mathbf{P} | g \rangle = \sqrt{\frac{8}{\pi^2}} \sqrt{\left(\frac{L^*}{a_B}\right)^2} \boldsymbol{\mu}_{cv} \tag{46}$$

in contrast to that of the perfectly coherent exciton

$$\langle \Psi_K | \mathbf{P} | g \rangle = \sqrt{\frac{2}{\pi}} \sqrt{\left(\frac{L}{a_B}\right)^2} \boldsymbol{\mu}_{cv} \tag{47}$$

Then the radiative decay rate is reduced by a factor of $(2L^*/\lambda)^2$ from the perfectly coherent case:

$$2\gamma^* = 2\gamma \left(\frac{2L^*}{\lambda}\right)^2 = 48\pi \left(\frac{L^*}{a_B}\right)^2 \gamma_s \tag{48}$$

$$2\gamma = 12\pi \left(\frac{\lambda}{a_B}\right)^2 \gamma_s \tag{49}$$

where γ_s denotes a radiative rate resulting from band-to-band transition and is given by

$$\gamma_s \equiv \frac{4n_o|\mu_{cv}^2|}{3\hbar\lambda^3} \quad (50)$$

where a_B is the exciton Bohr radius and n_o is the refractive index of the well for the wavelength λ. When the coherent length $L^* = 900$ Å, corresponding to $\hbar\Gamma = 0.22$ meV for a GaAs quantum well, the radiative decay time $T_1 \equiv 1/(2\gamma^*)$ is evaluated to be 100 ps whereas the observed lifetime is 180 ps (Feldman et al., 1987). This is in contrast to the case of a perfectly coherent exciton: $T_1 = 1/(2\gamma) = 2.8$ ps.

The spectrum width Γ is the sum of half the radiative decay rate $2\gamma^*$ and the dephasing rate γ'. Note that the spectrum width is described by the decay of the wavefunction ψ in the excited state whereas the radiative decay rate $2\gamma^*$ comes from the population decay $\psi^*\psi$ in the excited state. Therefore, only γ^* contributes to the spectrum width Γ — i.e.,

$$\Gamma = \gamma' + \gamma^* \quad (51)$$

As a result, γ^*, L^*, and Γ should be determined self-consistently from Eqs. (45), (48), and (51) — i.e.,

$$(L^*)^2 \left[1 + \frac{24\pi\gamma_s}{\gamma'}\left(\frac{L^*}{a_B}\right)^2\right] = \frac{\pi^2\hbar}{2M\gamma'} \quad (52)$$

Let us consider a quantum well of CdS. First, when $\gamma^* \gg \gamma'$,

$$L^* = \left(\frac{\pi\hbar a^2}{48M\gamma_s}\right)^{1/4} \sim 2000 \text{ Å} \quad (53)$$

When $\hbar\Gamma = 1$ meV and $M = m_o$, $\gamma' \gg \gamma^*$ and

$$L^* = \sqrt{\frac{2\hbar}{M\Gamma}} \sim 300 \text{ Å} \quad (54)$$

The result of Eq. (53) means that the coherent length of the 2D exciton is limited by its superradiant decay rate. As a result, the figure of merit comes to have the same limitation. The excitonic enhancement of the figure of merit defined by the left side of Eq. (55) is derived under the resonant excitation of the 2D exciton:

$$\frac{|\chi^{(3)}|}{\alpha\tau} = \frac{n_o^2}{\pi^2\lambda^2}\frac{|\mu_{cv}|^4}{\hbar^3\Gamma^2}\left(\frac{\gamma^* + \Gamma}{\gamma^*}\right)\left(\frac{L^*}{a_B}\right)^4 \qquad (55)$$

where α is the linear absorption coefficient at the exciton frequency and τ is the switching time (see Subsection 2 in Section II). When $\gamma' \gg \gamma^*$,

$$\left|\frac{\chi^{(3)}}{\alpha\tau}\right| = \frac{n_o|\mu_{cv}|^2}{32\pi^3\hbar^2\gamma'}\left(\frac{L^*}{a_B}\right)^2 \qquad (56)$$

and, in the ideal limit $\gamma^* \gg \gamma'$,

$$\frac{|\chi^{(3)}|}{\alpha\tau} = \frac{\lambda^4}{4^5\pi^4 \hbar a^2} \sim 10^{17}\left(\frac{\text{esu}\cdot\text{cm}}{\text{s}}\right) \qquad (57)$$

Even under the resonant pumping of a top-surface exciton in an anthracene crystal at low temperature (2 K), the figure of merit $|\chi^{(3)}|/\alpha\tau$ is on the order of 10^8 to 10^{10} esu·cm/s, because $|\chi^{(3)}| \sim 1$ esu (Kuwata-Gonokami, 1987), $T_1 \sim 10^{-12}$ s (Aaviksoo et al., 1987), and the linear absorption coefficient $\alpha \sim 10^2$ to 10^4 cm^{-1}. This is in contrast to the figure of merit 5×10^2 esu·cm/s for a GaAs quantum well system, where $|\chi^{(3)}| \sim 10^{-1}$ esu, $T_1 \sim 2 \times 10^{-8}$ s, and $\alpha \sim 10^4$ cm^{-1} (Feldman et al., 1987). As a conclusion, the top value of the observed figure of merit of 10^8 to 10^{10} esu·cm/s for the anthracene crystal is still much smaller than the limit value of 10^{17}. The first-surface exciton of the anthracene crystal exhibited superradiant decay, with a lifetime shorter than 2 ps. On the other hand, the second- and third-surface excitons had longer lifetimes of 15 ± 2 ps and 200 ps, respectively (Dan and Hanamura, 1994). The eigenfunctions of excitonic surface states have been solved exactly for multilayer organic quantum wells by taking into account the dipole-dipole interactions — i.e., the propagation effects — not only within two-dimensional networks but also among the multilayers. Several surface states are formed by image forces as a result of the interface between the crystal and the vacuum. The deepest surface state of the exciton is found to be localized almost at the top surface, so that it is called a top-surface exciton. The second-surface exciton state has smaller binding energy. It has the largest wavefunction amplitude at the second layer from the surface, but decays oscillating in the deep direction.

As a result, the top-surface exciton has the largest transition dipole moment because the molecular dipole moments are summed up with the same sign. On the other hand, the molecular dipole moments of the second-surface exciton are partially canceled out between the even and odd layers. Therefore, we have obtained radiative lifetimes of 1 and 10 ps for the top- and second-surface excitons of the anthracene crystal by using the material constants of this crystal (Dan and Hanamura, 1994).

4. COHERENT LIGHT EMISSION FROM QUANTUM WIRES

There are two well-known mechanisms for coherent light emission: laser emission and superradiance. The laser uses stimulated emission from population-inverted systems of atoms, molecules, or solids. Here the relaxation times of material systems are so short that they can follow the radiation field adiabatically. Conversely, in the case of superradiance, the cavity is so small that the emitted light escapes rapidly from the cavity and the radiation field follows the atomic system adiabatically. In this case, the equation of motion of the atomic system determines the emitted light profile. Nowadays, development of semiconductor fabrication technology has made possible the preparation of mesoscopic systems in which the electronic excitation is well quantized, while the microcavity can now be designed to enhance or quench spontaneous emission at will. Here arises a dream in which we may control quantum mechanically and mutually both the electronic system and the radiation field on the same footing within the same space of the microcavity. In this new regime, the conventional notions of laser and superradiance as they now stand are inapplicable to our system. In this subsection, we discuss a mechanism of coherent light emission from such a mesoscopic system that is quite different from laser radiation or superradiance (Yura and Hanamura, 1994).

We choose a one-dimensional (1D) system of Wannier excitons—e.g., semiconductor quantum wires (QWRs) or conjugated polymers, whose length is several times greater than the wavelength of the relevant light within the medium composing the microcavity. The characteristic feature of this system can be stated as follows. The relevant excitation here is the lowest-energy exciton, because it has the dominant oscillator strength and a very large binding energy. Among the various interactions present, the largest is the exciton–radiation field coupling, which strongly hybridizes this exciton into an exciton-polariton (EP) (Hopfield, 1958). The exciton and the photon within this system are well described as harmonic oscillators—i.e., bosons—and their interaction is well presented in bilinear form. Therefore, this interaction is diagonalized, and the upper- and lower-branch eigenmodes are obtained. The upper mode and the longitudinal mode become degenerate in the limit of long wavelength. Thus the longitudinal-transverse splitting of the EP is larger than the exciton-phonon interaction and the decay rate of the photon from the cavity. At the present stage of semiconductor fabrication technology, homogeneous as well as inhomogeneous broadenings of 1D excitons are inevitable owing to nonuniform cross sections of the QWRs and interface roughness. However, the homogeneous linewidth, 0.2 meV, of the lowest 1D exciton (Wegscheider et al., 1993), is smaller than the polariton splitting, 0.6 to 6.0 meV, reported for 2D excitons in quantum wells, which is in contrast to the splitting, 0.08 meV, of bulk GaAs.

In this subsection, we confine ourselves to the ideal EP in the QWRs. For simplicity, we neglect the inhomogeneous broadening and scattering of the EPs at the interface and solve the population dynamics of polaritons, taking into account only the exciton-phonon scattering and the leakage of polaritons at the end of the cavity. The EP with higher energy is scattered by 3D phonons, whereas the EP with energy lower than that of the bottleneck leaks more easily from the ends of the cavity. Then, reflecting this characteristic and the bosonic nature of polaritons, the EPs dominantly occupy the specified mode above a critical pumping rate. Here we are using the second advantage of a 1D exciton system with mesoscopic size — i.e., that the EP of this system has discrete energy levels, enabling single-mode emissions. As mentioned above, the lowest-energy 1D exciton interacts very strongly with the radiation field, and the EP picture becomes valid. The other interactions are treated perturbatively, which we shall now discuss.

First, the exciton-phonon scattering can be taken into account in the population dynamics of the quantized polaritons. Second, the relevant EP mode is, in a strict sense, a linear combination of the polariton inside a microcavity and the radiation field outside it. Here, however, such an effect is described only by the leakage rate of this EP at one end of the sample. The photon-like part of the EP is well quantized and its characteristic wavelength becomes of the same order of magnitude as the system size. Third, the EP-EP scattering is negligible to lowest order when the quantization energy is larger than the level broadening, because of the nonexistence, at this level of approximation, of scattering processes that simultaneously satisfy energy and momentum conservation laws. We have also neglected the higher-order scatterings associated with phonon scatterings. In these approximations the population dynamics of EPs is written, at sufficiently low lattice temperature, as follows:

$$\frac{d}{dt} n_j = \sum_{i>j} A_{ij} n_i (n_j + 1) - \sum_{k<j} A_{jk} n_j (n_k + 1) - C_j n_j + N_j^{in} \quad (58)$$

where A_{ij} is the phonon-scattering rate from the ith to the jth EP level, C_j

is the radiation leakage rate at the end face, and N_j^{in} is the supply rate of EPs at the jth mode from outside. The phonon-scattering rate A_{ij} is evaluated by the first Born approximation at zero lattice temperature:

$$A_{ij} = \frac{2\pi}{\hbar^2} \sum_q |\Xi_{ij}|^2 \delta(\omega_i - \omega_j - \omega_q) \tag{59}$$

Extension of Eqs. (58) and (59) to finite lattice temperature is easily done. Here, the EP-phonon scattering matrix element Ξ_{ij} is evaluated in terms of deformation potentials of the conduction and valence bands and the exciton-EP transformation coefficients. Three-dimensional acoustic phonons — e.g., corresponding to the QWR made of GaAs-AlGaAs — are used, but the main features of the dynamics are not changed for the 1D phonon model. The radiation leakage rate $C(\omega) = C_j$ is given by

$$C(\omega) = \frac{v_g(\omega)}{2L} T(\omega) \tag{60}$$

in terms of the transmissivity $T(\omega)$ at the end face, the EP group velocity $v_g(\omega)$, and the sample length L. The notations $i > j$ and $k < j$ in Eq. (58) denote summations over i and k under the conditions $\omega_i > \omega_j$ and $\omega_k < \omega_j$, respectively, at $T = 0$. For high lattice temperatures, we have modified Eq. (58) so that the EP flows, not only to the lower-energy side but also to the higher-energy side by phonon absorption, are taken into account similarly.

The EP leakage rate on the dispersion curve shown in Fig. 7(a) is evaluated from Eq. (60), as shown in Fig. 7(b). Here we used the well-known material constants of GaAs. The microcavity consists of a QWR of length $L = 0.5$ to $10\,\mu\text{m}$ having one free end and the other end with 100% reflection. Then we have discrete EP levels, as shown in Fig. 7(a) for $L = 0.5\,\mu\text{m}$. Note that the leakage rate becomes minimum near the polariton bottleneck and increases very rapidly on the lower-energy side, because the photon-like branch has both a large group velocity v_g and a large transmissivity T. The minimum leakage rate $C_{\min} = 1.13 \times 10^{10}\,\text{s}^{-1}$ will be used to normalize the timescale.

Next we will show how rapid growth of the EP number on the specified single mode can be expected by making use of the bosonic nature of polaritons. Let us choose a specified mode of EP near $C_{\min}$ — i.e., near the bottom of $C(\omega)$ and on the side of the photon-like branch — and call it the c-mode. For a rough discussion, we keep only three modes — b, c, and d — with modes b and d lying just below and above the c-mode (see Fig. 7). Limiting ourselves to the case where the polaritons are scattered between

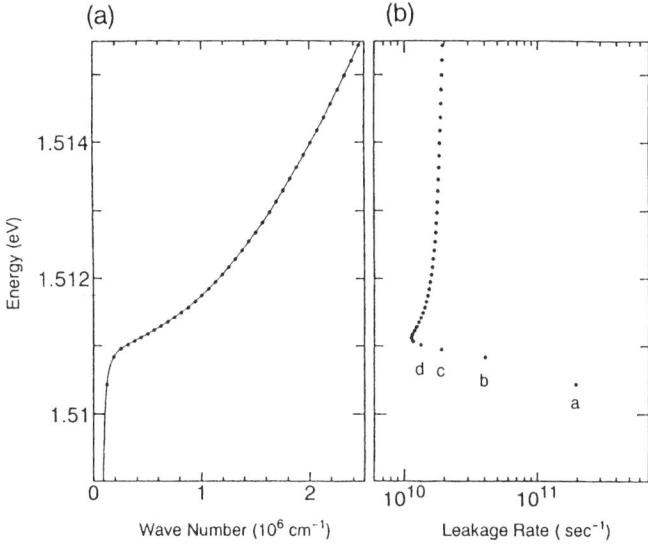

FIG. 7. Dispersion curve (a) and leakage rate (b) for the polariton in a quantum wire. On the dispersion curve, the continuous line shows the case of the long QWR. The dots correspond to the quantized modes for $L = 0.5\,\mu$m. We define a, b, c, and d modes, from the low-energy side (Yura and Hanamura, 1994).

neighboring modes by one-phonon processes, Eq. (58) gives

$$\frac{d}{dt}n_c = [A(n_d - n_b - 1) - C_c]n_c + An_d \qquad (61)$$

As shown in Fig. 7(b), n_b decays more rapidly than n_d in the photonic branch near $C_{\min}$ so that the coefficient of n_c on the right side of Eq. (61)—i.e. $[A(n_d - n_b - 1) - C_c]$—becomes positive when the total population of EPs increases, and then n_c grows exponentially in time.

Third, we have solved the coupled differential nonlinear equations (Eqs. 58) for every j numerically. As the first characteristics of the 1D Wannier exciton, the oscillator strength distributes dominantly on the lowest-energy state, and its binding energy is also as large as 20 meV—e.g., in a GaAs QWR of cross section 70×70 Å. For the EP scattering by phonons, we have evaluated the intraband scattering of an electron and a hole for which the EP envelope function of the exciton center-of-mass motion is approximated by

$$\psi_n(r, z) = N_n \sin(k_n z) J_0(\alpha r) \qquad (62)$$

where the normalization constant $N_n = (\pi L R_0^2/2)^{-1/2} J_1(\alpha_0)^{-1}$, with $\alpha_0 = 2.4048\ldots$, $\alpha = \alpha_0/R_0$, and $k_n = n\pi/2L$ ($n = 1, 2, \ldots$). Here the Bessel function J_0 indicates the lowest energy state in the lateral section of the cylinder with radius R_0, and n is the quantum number of the z-direction along the QWR with length L. We have assumed $L \gg a_B > R_0$, where a_B is the exciton Bohr radius, so that the dynamics of 1D excitons is justified. We used the deformation potential 16 eV of the exciton, the mass density, and the sound velocity of bulk GaAs, assuming that the electronic structure is well quantized but that the phonon structure is not substantially modified in the QWR of GaAs embedded in AlGaAs. Using these relations, the matrix A_{ij} is numerically calculated. The supply of EPs from the outside, described by N_j^{in}, are distributed uniformly over the range 1.526 to 1.530 eV—i.e., about 17 meV above the bottleneck. The total supply rate N is also normalized by C_{min} as $N = \tilde{N} C_{min}$.

The occupation numbers on each mode are plotted in Fig. 8 for the stationary flow state as a function of pump power. Below $\tilde{N} = 100$ for the pump power, the occupation numbers increase monotonically. The monopoly characteristic of the bosons begins to appear beyond $\tilde{N} = 100$—i.e.,

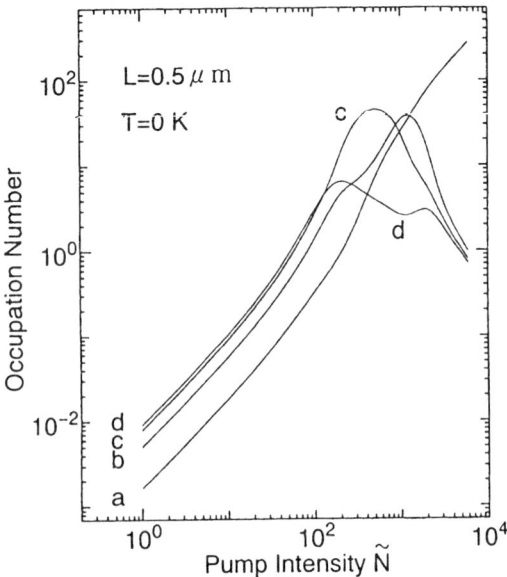

FIG. 8. Occupation number of each mode versus pump power for the QWR in Fig. 7. This indicates the nonlinearity of Eq. (58), reflecting the bosonic nature of the exciton polaritons (Yura and Hanamura, 1994).

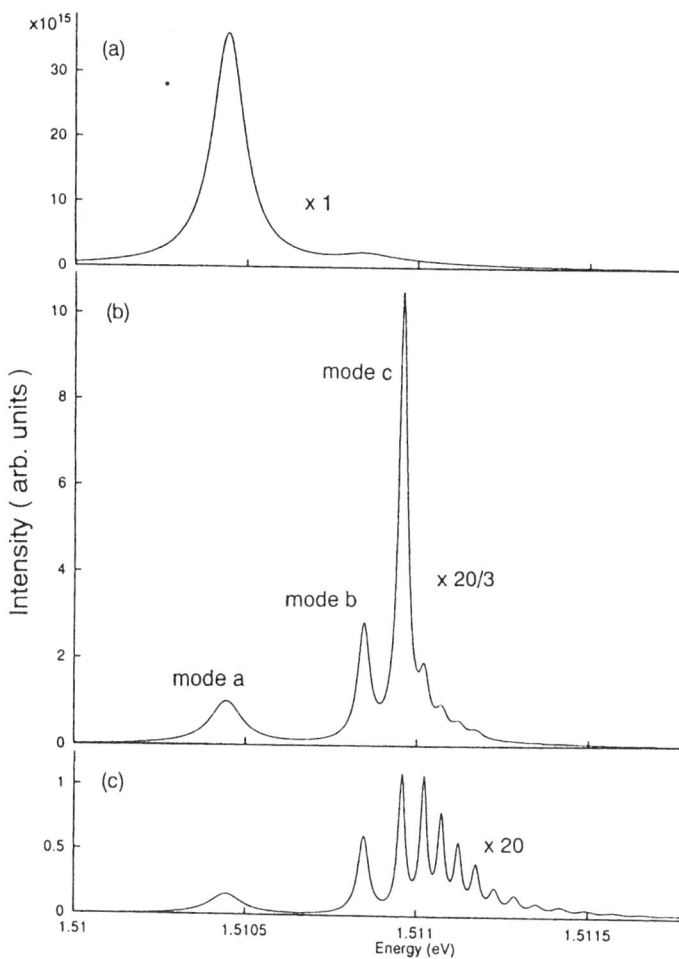

FIG. 9. Photoemission spectra of the QWR in Fig. 7 for various values of pumping power at 0 K. The characteristics of condensing on a single mode become observable by increasing the pump power $\tilde{N}$. (a) $\tilde{N} = 2000$, (b) $\tilde{N} = 300$, and (c) $\tilde{N} = 100$. (Yura and Hanamura, 1994).

the c-mode (see also Fig. 9) grows rapidly, swallowing the polaritons of the higher-energy side (e.g., the d-mode). On the other hand, the c-mode is overwhelmed by the b-mode beyond $\tilde{N} = 600$ for the pump power, and finally the b-mode is overwhelmed by the a-mode. In any case, once the rate of feeding into the mode of interest overcomes the decay rate through the sample end, the polariton of this mode begins to monopolize the population.

The emission spectrum is evaluated in terms of the calculated populations as a superposition of lorentzian forms with the broadening resulting from phonon scattering and decay through the cavity end. This is shown in Fig. 9 for several values of pumping at 0 K. We obtained a coherent emission — i.e., almost a single-mode emission — beyond the critical pumping, as shown in Fig. 9(a) and (b) for $\tilde{N} = 2000$ and 300. Below the critical pumping — e.g., at $\tilde{N} = 100$ — we have the multimode emission shown in Fig. 9(c). It is notable that 18% of the total supplied electron-hole pairs can be emitted through the c-mode EP at the normalized pump power of $\tilde{N} = 300$. This mode of EP is connected to the multimodes with the same photon energy but different wavevector of the radiation field outside the microcavity. Here we can choose between two cases. (1) When the single-mode emission of radiation is necessary, our 1D system is put in one of the vacuum cylinders with the 1D microcavity axis parallel to the cylinder axis in a 2D photonic bandgap (PBG) system. Then the emitted light is coherent in the sense that a large number of polaritons occupy a single mode. Physically, this originates in the single mode of the radiation field that propagates along the cylinder axis of the 2D PBG system. However, the coherent time of the emitted light is limited by the decay rate $C(\omega_j)$ of the polariton. (2) In the second case, without the PBG system, the radiation is emitted in any direction within the angle determined by the cross section and length of the microcavity but with a well-defined photon energy. We have also studied the temperature effect on the polariton dynamics. The phonon scatterings, at temperatures on the order of 50 K, are so frequent that the lowest-energy polaritons are dominantly populated. However, the growth of this mode c is also brought about beyond $\tilde{N} \geq 100$, the same value as at 0 K.

The exciton supply density at the threshold for condensation is approximately 10^{12} e-h pairs per second per single wire, and the stationary density is estimated to be 50 e-h pairs per wire for $L = 0.5$ μm — i.e., 10^6 cm^{-1}. This threshold is of the same order of magnitude as for the laser threshold due to excitons in a GaAs QW. As realized from the fact that the 1D exciton has a binding energy five times as large as in bulk GaAs, the 1D exciton is very stable, and the exciton Bohr radius $a_B^{(1)}$ is expected to be reduced to less than half the 3D value $a_B = 100$ Å. Therefore, the exciton density $n = 10^6$ cm^{-1} may correspond to $na_B^{(1)} < 0.5$, so that we expect the 1D excitons to be stable sufficiently near the threshold. In the present paper, we use the free end for the microcavity. When we introduce the mirror at the end of the cavity and can assume a uniform reduction of the leakage rate $C(\omega)$, the condensation of EPs is realized at a much lower total concentration of electron-hole pairs. This exciton stability at these concentrations is also attributable to the fact that the first-order exciton-exciton scattering is prohibited, as mentioned at the beginning of this subsection. The state filling and other many-body effects work but are expected not to induce serious modification of the present results near the threshold density of condensation in this 1D system. Therefore we expect that it is possible to observe this condensation effect of EPs in a GaAs QWR. It is also interesting to study the effect of exciton lattice formation as a result of the higher-order many-body effect. Note also the difference between our model and the system of Wegscheider *et al.* (1993). They used QWRs of $L = 600\,\mu$m, whereas ours is as short as 10 μm. Here the quantization of EPs plays an important role.

These results are almost independent of the manner of externally supplying the polaritons because the thermalization process of electron-hole pairs or excitons is very rapid on the exciton-like branch and the dynamics of polaritons is almost determined near the bottleneck of the polariton. This means it is possible to supply polaritons externally by electron-hole injection—e.g., by current injection.

II. Experimental Results

As discussed in Section I, excitons look promising as elementary excitations responsible for the nonlinear optical response (Hanamura, 1988a, 1988b). They seem promising because the third-order optical susceptibility can show mesoscopic enhancement under resonant excitation of excitons at low temperature (see Subsection I.2) and because the excitons can radiatively decay so rapidly through the superradiant process (see Subsection I.1). Therefore, we can expect a large figure of merit for nonlinear optical responses, which is defined by $|\chi^{(3)}(\omega)|/\alpha(\omega)\tau$. Here, $\chi^{(3)}(\omega) \equiv \chi^{(3)}(\omega;\omega,-\omega,\omega)$ is the third-order susceptibility defined in Subsection I.2, $\alpha(\omega)$ is the linear absorption coefficient at the relevant frequency ω, and τ is the switching time that we assume the superradiative decay to determine. In this section, we discuss some of the experimental observations of these theoretical results.

1. SUPERRADIANCE OF EXCITONS

The superradiant decay of Wannier excitons was confirmed by picosecond time-resolved spectra of the exciton luminescence from CuCl microcrystallites embedded in an NaCl matrix crystal and glass (Itoh *et al.*, 1990; Nakamura *et al.*, 1989). These microcrystallites are observed to be somewhat oblate but are estimated to be almost spherical, with the radius R

ranging from 1.7 to 15 nm. The Wannier exciton of the CuCl crystal has an eigenenergy of 3.2 eV, while the absorption edge of the NaCl crystal is 7 eV. Therefore, we may approximate an exciton in a CuCl microcrystallite to be confined by a potential barrier with infinite activation energy. Then the theoretical result in Subsection I.1 is applicable to the radiative decay rate of these excitons.

CuCl microcrystallites with mean radius of 3.8 nm were grown inside an NaCl crystalline matrix with 1 mol % of CuCl. The absorption and luminescence spectra of the CuCl microcrystallites were measured at 77 K, as shown, respectively, by the solid and dotted curves at the top of Fig. 10. No

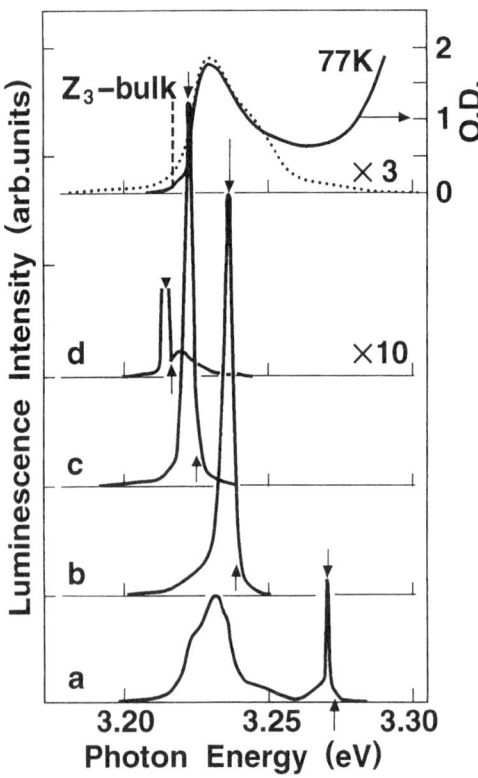

FIG. 10. Exciton luminescence spectra of CuCl microcrystallites of 3.8-nm mean effective radius, measured at 77 K under selective excitation at different energies, indicated by the upward-pointing arrows. The exciton absorption and luminescence spectra under nonresonant excitation are also shown (at top) by the solid and dotted curves, respectively. For an explanation of the downward-pointing arrows, refer to Fig. 11 (Itoh et al., 1990).

Stokes shift is observed between the exciton absorption and luminescence bands. Their peak energy is blue-shifted by 11.3 meV from the bulk exciton, and this results from the quantization of the center-of-mass motion for the confined exciton. As long as we assume a spherical microcrystallite with radius R, the blue shift δE from the bulk exciton energy is well represented by

$$\delta E = \frac{\hbar^2}{2M}\left(\frac{\pi}{R^*}\right)^2 \tag{63}$$

with the effective radius $R^* = R - a_B/2$. Here, the effective Bohr radius a_B of the exciton is 0.68 nm and the center of mass M is $2.3m$, with m being the electron mass in vacuum. This large mass M comes from heavy hole mass composing the excitons (Mita et al., 1980; Onodera, 1980).

The exciton absorption and luminescence spectra under nonresonant excitation are inhomogeneously broadened, as shown at the top of Fig. 10. This results from the wide distribution of the size of microcrystallites R and the size-dependent blue shift given by Eq. (63). By pumping this system with monochromatic laser light at a photon energy shifted by δE from the bulk exciton, we can observe selectively the luminescence of the microcrystallites with the specified radius R that satisfies Eq. (63). The luminescence lines are observed at nearly the same frequency as the pump, as shown in Fig 10(a), (b), (c), and (d).

The time-profile of the luminescence intensity depends on the pump frequency—i.e., the size R of the microcrystallites—as shown in Fig. 11. These time dependencies were analyzed by taking into account the reabsorption of emitted photons and thereby obtaining the size dependence of the radiative decay time τ in Fig. 12. Not only the R^3 dependence of the observed radiative decay rate $1/\tau$ but also the absolute value of τ have been found to be in good agreement with the theoretical result (Eq. 7)—i.e., $\tau = T_1$ for $R < 7$ nm. When the size R increases beyond the critical value L^*, which depends on the lattice temperature, the observed value of $1/\tau$ deviates from the R^3 dependence, because the coherent length L^* of the exciton becomes shorter than the size R of the microcrystallite for $R > L^*$, so that τ becomes independent of the size R but dependent on the lattice temperature. This gives evidence that the interaction of the exciton with phonons determines the coherent length of the exciton in this region.

2. FIGURE OF MERIT FOR NONLINEAR OPTICAL RESPONSES

The possibility of optical information processing using nonlinear optics has been discussed, but a single conclusive form has not yet been obtained.

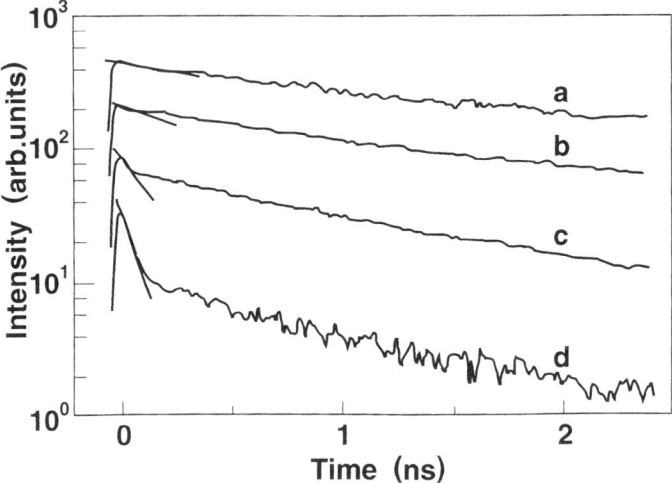

FIG. 11. Picosecond time response of the exciton luminescence at the energies indicated by the downward-pointing arrows in Fig. 10. (a), (b), (c), and (d) correspond to energies indicated in Fig. 10. The ordinate for each curve is shifted by an arbitrary amount. Solid lines indicate the transient decay with τ_{tr} (Itoh et al., 1990).

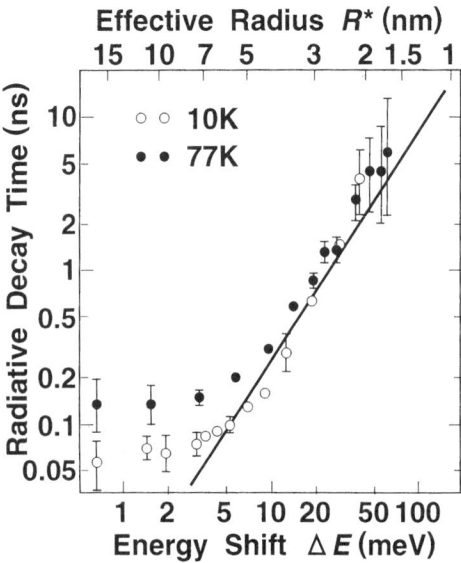

FIG. 12. Dependencies of the exciton radiative decay time on the effective radius for both 10 K and 77 K, indicated by open and closed circles, respectively. The solid line represents the theoretical radiative decay time calculated from Eq. (7) (Itoh et al., 1990).

Pattern processing by optical methods is one of the most promising subjects, where the two-dimensional information is treated by decomposing the picture at time t into a great number of elements. For example, we may use an optical bistable response for one of these elements. The optical bistable response was observed as shown in Fig. 13 for a CdS crystal with a thickness of 11 μm doped by neutral donors to a value on the order of $10^{15}\,\mathrm{cm}^{-3}$ (Dagenais and Sharfin, 1985). When we irradiated this crystal by laser light at the frequency resonant to a bound exciton, the transmitted light intensity I_t showed hysteresis against changing incident laser intensity, as shown in Fig. 13. This optical bistability comes from a combined effect of the optical nonlinearity and reflective feedback from both sides of the crystal. We make the two states U and L of the high and low transmissivity correspond to 1 and 0 in the bit signal, respectively, and perform information processing. This is restricted to incident laser intensities I_i that satisfy

$$I_s < I_i < I_{cr} \tag{64}$$

Here I_s and I_{cr} are called the sustained (or holding) and critical incident intensity, respectively, and are shown in Fig. 13.

Nonlinear optical materials with larger optical nonlinearity and shorter switching times are required in order to obtain more efficient information

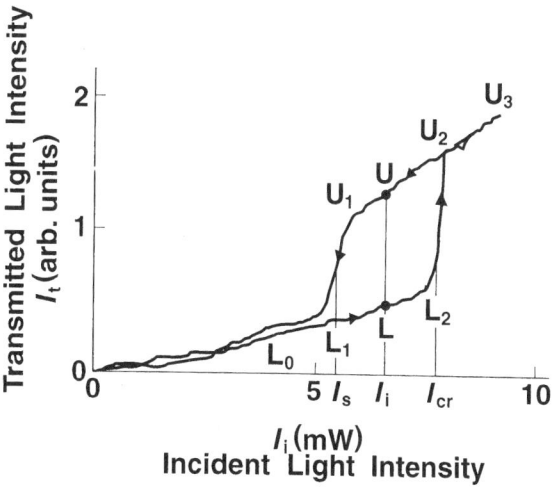

FIG. 13. Optical bistability at 4.2 K due to bound excitons in CdS with thickness 11 μm and neutral-donor density $10^{15}\,\mathrm{cm}^{-3}$. Two stable states U and L coexist for the incident laser intensity I_i, which satisfies $I_s < I_i < I_{cr}$ (Dagenais and Sharfin, 1985).

processing. A smaller linear absorption coefficient at the relevant frequency is also more advantageous, to avoid attenuating the incident as well as the transmitted signals. Therefore, the figure of merit for the nonlinear optical response is defined by

$$F = \frac{|\chi^{(3)}(\omega;\omega,-\omega,\omega)|}{\tau\alpha(\omega)} \quad (65)$$

where τ and $\alpha(\omega)$ are, respectively, the switching time and the linear absorption coefficient at frequency ω. Here we are using the optical Kerr effect and/or the absorption saturation as the nonlinear optical response, both of which can be described by the third-order susceptibility $\chi^{(3)}(\omega;\omega,-\omega,\omega)$.

In 1987, 50 distinguished researchers reported the experience of roughly constant figure of merit, independent of materials and frequency (Auston et al., 1987). This means a trade-off relation between the magnitude of the nonlinearity $|\chi^{(3)}|$ and the switching time τ, as shown by the shaded zone in Fig. 14. This report brought about pessimistic attitudes regarding the future of optical information processing, but studies done in 1988 (Hanamura, 1988a, 1988b) pointed out that the figure of merit can be increased beyond the restriction when a coherent exciton is resonantly pumped. As discussed

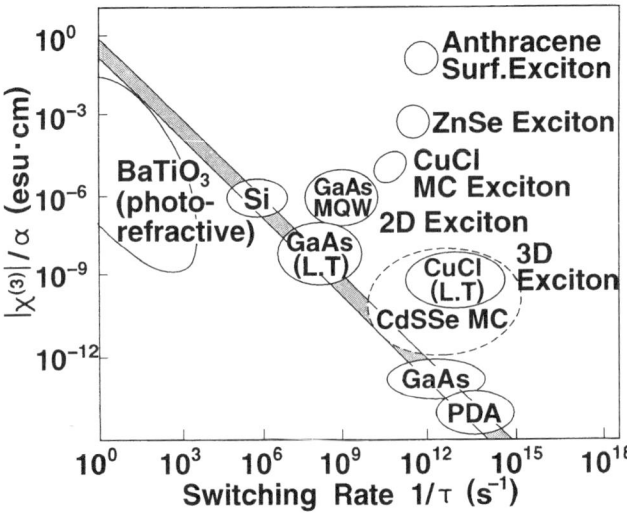

FIG. 14. Figure of merit for nonlinear optical response $F \equiv |\chi^{(3)}|/\alpha\tau$. The law F = const found by experience in bulk semiconductors is denoted by the shaded zone.

in Section I, the coherent exciton has large oscillator strength, so that $|\chi^{(3)}|$ increases in proportion to the coherent volume of the exciton, and the switching time τ becomes shortened by the superradiant decay. Here we assumed that the switching time is almost determined by the superradiant decay of the exciton. The first experimental confirmation of these predictions was made by using a system of CuCl microcrystallites embedded in an NaCl crystal or glass matrix (Masumoto et l., 1988; Kataoka et al., 1993). The switching time was measured by the time dependence of luminescence intensity, as shown in the preceding section, and the optical nonlinearity $|\chi^{(3)}|$ was measured by absorption saturation (Masumoto et al., 1988) and four-wave mixing (Kataoka et al., 1993). We notice that the observed values ($|\chi^{(3)}|/\alpha$ and $1/\tau$) are located above the line of constant F in Fig. 14.

At present, the surface exciton that is well localized on the top surface of the anthracene crystal gives the highest value of the figure of merit at low temperature.

The decay times of the top-, second-, and third-surface excitons were observed, respectively, to be shorter than 2, 15, and 200 ps (Aaviksoo et al., 1987). The magnitude of $|\chi^{(3)}|$ was observed to be on the order of 1 esu by reflection-type photo-induced polarization spectroscopy (Kuwata-Tonokami, 1987).

The coherent length of excitons is much longer than the size of a unit cell only in good crystals at low temperature. Even in microcrystallites, the quantization energy for center-of-mass motion is smaller than room temperature as long as we confine ourselves to weak confinement in which the enhancement of figure of merit is expected. Therefore, we have no idea how to get the enhanced figure of merit at room temperature.

3. BIEXCITONS AS A NONLINEAR MEDIUM

While a single exciton has large oscillator strength, the oscillator strength for the conversion of this single exciton into a biexciton is also much larger than that for a single molecule or for a unit cell in a good crystal at low lattice temperature. This is the case because we may choose any two valence electrons within a large orbital of a biexciton, which is a loosely bound state of two Wannier excitons (Akimoto and Hanamura, 1972; Hanamura, 1973). This is called the giant oscillator strength of a biexciton, and has been observed to enhance several nonlinear optical responses involving excitons and biexcitons. First of all, giant two-photon absorption due to the biexciton was proposed in 1973 (Hanamura, 1973), and this coherent two-photon absorption was experimentally confirmed in the next year (Gale and Mysyrowicz, 1974). Second, strong induced absorption is expected to result from optical conversion of a single exciton into a biexciton. Third, optical bistability was observed under two-photon resonant excitation of a biexciton (Peyghambarian et al., 1983; Hönerlage et al., 1984), and finally laser action was observed at the transition frequency between a biexciton and a single exciton in microcrystallites of CuCl (Masumoto et al., 1993). These topics are introduced in the subsections that follow.

a. Giant Two-Photon Absorption Resulting from Formation of Biexcitons

The two-photon absorption rate resulting from formation of a biexciton is evaluated by the following second-order perturbation (Hanamura, 1973):

$$W^2(\Gamma_1;\omega) = \frac{2\pi}{\hbar}\left|\left\langle \Gamma_1^{\text{bix}}(\mathbf{K})\left|H'_-\sum_i\frac{|i\rangle\langle i|}{E_{ig}-\hbar\omega}H'_-\right|g\right\rangle\right|^2 \delta[2\hbar\omega - E_{\text{bix}}(\mathbf{K})] \quad (66)$$

where $E_{\text{bix}}(\mathbf{K})$ is the energy of a biexciton with wavevector $\mathbf{K}$ for its center-of-mass motion, $|g\rangle$ and $|\Gamma_1^{\text{bix}}(\mathbf{K})\rangle$ denote the ground state and the excited state with a biexciton of wavevector $\mathbf{K}$ and symmetry Γ_1, respectively, and $H'_- = -\Sigma_\mathbf{k} \mathbf{P}_\mathbf{k} \cdot \mathbf{E}(\omega, \mathbf{k})$ describes the photon annihilation component. The binding energy $E_{\text{bix}}^{\text{b}}$ of a biexciton is much smaller than the excitonic binding energy, so that the single photon energy, which is equal to half the biexciton energy

$$\hbar\omega = \frac{E_{\text{bix}}}{2} = E_{1s} - \frac{E_{\text{bix}}^{\text{b}}}{2} \quad (67)$$

is very close to the lowest exciton level E_{1s}. Therefore, the energy denominator in Eq. (66) is the smallest for $i = 1s$, and the matrix elements $\langle 1s|H'_-|g\rangle$ and $\langle \Gamma_1^{\text{bix}}|H'_-|1s\rangle$ are the largest. As a result, we choose only the lowest exciton state, with the electron-hole relative state $|1s\rangle$ as the intermediate state $|i\rangle$ in Eq. (66). The transition dipole moment $\langle \Gamma_1^{\text{bix}}|\mathbf{P}_\mathbf{k}|1s\rangle$ has the following enhancement in comparison with the unit cell values μ_{cv}:

$$\frac{1}{\sqrt{\pi a_B^3}}\int g(\mathbf{R})\,d\mathbf{R} \quad (68)$$

where the envelope function of a biexciton is approximated by

$$\phi_{1s}(\mathbf{r})\phi_{1s}(\mathbf{r}')g(\mathbf{R})$$

and

$$g(\mathbf{R}) = \frac{1}{\sqrt{\pi a_{\text{bix}}^3}} \exp\left(-\frac{R}{a_{\text{bix}}}\right) \tag{69}$$

Here $\mathbf{R}$ describes the relative motion of two excitons within a biexciton, a_{bix} which measures the average distance between two excitons composing a biexciton, and the exciton envelope function $\phi_{1s}(\mathbf{r})$ for the electron-hole relative motion is assumed not to be modified in the biexciton formation. See Akimoto and Hanamura (1972) for the validity of this approximation.

When we compare the two-photon absorption rate of Eq. (66) to the one-photon absorption rate $W^{(1)}(\omega)$ due to a single excitation we obtain

$$\frac{W^{(2)}[\Gamma_1; \hbar\omega = (E_{1s} - E_{\text{bix}}^b/2)]}{W^{(1)}(\hbar\omega = E_{1s})} = 3 \times 10^{-15} \left(\frac{N}{V}\right) \tag{70}$$

for the exciton-biexciton system of CuCl, where N/V is the photon density of incident laser light and is 10^{15} cm^{-3} for an incident power of 1 MW/cm^{-2}. The factor N/V comes from the fact that the rate of the single exciton excitation $W^{(1)}$ is linearly proportional to the incident power, while the biexciton excitation rate $W^{(2)}$ is proportional to the square of the incident power. Therefore, we can expect a larger absorption coefficient at $\hbar\omega = E_{1s} - E_{\text{bix}}^b/2$ than at $\hbar\omega = E_{1s}$ when the incident laser power becomes larger than 1 MW/cm^2. This giant two-photon absorption due to the biexciton formation originates from the following two factors: (1) the giant oscillator strength $64(a_{\text{bix}}/a_B)^3 \simeq 10^3$ coming from the square of Eq. (68) for the electric-dipole transition between a single exciton and a biexciton, and (2) the resonant enhancement coming from the small denominator in Eq. (66). For example, this energy denominator becomes as small as $E_{\text{bix}}^b/2 \simeq 15$ meV, in contrast to a few electron volts in conventional two-photon absorption. This prediction was confirmed not only qualitatively but also quantitatively by Gale and Mysyrowicz (1974).

We can make the best use of two-photon absorption in identifying the symmetry of the biexciton state (Inoue and Toyozawa, 1965). When we denote the polarization directions of two laser beams by $\mathbf{e}_1$ and $\mathbf{e}_2$, the two-photon absorption probability depends on these directions relative to the crystal axes. For example, the valence band of a CuBr crystal consists of $J = \frac{3}{2}$, and we have three biexciton levels, Γ_1, Γ_3, and Γ_5, that are two-photon allowed (Hanamura, 1975a, 1975b). These levels have different dependencies of two-photon absorption coefficients on $\mathbf{e}_1$ and $\mathbf{e}_2$, so that they can be identified by these measurements (Vu Duy Phach and Lévy, 1979).

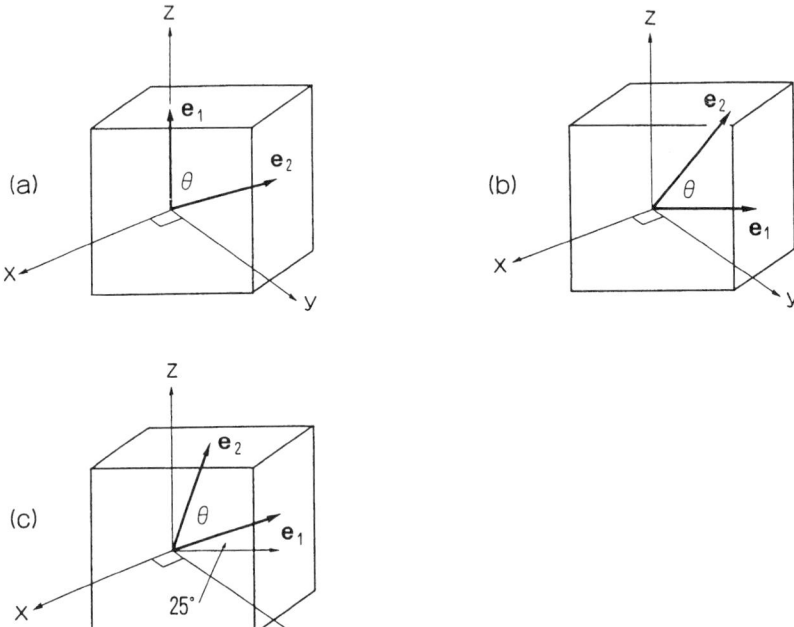

FIG. 15. Polarization configurations of two incident laser beams in two-photon measurements of biexcitons: (a) $e_1 = [0, 0, 1]$ and $e_2 = [-(\sin\theta)/\sqrt{2}, (\sin\theta)/\sqrt{2}, \cos\theta]$: (b) $e_1 = [-1/\sqrt{2}, 1/\sqrt{2}, 0]$ and $e_2 = [-(\cos\theta)/\sqrt{2}, (\cos\theta)/\sqrt{2}, \sin\theta]$; (c) $e_1 = [-(\cos 25°)/\sqrt{2}, (\cos 25°)/\sqrt{2}, \sin 25°]$ and $e_2 = [-[\cos(\theta + 25°)]/\sqrt{2}, [\cos(\theta + 25°)]/\sqrt{2}, \sin(\theta + 25°)]$.

Three directions of e_1, the polarization of the first pump field, were chosen and fixed in the frame of the cubic crystalline axes, as shown in Fig. 15. The angle θ of the second pump beam is changed in the plane, as described in the legend of Fig. 15. Two-photon absorption coefficients at three peaks of sum-frequencies $\hbar(\omega_1 + \omega_2) = 5.906$, 5.910, and 5.913 eV were plotted as a function of θ, as shown in Fig. 16, corresponding to the three configurations in Fig. 15. In comparison with the theoretical curves by Inoue and Toyozawa (1965), these three peaks are identified as Γ_1 (A_1), Γ_3 (E), and $\Gamma_5(T_2)$, in this order. See also, Ueta et al. (1986).

b. Transition Between Single Excitons and Biexcitons

In analyzing many nonlinear optical responses of an excitonic system, the crystal ground state, the single exciton state, and the biexciton state can be

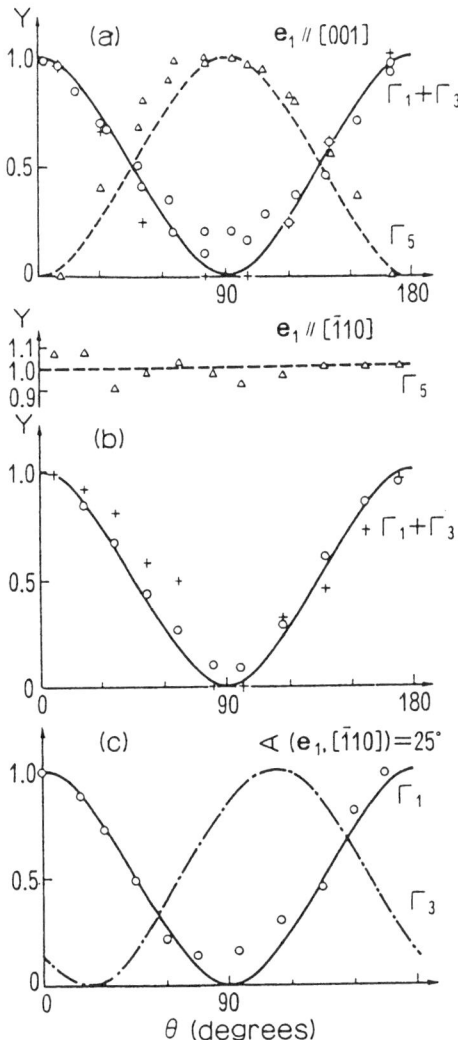

FIG. 16. Dependence of the giant two-photon transition intensity on the angle θ between the polarizations $\mathbf{e}_1$ and $\mathbf{e}_2$ of the two incident laser beams, corresponding to the three configurations in Fig. 15. ○: two-photon absorption at low energy ($\hbar\omega_1 + \hbar\omega_2 = 5.9061$ eV, Γ_1 symmetry). △: two-photon absorption at intermediate energy (5.9103 eV, Γ_5). +: two-photon absorption at high energy (5.9128 eV, Γ_3) (Vu Duy Phach and Lévy, 1979).

treated approximately as a system of three levels. Because of the giant dipole moment $\langle \Gamma_1^{\text{bix}} | \mathbf{P}_k | 1s \rangle$ (Rashba, 1975), strong induced absorption under the presence of $1s$ excitons is observed at $\hbar\omega = E_{\text{mol}} - E_{1s}$. The induced absorption rate $W_{\text{ind}}^{(1)}$ in the presence of excitons with density ρ_{exc} is also enhanced by the giant oscillator strength. The ratio of $W_{\text{ind}}^{(1)}$ to the single exciton excitation rate $W_{\text{exc}}^{(1)}$ is evaluated in terms of Eq. (68) as

$$\frac{W_{\text{ind}}^{(1)}}{W_{\text{exc}}^{(1)}} = \left[\int g(\mathbf{R}) \, d\mathbf{R} \right]^2 \rho_{\text{exc}} = 64\pi a_{\text{bix}}^3 \rho_{\text{exc}} \qquad (71)$$

For CdS, the size of a biexciton a_{bix} is about 60 Å, so that the induced absorption coefficient at $\hbar\omega = E_{\text{bix}} - E_{1s}$ is of the same order of magnitude as the linear absorption peak $\hbar\omega = E_{1s}$, for exciton density $\rho_{\text{exc}} = 10^{17}\,\text{cm}^{-3}$. The reverse process is also observable as high-efficiency luminescence from the biexciton at $\hbar\omega = E_{\text{bix}} - E_{1s} = E_{1s} - E_{\text{bix}}^{\text{b}}$, because, in a good crystal at low temperature, the biexciton decays radiatively into a single exciton and a photon, by using the giant oscillator strength, before nonradiative decay of the biexciton begins. Strictly speaking, a biexciton decays into two exciton-polaritons (EPs) (Ueta et al., 1979), but EPs at $\hbar\omega = E_{1s}$ and $\hbar\omega = E_{1s} - E_{\text{bix}}^{\text{b}}$ can be treated as an exciton and a photon, respectively. By virtue of this property, this transition is used in laser oscillation, as shown in the following subsection.

c. *Optical Bistability and Laser*

Under two-photon resonance of the biexciton, the third-order susceptibility $\chi^{(3)}(\omega;\omega,-\omega,\omega)$ is expected to become very large, so that optical bistability is expected (Hanamura, 1981; Koch and Haug, 1981). Peyghambarian et al. (1983) succeeded in observing optical bistable response under two-photon resonance with the biexciton of a CuCl crystal thickness 10 μm, polished and coated to a reflectivity of 0.9. The response time was observed to be shorter than 100 ps for the incident power of 15 to 30 MW/cm^{-2}.

As discussed in Section I, excitons in semiconductor microcrystallites can show large optical nonlinearity and rapid switching. In addition to these advantages of semiconductor microcrystallites as a nonlinear optical medium, another advantage is that the excitations do not diffuse but are confined to the irradiated region. Making use of these advantages, Masumoto et al. (1993) succeeded in observing the laser action of the exciton-biexciton levels in CuCl microcrystallites embedded in an NaCl matrix. They made a Fabry-Perot resonator by providing dielectric coatings on both sides of a 0.52-mm-thick NaCl matrix containing 0.1 mol % of CuCl

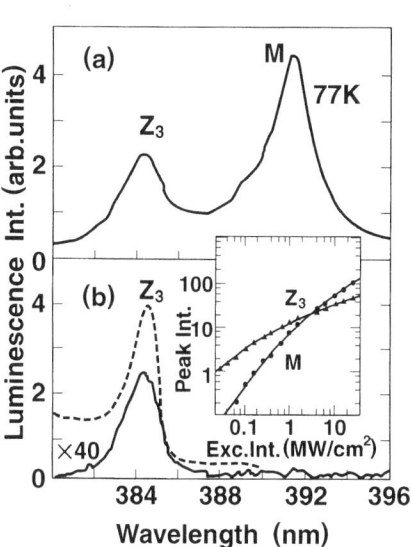

FIG. 17. Luminescence spectra from CuCl microcrystallites at pump powers of (a) 24 MW/cm^2 and (b) 39 kW/cm^2. The dotted line describes the absorption spectrum. The inset shows the pump power dependence of emission peak intensity of the M line due to the transition of a biexciton into a single exciton and of the Z_3 line due to a single exciton annihilation (Masumoto et al., 1993).

microcrystallites. When the pump power was 39 kW/cm^2, only the exciton luminescence was observed, as shown in Fig. 17(b). When the pump power increased to 24 MW/cm^2, the luminescence of biexcitons, labeled M in Fig. 17(a), was more intense than the exciton line, labeled Z_3.

The emission intensity was also observed to increase rapidly for pump powers beyond the threshold of 2.1 MW/cm^2, as shown in Fig. 18. Just above the threshold—i.e., at $I = 1.08 I_{th}$—five laser modes with wavelength separation of 0.08 nm, corresponding to the Fabry-Perot resonator length of 0.62 mm, were observed, as shown by the solid lines in Fig. 19. These signals disappeared when the pump power I was reduced to $0.86 I_{th}$, as shown by the dotted lines in Fig. 19. These emission lines are located at the wavelengths of the emission bands of the biexciton-to-exciton luminescence.

4. WEAK LOCALIZATION AND NONLINEAR OPTICAL RESPONSES

In this subsection, we discuss novel nonlinear optical responses induced and/or enhanced by weak localization of exciton-polaritons (EPs). First, the

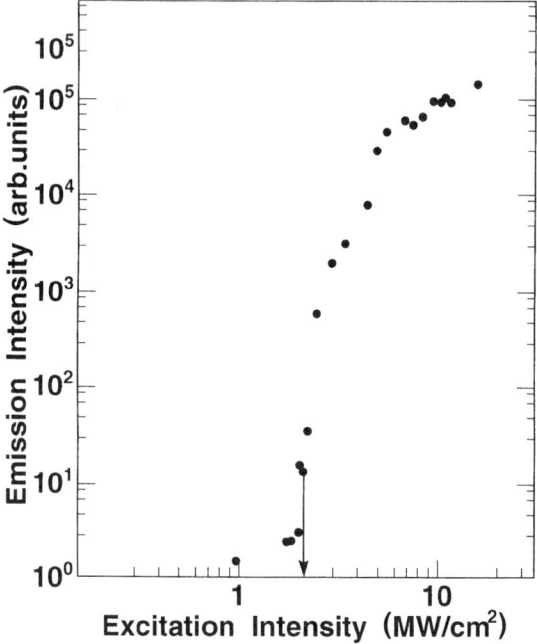

FIG. 18. Luminescence intensity of a biexciton decay as a function of pump power. The lasing threshold is at 2.1 MW/cm² (Masumoto et al., 1993).

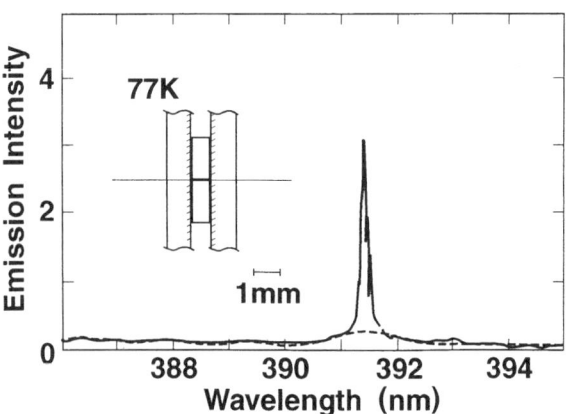

FIG. 19. Luminescence spectra for pump power values of $I = 1.08 I_{th}$ beyond the lasing threshold (solid line) and $I = 0.86 I_{th}$ below it (dotted line). The inset describes Fabry-Perot cavity (Masumoto et al., 1993).

phase-conjugated wave (PCW) is theoretically shown to be generated by scattering of excitation and to be enhanced by weak localization of excitons or EPs (Hanamura, 1989a). Weak localization is the enhancement of backward scattering of elementary excitations in a dense distribution of elastic scatterers, and comes from constructive interference of its multiple scatterings and their time-reversal processes, as shown in this section. On the other hand, conventional PCW is a time-reversal propagation of a test beam under application of forward and backward pump beams. That is, PCW is one of the third-order optical processes and is called a four-wave mixing — i.e., the fourth wave generation under application of three external beams.

The third-order polarization describing generation of the PCW under three-beam irradiation is written phenomenologically as follows:

$$\mathbf{P}^{(3)} = \alpha(\omega)[(\mathbf{E}_t^* \cdot \mathbf{E}_f)\mathbf{E}_b + (\mathbf{E}_t^* \cdot \mathbf{E}_b)\mathbf{E}_f] + \beta(\omega)(\mathbf{E}_f \cdot \mathbf{E}_b)\mathbf{E}_t^* \qquad (72)$$

The first (second) term of Eq. (72) describes the process in which a population grating of excitons is formed by the test field $\mathbf{E}_t^*$ and the pump field $\mathbf{E}_f(\mathbf{E}_b)$, and the other pump wave $\mathbf{E}_b(\mathbf{E}_f)$ is diffracted by the population grating. For the process described by the third term in Eq. (72), two pump beams $\mathbf{E}_f$ and $\mathbf{E}_b$ create the spatially uniform polarization oscillating with 2ω, and the PCW with $-\mathbf{k}_t$ and ω is produced by dividing the polarization at 2ω into $\omega_t = \omega(\mathbf{k}_t)$ and $2\omega - \omega_t = \omega(-\mathbf{k}_t)$. Since PCW through the first $\alpha(\omega)$ process was shown microscopically to be generated by elastic scattering of excitations and to be enhanced by weak localization (Hanamura, 1989), several theoretical papers on the effects of weak localization on optical nonlinearities have been published (Krabtsov et al., 1990; McGurn et al., 1991; Yudson and Reineker, 1992; Taniguchi and Hanamura, 1993). No experimental observation of these effects, however, has been reported. For the case of the conventional phase-conjugated process, three beams are necessary, and forward and backward pump beams are required to irradiate the front and rear surfaces at a frequency with a finite absorption coefficient. This configuration makes these experiments difficult.

The novel nonlinear optical phenomena that may come from the weak localization of EPs were observed in a semiconductor microcavity by Rhee et al. (1996). They observed the emission of light in a direction normal to the surface with a small divergence angle, even though the EP states were excited at an angle of 3 degrees from the normal. It was concluded that (1) the emitted light is coherent, as confirmed by interfering the emission with the incident pump beam; (2) the emitted light intensity increases more rapidly than the square of the pump power; and (3) the coherent light emission has a delay of 0.5 ps after the pumping pulse. These three experimental facts are explainable by introducing the weak localization of two-dimensional (2D) EPs and the bosonic character of EPs into the scattering and exciton-exciton collisional processes. In this subsection, we describe theory and experiment for this novel nonlinear process (Hanamura and Norris, 1996b).

The pump excites EPs with an initial in-plane wavevector $\mathbf{k}_o$. These EPs may be coherently backscattered, owing to disorder in the quantum well confinement potential, into the state with wavevector $-\mathbf{k}_o$. Collisions between EPs with momentum $+\mathbf{k}_o$ and $-\mathbf{k}_o$ result in generation of a population of EPs at exactly $\mathbf{k} = 0$, giving rise to coherent emission of light in the normal direction. This occurs in the lowest order of the external fields. Under much stronger pumping, the bosonic nature of EPs is well reflected by this nonlinear process. The EP with $\mathbf{k} = 0$ has the longest lifetime in the microcavity, so that the highest population is in this state. The induced scattering and reverse process among these EPs with $\mathbf{k}_o$, $-\mathbf{k}_o$, and $\mathbf{k} = 0$ are enhanced by the boson factors. That is, the highly populated states of EPs are attracting EPs to these states because of the boson enhancement factors in the matrix elements of the scattering and the EP-EP collision. As a result, the populations in the backscattered state $-\mathbf{k}_o$ and the EP with $\mathbf{k} = 0$ are much more populated than irrelevant EPs. We will now describe these processes in more detail.

The emission intensity in the normal direction is proportional to the generation rate of EPs $c_s [\equiv c(\mathbf{k}=0)]$ and $c_s^\dagger$, and is evaluated under cw excitation for theoretical simplicity as

$$I_s = \lim_{t \to \infty} \frac{\partial}{\partial t} \left\langle\!\!\left\langle c_s^\dagger c_s \rho(t) \right\rangle\!\!\right\rangle \tag{73}$$

The double angle brackets in Eq. (73) signify both the quantum-mechanical and ensemble averages over the distribution of scattering potentials, which come mainly from the fluctuation in quantum well width in the present problem. An irrelevant factor resulting from transmissivity, which measures the coupling strength of the polaritons to the external electric field, has been omitted. The density operator $\rho(t)$ of the total system is described as follows:

$$\rho(t) = \exp(-iH_T t)\rho_o \exp(iH_T t) \tag{74}$$

Here,

$$H_T = H + H' + H'' \tag{75}$$

where

$$H = \sum_{\mathbf{k}} \omega(k) c_{\mathbf{k}}^{\dagger} c_{\mathbf{k}} + \sum_{\mathbf{k}q,j} V_o(\mathbf{q}) e^{i\mathbf{q}\cdot\mathbf{r}_j} c_{\mathbf{k}+\mathbf{q}}^{\dagger} c_{\mathbf{k}} \tag{76}$$

$$H' = \sum_{i=\alpha,s,d} (c_i + c_i^{\dagger}) \boldsymbol{\mu}_i \cdot \mathbf{E}_i \tag{77}$$

$$H'' = \sum_{\mathbf{k}} V_{\mathbf{k}} c_s^{\dagger} c_d^{\dagger} c_{\alpha} c_{\mathbf{k}} + h.c. \tag{78}$$

Here and hereafter we set $\hbar = 1$. The radiation field has been divided into two parts: inside and outside the microcavity. The former has been taken into account by forming EP $(c_{\mathbf{k}}, c_{\mathbf{k}}^{\dagger})$ — i.e., the hybrid exciton photon in the cavity. The external field $\mathbf{E}_{\alpha}$ excites the EP with wavevector $\mathbf{k}_0$ coherently through Eq. (77) — i.e., $c_{\alpha}|\alpha\rangle = \alpha|\alpha\rangle$. Here, $\mathbf{k}_0$ is the two-dimensional wavevector of the EP in the quantum well plane. This EP suffers from scattering by the random potential $V_o(\mathbf{r}_i)$, and $V_o(\mathbf{q})$ in Eq. (76) is the Fourier component of $V_o(\mathbf{r})$. The interaction H'' in Eq. (78) describes the scattering of EPs with $\mathbf{k}_0$ in $|\alpha\rangle$ and $\mathbf{k}$ into the signal $c_s^{\dagger}(\mathbf{k}=0)$ and the idler $c_d^{\dagger}$. This idler EP has 2D wavevector sums of $\mathbf{k}_o$ and $\mathbf{k}$. Therefore, this idler becomes coincident to the signal wave when 2D EP with $\mathbf{k}$ is the backscattered EP with $\mathbf{k} = -\mathbf{k}_o$, as will be discussed soon.

The scattering of EPs by $V_o(\mathbf{q}) \exp(i\mathbf{q}\cdot\mathbf{r}_j)$ must be taken into account to infinite order to evaluate the effects of the weak localization of EPs. Therefore, we keep the Hamiltonian H of Eq. (76) in its form to the final stage of the calculation, where the ensemble average of observed physical variables is taken. The present signal appears only in the second and higher orders in H' (the interaction with the external field) and also in H'' (the EP-EP collision). Therefore, we expand the density operator $\rho(t)$ in Eq. (74) in H' and H'', as shown in Fig. 20:

$$I_s = A G^R(\omega_s + \omega_d - \omega_o, \mathbf{k}) \Xi(\mathbf{k}, \mathbf{k}_0)[G^R(\omega_0, \mathbf{k}_0) - G^A(\omega_0, \mathbf{k}_0)] + c.c. \tag{79}$$

where $G^{R,A}(\omega, \mathbf{k})$ and $\Xi(\mathbf{k}, \mathbf{k}_0)$ are the retarded (advanced) Green's function and the double Green's function describing the shaded parts in Fig. 20, respectively. Here,

$$A = |\langle n_0 + 2, \alpha | H'' | n_0, \alpha \rangle \langle \alpha | c_{\alpha} | \alpha \rangle \boldsymbol{\mu} \cdot \mathbf{E}_{\alpha}|^2 \tag{80}$$

where n_0 is the number of the signal EP with $\mathbf{k} = 0$. We can evaluate $\Xi(\mathbf{k}, \mathbf{k}_0)$ following the procedure of Eq. (16) in Hanamura (1989a), which comes from summation of the maximally crossed diagrms Γ_c, shown in

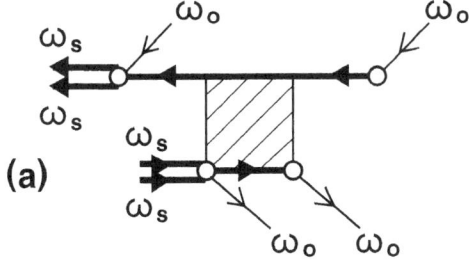

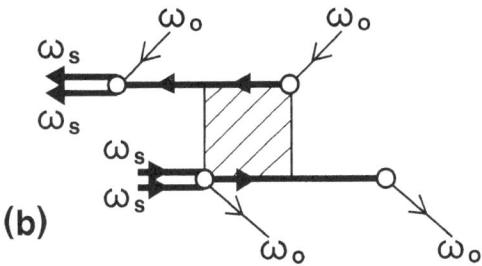

FIG. 20. Two diagrams that contribute to the emission into the direction normal to the surface under the laser pumping tilted from normal incidence. Solid lines describe propagation of the exciton polariton and the signal (ω_s) polariton, and slanted lines describe the coherent incident polariton ($\omega_0, \mathbf{k}_0$).

Fig. 21(c), and the ladder diagrams Γ_l, shown in Fig. 21(b):

$$\Gamma_c(\mathbf{k} + \mathbf{k}_0) = \frac{2(\gamma_o + \gamma)U_o}{2\gamma_o + D(\mathbf{k} + \mathbf{k}_0)^2} \tag{81}$$

$$\Gamma_l(\mathbf{k}_0 - \mathbf{k}'_0) = \frac{2(\gamma_o + \gamma)U_o}{2\gamma_o + D(\mathbf{k}_0 - \mathbf{k}'_0)^2} \tag{82}$$

$$\Xi(\mathbf{k}, \mathbf{k}_0) = \frac{-1}{2\gamma_o(\gamma_o + \gamma)^2}(\Gamma_l + \Gamma_c) \tag{83}$$

Here, γ_o describes the rate of inelastic scattering, $\gamma \equiv \pi N(\omega)n_i|V_o(0)|^2$, with $N(\omega)$ the EP state density and n_i the number density of scatterers, $U_o \equiv n_i|V_o(0)|^2$, and $D \equiv v_g^2/6(\gamma_o + \gamma)$ is the diffusion constant of EP, with v_g the group velocity of EP. Figure 21(a) describes the Born scattering rate γ.

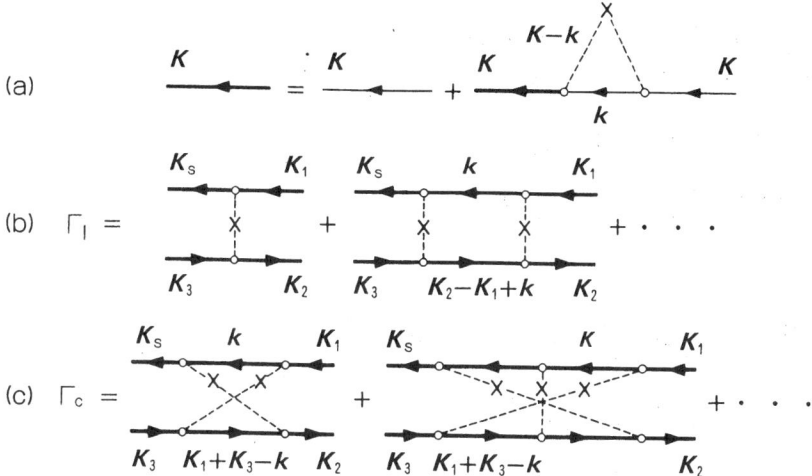

FIG. 21. (a) Diagram describing the lowest-order contribution to the self-energy of polariton K. The cross and dotted lines mean a scattering center and the interaction of polariton with the scattering center, respectively. (b) Ladderlike scatterings of the leftward and rightward polaritons. (c) The maximally crossed scattering mechanism of polaritons at impurity centers.

Under single beam pumping, $\mathbf{k}'_0 = \mathbf{k}_0$ in Eq. (82), so that I_s is given by

$$I_s = \frac{4AU_o}{\gamma_o^2(\gamma_o + \gamma)^2}\left[1 + \frac{1}{1 + D(\mathbf{k}_0 + \mathbf{k})^2/2\gamma_o}\right] \quad (84)$$

This intensity I_s is proportional to $|\alpha|^4$—i.e., the square of the incident power I_o—in the lowest order. To this order, the ladder-type contribution Eq. (82) results in a uniform distribution of scattered EPs for $\mathbf{k}'_0 = \mathbf{k}_0$, i.e., a contribution one in the square brackets in Eq. (84). Therefore, the signal in the normal direction is only four times as strong as in other directions by choosing $\mathbf{k} = -\mathbf{k}_0$. Here the factor 2 comes from the enhancement of backward scattering, as shown in the square brakets in Eq. (84). Another factor 2 comes from the fact that both the signal and idler become coincident for the case $\mathbf{k} = -\mathbf{k}_0$. When higher-order processes are taken into account, the bosonic factor begins to work and enhance the induced scatterings into $\mathbf{k} = 0$ and $-\mathbf{k}_0$ state. This is because the EP with $\mathbf{k} = 0$ has the longest lifetime in the microcavity and the EP with $-\mathbf{k}_0$ is also highly populated due to the enhancement of backward scattering by the weak localization effect. Therrefore, the sharp peak in the normal direction is expected to originate from a combined effect of the weak localization and the induced scattering to the signal mode $\mathbf{k} = 0$. The angle of signal enhancement is within $\theta_c \equiv 2\gamma_o/Dk_o^2 = 12\gamma_o(\gamma_o + \gamma)/(k_o v_g)^2$. First, the signal light has the same frequency and phase as the incident pulse, because the EP suffers only from elastic scatterings within the lifetime $1/2\gamma_o$, so that it is coherent with the pump beam. Second, this interference persists for a time $1/2\gamma_o$ before the EP suffers from inelastic scattering. Third, the sharp signal increases more strongly than by the square of I_o. These results are consistent with the observation of Rhee et al. (1996). Finally, a clearer method of confirming the weak localization was proposed in terms of two-beam configuration (Hanamura, 1996a).

List of Acronyms

EP	exciton-polariton
HOMO	highest occupied molecular orbital
LUMO	lowest unoccupied molecular orbital state
PBG	photonic bandgap
PCW	phase-conjugated wave
PMDA	pyromellitic acid dianhydride
QWR	quantum wire

References

Aaviksoo, Y., Lippmaa, Y., and Reinot, T. (1987). *Opt. Spectrosk. (USSR)* **62**, 419.
Akimoto, O. and Hanamura, E. (1972). *J. Phys. Soc. Jpn.* **33**, 1537. *Solid State Commun.* **10**, 253.
Auston, D. H. et al. (1987). *Appl. Opt.* **26**, 211.
Baxter, R. J. (1982). *Exactly Solved Models in Statistical Mechanics*. Academic Press.
Dagenais, M. and Sharfin, W. (1985). *J. Opt. Soc. Am. B* **2**, 1179.
Dan, N. T. and Hanamura, E. (1996). *Phys. Rev. B* **54**, 2739.
Dicke, R. H. (1954). *Phys. Rev.* **93**, 99.
Feldman, J. Peter, G., Goebel, E. O., Dawson, P., Moore, K., Foxon, C., and Elliott, R. J. (1987). *Phys. Rev. Lett.* **59**, 2337.
Gale, G. M. and Mysyrowicz, A. (1974). *Phys. Rev. Lett.* **32**, 727.
Gross, M. and Haroche, S. (1982). *Phys. Rep.* **93**, 301.
Hanamura, E. (1973). *Solid State Commun.* **12**, 951.
Hanamura, E. (1975a). *J. Phys. Soc. Jpn.* **39**, 1506.
Hanamura, E. (1975b). *J. Phys. Soc. Jpn.* **39**, 1516.
Hanamura, E. (1981). *Solid State Commun.* **38**, 939.
Hanamura, E. (1988a). *Phys. Rev. B* **37**, 1273.
Hanamura, E. (1988b). *Phys. Rev. B* **38**, 1228.
Hanamura, E. (1989a). *Phys. Rev. B* **39**, 1152.
Hanamura, E. (1989b). In *Proceedings of a NATO Advanced Research Workshop on Optical Switching of Low-Dimensional Systems* (H. Haug and L. Banyai, eds.). Plenum, pp. 203–210.

Hanamura, E. (1996a). *Phys. Rev. B* **54**, 11219.
Hanamura, E. and Norris, T. (1996b). *Phys. Rev. B* **54**, R2292.
Hönerlage, B., Bigot, J. Y., and Levy, R. (1983). In *Optical Bistability* (H. Gibbs, S. McCall, and C. Boden, eds.). Plenum, p. 253.
Hopfield, J. J. (1958). *Phys. Rev.* **112**, 1555.
Inoue, M. and Toyozawa, Y. (1965). *J. Phys. Soc. Jpn.* **20**, 363.
Itoh, T., Furumiya, M., and Ikehara, I. (1990). *Solid State Commun.* **73**, 271.
Kataoka, T., Tokizaki, T., and Nakamura, A. (1993). *Phys. Rev. B* **48**, 2815.
Knox, R. S. (1963). *Theory of Excitons.* Solid State Physics Suppl. Vol. 5 (F. Seitz and D. Turnbull, eds.) Academic Press.
Koch, S. W. and Haug, H. (1981). *Phys. Rev. Lett.* **46**, 450.
Krabtsov, V. E., Yudson, V. I., and Agranovich, V. M. (1990). *Phys. Rev. B* **41**, 2794.
Kuwata-Gonokami, M. (1987). *J. Lumin.* **38**, 247.
Manabe, Y., Tokihiro, T., and Hanamura, E. (1993). *Phys. Rev. B* **48**, 2773.
Masumoto, Y., Kawamura, T., and Era, K. (1993). *Appl. Phys. Lett.* **62**, 225.
Masumoto, Y., Yamazaki, M., and Sugawara, H. (1988). *Appl. Phys. Lett.* **53**, 1527.
McGurn, A. R., Leskova, T. A., and Agranovich, V. M. (1991). *Phys. Rev. B* **44**, 11441.
Mita, T., Sôtome, K., and Ueta, M. (1986). *J. Phys. Soc. Jpn.* **48**, 496.
Nakamura, A., Yamada, H., and Tokizaki, T. (1989). *Phys. Rev. B* **40**, 8585.
Onodera, Y. (1980). *J. Phys. Soc. Jpn.* **49**, 1845.
Peyghambarian, N., Gibbs, H. M., Weinberger, D. A., Rushford, M. C., and Sarid, D. (1983). *Phys. Rev. Lett.* **51**, 1692.
Rashba, E. I. (1975). In *Excitons at High Density* (H. Haken and S. Nikitine, eds.). Springer Tracts Mod. Phys., Vol. 73, p. 150.
Rhee, J.-K., Citrin, D. S., Norris, T. B., Hanamura, E., Nishioka, M., and Arakawa, Y. (1966). In *Ultrafast Phenomena*, 1996 OSA Technical Digest Series, Vol. 8, Optical Society of America, Washington, D.C., p. 220.
Shen, Y. R. (1984). *The Principles of Nonlinear Optics.* John Wiley & Sons.
Suzuura, H., Tokihiro, T., and Ohta, Y. (1994). *Phys. Rev. B* **49**, 4344.
Taniguchi, N. and Hanamura, E. (1993). *Phys. Rev. B* **47**, 12470.
Tokihiro, T., Manabe, Y., and Hanamura, E. (1995). *Phys. Rev. B* **51**, 7655.
Tokihiro, T., Manabe, Y., and Hanamura, E. (1993). *Phys. Rev. B* **47**, 2019.
Ueta, M., Kanzaki, H., Kobayashi, K., Toyozawa, Y., and Hanamura, E. (1986). *Excitonic Processes in Solids.* Springer-Verlag, Ch. 2 and 3.
Ueta, M., Mita, T., and Itoh, T. (1979). *Solid State Commun.* **32**, 43.
Vu Duy Phach and Lévy, R. (1979). *Solid State Commun.* **29**, 247.
Wegscheider, W., Pfeiffer, L. N., Dignam, M. M., Pinczuk, A., West, K. W., Mccall, S. L., and Hull, R. *Phys. Rev. Lett.* **71**, 4071.
Yudson, V. I. and Reineker, P. (1992). *Phys. Rev. B* **45**, 2073.
Yura, F. and Hanamura, E. (1994). *Phys. Rev. B* **50**, 15457.

CHAPTER 4

Semiconductor Nonlinearities for Solid-State Laser Modelocking and Q-Switching

Ursula Keller

INSTITUTE OF QUANTUM ELECTRONICS, PHYSICS DEPARTMENT
SWISS FEDERAL INSTITUTE OF TECHNOLOGY (ETH)
ZÜRICH, SWITZERLAND

- I. INTRODUCTION 211
  1. Motivation 211
  2. Semiconductor Saturable Absorbers: Historical Overview 214
- II. SEMICONDUCTOR SATURABLE ABSORBER MIRRORS (SESAMs) 221
  1. Macroscopic Properties of SESAMs 221
  2. Requirements for Passive Modelocking 223
  3. Requirements for Passive Q-Switching 243
  4. Microscopic Properties 245
- III. SESAM DESIGNS 249
  1. Antiresonant Fabry-Perot Saturable Absorber (A-FPSA) 249
  2. Dispersive SESAMs 259
  3. Antiresonant Fabry-Perot p-i-n Modulator 259
  4. General SESAM Design 260
- IV. PASSIVELY MODELOCKED SOLID STATE LASERS USING SESAMs 260
  1. Laser Design and Diode Pumping 260
  2. Approaching Limits: Ti:Sapphire Lasers 267
- V. PASSIVELY Q-SWITCHED SOLID STATE LASERS USING SESAMs 272
  1. Laser Design d Diode Pumping 272
  2. Approaching Limits: Microchip Lasers 272
- VI. CONCLUSIONS AND OUTLOOK 274
  LIST OF ABBREVIATIONS AND ACRONYMS 275
  REFERENCES 276

I. Introduction

1. MOTIVATION

Semiconductor materials are well established for use in electronic and optoelectronic devices. Remarkable advances in semiconductor growth and

List of Abbreviations and Acronyms can be located preceding the references to this Chapter.

processing technology have allowed for the fabrication of artificial microstructures approaching atomic dimensions. Epitaxial growth techniques, such as molecular beam epitaxy (MBE) and metal-organic chemical vapor deposition (MOCVD), can be used with different growth parameters to tailor linear and nonlinear optical properties in ultrathin layer structures. This precise control of optical nonlinearities, combined with the large bandgap variations of different semiconductor materials from the visible to the infrared, make these materials very attractive as saturable absorbers in solid state lasers. This specialized application can benefit from a technology base that is driven by a large worldwide market in optoelectronics.

Why combine semiconductor materials with solid state lasers? Solid state lasers still provide some of the best laser qualities in terms of high-quality spatial modes, high output power, energy storage and large pulse energy when Q-switched, and the large optical bandwidth necessary for ultrashort pulse generation. Combined with the recent advances in semiconductor lasers (diode lasers) as pump sources, passive modelocking or Q-switching of solid state lasers with semiconductor saturable absorbers forms a new and promising basis for an all-solid-state ultrafast laser technology with the potential for "turn-key" operation. Semiconductor saturable absorbers can offer significant improvements in performance, size, robustness, and cost of ultrafast lasers, compared with the saturable absorbers used in the past, such as organic dyes, color filter glasses, dye-doped solids, and ion-doped crystals. Many of these absorbers provided saturation for only a limited range of wavelengths and with only a limited range of recovery times and saturation fluences. In addition, organic dyes suffer from well-known problems such as short lifetime and complicated handling (pumps, dye jets, flow tubes, and dye toxicity).

Epitaxial growth techniques also allow us to design devices at the optical wavelength level. We typically integrate a semiconductor saturable absorber directly into a mirror structure, resulting in a device whose reflectivity increases as the incident optical intensity increases. We call this general class of devices semiconductor saturable absorber mirrors (SESAMs) (Keller *et al.*, 1992a, 1996). SESAMs can be applied to modelocked solid state lasers from the sub-10-fs range through the picosecond regime, and then can be extended to Q-switched pulses from the sub-100-ps range through the nanosecond regime and even higher.

The SESAM also helped to establish the feasibility of pure continuous wave (cw) passive modelocking in solid state lasers with upper state lifetimes (microseconds to milliseconds) much longer than the cavity round trip time (typically nanoseconds). Earlier, most solid state lasers also Q-switched during passive modelocking (Haus, 1976) — an undesirable condition in many applications. In Q-switched modelocking (Fig. 1), the modelocked

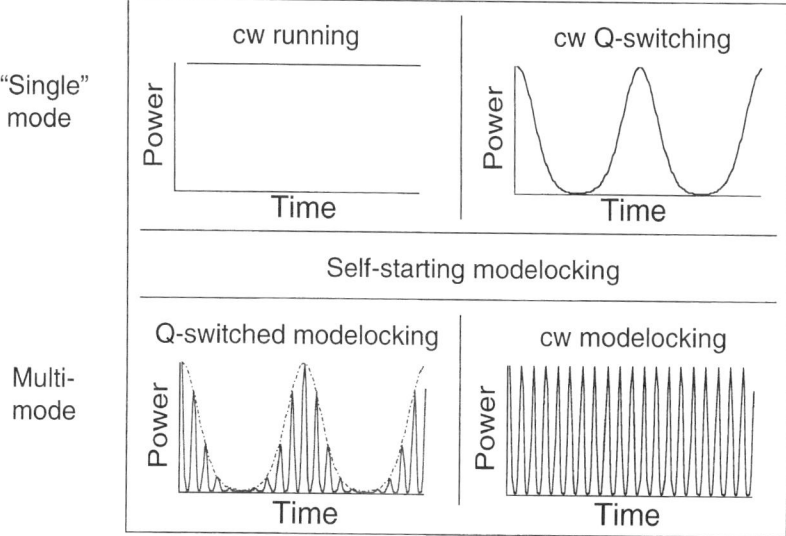

FIG. 1. Different modes of operation of a laser with a saturable absorber. Continuous wave (cw) Q-switching typically occurs with much longer pulses and lower pulse repetition rates than cw modelocking.

pico- or femtosecond pulses are inside much longer Q-switched pulse envelopes (typically in the microsecond regime) at much lower repetition rates (typically in the kilohertz regime). The combination of adjustable device structure and the material parameters of the SESAM provided sufficient design freedom to choose the key macroscopic parameters such as modulation depth, recovery time, saturation intensity, and saturation fluence to suppress Q-switched operation. The design freedom of SESAMs has also allowed us systematically to investigate the stability regimes of passive Q-switching (Spühler et al., 1998) and passive cw modelocking with an improved understanding and modeling of Q-switching (Kärtner et al., 1995a; Hönninger et al., 1998b) and multiple pulsing instabilities (Aus der Au et al., 1997; Kärtner et al., 1998).

Additionally, SESAMs have allowed us to push several key limits in laser performance. Presently, we have demonstrated pulses as short as 6.5 fs directly from a Ti:sapphire laser without further pulse compression (Jung et al., 1997c), the shortest cw modelocked pulses reported to date. In addition, we have demonstrated the shortest Q-switched pulses — 37 ps — from a solid state laser (Spühler et al., 1998). With optimized modematching, we have generated the highest reported average powers in diode-pumped, modelocked femtosecond Cr:LiSAF and Nd:glass lasers.

On the practical side, modelocked, diode-pumped solid state lasers using SESAMs are becoming commercially available, providing more compact, reliable, and "hands-off" ultrafast lasers. The pulse generation process is self-starting, and cavity stability adjustments are less critical with SESAMs than with Kerr lens modelocking (KLM). The design freedom of the laser cavity with SESAMs allow good performance with diode pumping. Picosecond and femtosecond pulses have been demonstrated from many solid state lasers, and thus we can expect to see further applications of these lasers to commercial products in many different wavelength and pulsewidth regimes.

The goal of this chapter is to summarize the current status of passively modelocked and Q-switched solid state lasers using semiconductor nonlinearities. In addition, we provide explicit guidelines for the macroscopic saturable absorber parameters and laser cavity design to obtain stable pulse generation. We show the approaching limits of pulse duration in both passive modelocking and Q-switching. Very little emphasis is placed on the microscopic properties of the optical nonlinearities in the semiconductor saturable absorber, which are covered in much more detail in other chapters of this book. However, our understanding of the microscopic semiconductor nonlinearities is limited, especially in the femtosecond regime, because we typically use a very high excitation level, often many times above the saturation fluence, which results in very complex dynamics. Although this complexity provides a rich and interesting field for further semiconductor spectroscopic studies, our modelocking or Q-switching approach fortunately does not critically depend on a full understanding of the detailed ultrafast recovery processes of the semiconductor saturable absorbers.

2. Semiconductor Saturable Absorbers: Historical Overview

The use of saturable absorbers in solid state lasers is practically as old as the solid state laser iteself (Mocker and Collins, 1965; De Maria et al., 1966; Smith, 1970). Early examples are ruby lasers (Mocker and Collins, 1965) using dye and color filter glass saturable absorbers and Nd:glass lasers (De Maria et al., 1966) using dye saturable absorbers. However, the first attempts at passive modelocking in solid state lasers with long upper state lifetimes (i.e., significantly longer than the cavity round trip time) always resulted in Q-switched modelocking. This limitation was attributable mostly to the parameter ranges of the available saturable absorbers (Haus, 1976). For many years, the success of passively modelocked dye lasers diverted the research interest away from solid state lasers (Shank, 1988; Diels, 1990).

Initially, semiconductor saturable absorbers were successfully used to modelock CO_2 and semiconductor diode lasers. As early as 1974, passive modelocking of a transversely excited atmospheric pressure (TEA) CO_2 laser was obtained with a p-doped germanium saturable absorber, producing pulses of several hundred picoseconds duration (Gibson *et al.*, 1974). Fast absorption bleaching of p-type Ge in the 9- to 11-μm-wavelength region results primarily from intravalence band transitions between the heavy and light hole bands. In the early 1980s, semiconductor diode lasers were passively modelocked for the first time with a semiconductor saturable absorber, where the recovery time was reduced by damage induced either during the aging process (Ippen *et al.*, 1980) or by proton bombardment (van der Ziel *et al.*, 1981). More reliable passive modelocking of diode lasers was obtained with excitonic nonlinearities in a GaAs/AlGaAs multiple quantum well saturable absorber (Silberberg *et al.*, 1984; Smith *et al.*, 1985). A color center laser was the first solid state laser that was cw modelocked with an intracavity semiconductor saturable absorber (Islam *et al.*, 1989). For both diode and color center lasers, the upper state lifetime of the laser medium is in the nanosecond regime, where dynamic gain saturation supports pulse formation and the large gain cross section strongly reduces the tendency for self-Q-switching. This is not the case for most other solid state lasers, which have upper state lifetimes in the microsecond to millisecond regime.

The first passively cw modelocked solid state lasers with longer upper state lifetimes were reported in 1990, and used SESAMs in coupled cavities, a technique termed RPM (resonant passive modelocking) (Keller *et al.*, 1990; Haus *et al.*, 1991). This work was motivated by previously demonstrated soliton lasers (Mollenauer and Nelson, 1988) and APM (additive pulse modelocking) lasers (Blow and Nelson, 1988; Kean *et al.*, 1989; Ippen *et al.*, 1989), in which a nonlinear phase shift in a fiber inside a coupled cavity provided effective saturable absorption. An RPM Nd:YLF laser produced pulses as short as 3.7 ps (Keller *et al.*, 1991b, 1992a), and an RPM Ti:sapphire laser generated picosecond pulses (Keller *et al.*, 1990). Femtosecond pulses were obtained only when RPM was used as a starting mechanism for Kerr lens modelocking (KLM) (Keller *et al.*, 1991a, 1992b). Currently, however, most uses of coupled cavity techniques have been supplanted by intracavity saturable absorber techniques based on either KLM or intracavity SESAMs, owing to their greater inherent simplicity. In 1992, we demonstrated stable, purely cw modelocked Nd:YLF and Nd:YAG lasers using an intracavity SESAM design, referred to as the antiresonant Fabry-Perot saturable absorber (A-FPSA) (Keller *et al.*, 1992c). This was the first intracavity SESAM device (Fig. 2) that allowed cw modelocking in solid state lasers with long upper state lifetimes without any Q-switching.

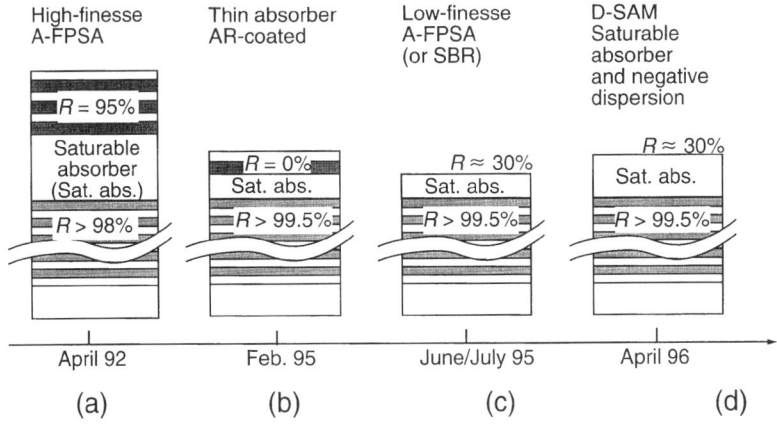

FIG. 2. Different semiconductor saturable absorber mirror (SESAM) devices in historical order: (a) high-finesse antiresonant Fabry-Perot saturable absorber (A-FPSA) (Keller et al., 1992c; Brovelli et al., 1995c); (b) AR-coated SESAM (Brovelli et al., 1995a); (c) saturable Bragg reflector (SBR) (Tsuda et al., 1995b) or, more generally, low-finesse A-FPSA (Hönninger et al., 1995; Jung et al., 1995a); and (d) dispersive saturable absorber mirror (D-SAM) (Kopf et al., 1996b).

Since then, many new SESAM designs have been developed that provide stable pulse generation in the pico- and femtosecond regimes for a variety of solid state lasers.

In the picosecond modelocking regime, Nd:YLF lasers (Keller, 1992c, 1994; Weingarten et al., 1993) and Nd:YAG lasers (Weingarten et al., 1993; Keller, 1994) have generated pulses as short as 2.8 and 6.8 ps at lasing wavelengths of about 1 μm using a low-temperature molecular beam epitaxy (MBE) grown 50 × InGaAs/GaAs multiple quantum well (MQW) saturable absorber, 0.6 μm thick, integrated within a Fabry-Perot formed by the lower GaAs/AlAs Bragg mirror and the higher evaporated TiO_2/SiO_2 Bragg mirror. The Fabry-Perot was operated at antiresonance mainly to prevent significant wavelength-dependent reflectivity and optical bistability effects. The low-temperature MBE growth provided a fast picosecond saturable absorber owing to the fast trapping of the electrons in the As antisites as well as good optical quality without any surface striations to scatter the laser light. The large lattice mismatch of InGaAs on GaAs causes such degradations at higher growth temperatures. The InGaAs/GaAs MQW bandgap — not a very critical parameter — is adjusted approximately to the lasing wavelength. The responsive optical nonlinearity providing the self-amplitude modulation for passive modelocking is based on band filling and the Pauli exclusion principle—no quantum well nonlinearities are required in principle. This has the advantage that we do not have to rely on narrow-band excitonic nonlinearities and can obtain broad wavelength tunability (Kopf et al., 1997b; Hönninger et al., 1998a; Sutter et al., 1998b) and modelocking of an extremely broad spectrum in the sub-10-fs pulse-width regime (Jung et al., 1997b). To achieve nearly all the available modulation depth, we typically operate the saturable absorber several times above the saturation fluence.

Today, we distinguish between different designs of the A-FPSA. As long as the laser mode incident on the A-FPSA can be freely adjusted, we can use different top reflectors ranging from an antireflection- (AR)-coated reflector to one with a high reflectivity of about 98%. We then refer to the device with the high top reflector as the high-finesse A-FPSA (Fig. 2a) and to the one with the lower top reflector as the low-finesse A-FPSA (Fig. 2c). The simplest low-finesse A-FPSA uses no additional coating, because the top reflector is formed by the semiconductor/air interface, which has a reflectivity of about 30% (Fig. 2c). For a lower top reflector on the A-FPSA, we typically have to reduce the thickness of the saturable absorber layer to minimize the insertion loss. A more detailed discussion of the different SESAM designs with their benefits and trade-offs is presented later in this chapter. Using a low-finesse A-FPSA with one or two InGaAs quantum wells, we demonstrated pulses as short as 1.6 ps from an Nd:LSB laser at a wavelength of 1.06 μm (Braun et al., 1996) and 1.9 ps from an Nd:YAlO$_3$ laser at 930 nm (Kellner et al., 1998). The first SESAM design at a longer wavelength of about 1.3 μm was used inside Nd:YLF and Nd:YVO$_4$ (Fluck et al., 1996b) lasers, producing pulses as short as 5.7 and 4.6 ps, respectively. In this case, we used both a high-finesse and a low-finesse A-FPSA with a low-temperature MBE-grown bulk InGaAs absorber, which experimentally confirmed that quantum well nonlinearities are not a necessity.

In the femtosecond modelocking regime, we first modelocked an Nd:glass laser with a high-finesse A-FPSA generating pulses as short as 130 fs (Keller et al., 1993). We used the same high-finesse A-FPSA as for the picosecond Nd:YLF laser described above. Initial experimental results were not well predicted by existing theory, leading us to further investigations in which we replaced the SESAM with an acousto-optic modelocker to provide regenerative active modelocking of the Nd:glass laser, producing 310-fs pulses (Kopf et al., 1994a). These pulses were much shorter than any previous theory of active modelocking would have predicted (Haus, 1975; Haus and Silberberg, 1986). The soliton modelocking model, discussed in more detail in Section II, was developed to explain this result (Kärtner et al., 1995c) and was then extended to the case with a slow saturable absorber, such as a SESAM (Kärtner and Keller, 1995b; Kärtner et al., 1996). This improvement in our understanding of the modelocking process resulted in shorter pulses from different diode-pumped Nd:glass lasers (Kopf et al., 1995b), with pulses as short as 60 fs (Aus der Au et al., 1997). We recently improved the average output power of a diode-pumped femtosecond Nd:glass laser by approximately two orders of magnitude, to more than 1 W (Aus der Au et al., 1998). For this result, we optimized the laser cavity design for high output power and adjusted the SESAM design to this specific cavity. A key feature of soliton modelocking is that it decouples the saturable absorber from the cavity. In addition to Nd:glass lasers, many other femtosecond solid state lasers have been passively modelocked using different SESAMs.

At shorter wavelengths, a Ti:sapphire laser was passively modelocked in 1993 with an intracavity SESAM generating 130-fs pulses at 10-mW average power and 7-W pump power for the first time (Mellish et al., 1993). However, the AR-coated proton-implanted MQW absorber mirror with 100 periods of GaAs/AlGaAs quantum wells was originally developed for modelocking external cavity semiconductor diode lasers, which can accept much higher insertion loss (Delfyett et al., 1992). We improved on this initial result with a SESAM design optimized for lower insertion loss, which is important for solid state lasers. This SESAM consisted of a single 15-nm-thick GaAs quantum well absorber embedded inside an AR-coated AlAs/AlGaAs Bragg reflector and produced pulses as short as 34 fs with an average power of 140 mW at 3-W pump power (Brovelli et al., 1995a) The modelocking mechanism was also based on soliton modelocking, and had a modelocking buildup time of only about 3 μs over the full cavity stability regime. Shortly afterwards, a similar saturable absorber mirror with an embedded 10-nm-thick GaAs quantum well absorber layer, but without the AR coating (termed a saturable Bragg reflector, or SBR), generated longer 90-fs pulses, most likely as a result of the lower modulation depth of this device (Tsuda et al., 1995b). Continued research efforts have ultimately resulted in sub-10-fs pulse generation using a broadband SESAM (Jung et al., 1997c). This is discussed in more detail in Section IV.

Diode-pumped Cr-doped crystals such as Cr:LiSAF (Payne et al., 1959), Cr:LiCAF (Payne et al., 1988), Cr:LiSCAF (Wang et al., 1954), and Cr:LiSGAF (Smith et al., 1992) have attracted much recent interest. These crystals have fluorescence linewidths similar to Ti:sapphire and can be pumped at wavelengths near 670 nm, where commercial high-brightness, high-power diode arrays are available. The first diode-pumped Cr:LiSAF laser that was cw modelocked with an intracavity SESAM used a device originally developed for external cavity semiconductor diode lasers (Delfyett et al., 1992), as described above for the Ti:sapphire laser. The high insertion loss of this device limited the performance to 220-fs pulses with only 10-mW average output power (Mellish et al., 1994). Currently, femtosecond pulses as short as 45 fs (Kopf *et al.*, 1997b) and with average output power as high as 500 mW (Kopf *et al.*, 1997c) have been obtained with improved SESAM and laser cavity design. A more detailed discussion is presented in Section IV.

Yb-doped materials are interesting as diode-pumped, tunable, and femtosecond high-power laser sources as a result of their broad absorption and emission bandwidths and their low thermal loading. Additional benefits are the long fluorescence lifetime (1 to 2 ms), which is desirable for ultrashort pulse amplification or Q-switching, and the simple spectroscopic structure, which avoids loss processes such as excited-state absorption, up-conversion, and concentration quenching (DeLoach *et al.*, 1993). Yb:YAG was passively modelocked for the first time with both high-finesse and low-finesse A-FPSAs, producing pulses as short as 570 fs (Hönninger *et al.*, 1995) and, more recently, 340 fs (Keller *et al.*, 1996). We generated pulses as short as 58 fs with the Yb:phosphate laser and 61 fs with the Yb:silicate glass laser at average output powers of 65 and 53 mW, respectively (Hönninger *et al.*, 1998a). Additionally, we demonstrated tunability of femtosecond pulses from 1025 to 1065 nm for the Yb:phosphate laser and from 1030 to 1082 nm for the Yb:silicate glass laser. For both cases, we used a low-finesse A-FPSA with an AlAs/GaAs Bragg mirror and a 20- or 30-nm-thick InGaAs absorber layer.

At longer wavelengths, the broadband Cr:YAG laser is tunable from 1.35 to 1.56 μm (Shestakov *et al.*, 1988), which is an interesting wavelength range for optical communications. Diode-pumped Nd:YAG or YVO$_4$ lasers can provide relatively compact, efficient pump sources for the Cr:YAG laser. An intracavity InGaAs SBR generated 110-fs pulses with 70-mW average output power (Collings *et al.*, 1996). It was assumed that the modelocking was based on a fast saturable absorber provided by excitonic nonlinearities in the two InGaAs quantum wells (Tsuda *et al.*, 1996b). Similar results with 114-fs pulses and average output power of about 90 mW have been obtained in the soliton modelocking regime using four InGaAs quantum wells in a low-finesse A-FPSA (Spälter *et al.*, 1997). Cr:forsterite is another broadband laser whose laser emission peaks at 1235 nm and that is tunable over the spectral range of 1167 to 1345 nm (Petricevic *et al.*, 1988). This laser has recently been passively modelocked with an intracavity low-finesse A-FPSA with a single InGaAs quantum well producing pulses as short as 40 fs (Zhang *et al.*, 1997). For both lasers, somewhat shorter pulses have been generated with KLM.

There has also been remarkable progress in ultrashort pulse generation with fiber lasers. Various semiconductor saturable absorber structures have been incorporated into these lasers to obtin stable modelocking. The requirements of SESAM design for fiber lasers are somewhat different from those for bulk solid state lasers. The much larger gain with fiber lasers normally can support SESAMs with much higher insertion loss, but we would expect stability requirements similar to those of bulk solid state lasers. However, only limited experimental data are available, and improved understanding will allow for further optimization, and scaling to higher powers, of such lasers. We restrict the focus of our discussion in this chapter to bulk solid state lasers. For general information, we provide a short historicl overview of cw modelocked fiber lasers using various semiconductor saturable absorbers.

An InGaAs/GaAs on GaAs superlattice, consisting of 120 periods of 7.4 nm-thick GaAs and 10.2-nm-thick $In_{0.63}Ga_{0.37}As$, was MBE grown at low-growth temperatures and was used as an intracavity saturable absorber for an Er-doped fiber laser at a center wavelength of 1.557 μm (Zirngibl et al., 1991). Transform-limited pulses as short as 1.2 ps were generated at a pulse repetition rate of 50 MHz with an average power of 15 to 22 mW. An InGaAs/InP MQW SESAM was contacted to one end of the polished Er,Yb-doped fiber laser and generated pulses as short as 7.6 ps at a center wavelength of approximately 1.54 μm (Loh et al., 1993). A 2-μm-thick AR-coated bulk InGaAsP saturable absorber grown on InP was directly mounted on the output coupler of an Er-doped fiber laser and generated transform-limited 320-fs pulses with 40-pJ pulse energy, an average power of 1.4 to 2.9 mW, and a pulse repetition rate of about 70 MHz (de Souza et al., 1993). Pulses of 2.3 nJ and 5.5 ps were generated by an Er,Yb-doped fiber laser that suppressed multiple-pulse behavior at high pump powers with bulk and MQW absorbers (Barnett et al., 1995). A 50-period, low-temperature-grown InGaAs/GaAs MQW low-finesse A-FPSA provided reliable self-starting of an Nd-doped fiber laser that was modelocked by the Kerr induced nonlinear polarization rotation and generated pulses as short as 260 fs with an average output power of 4 mW (Ober et al., 1993). It is interesting to note that the low-finesse A-FPSA has the same semiconductor structure as that of the high-finesse A-FPSA used for passive modelocking of Nd-doped bulk lasers described above. However, for fiber lasers we can accept more insertion loss and higher modulation depth without getting into Q-switching instabilities. Therefore, the top reflector was reduced to the Fresnel reflection at the semiconductor/air interface. More recently, Tm-doped silica fiber laser at a lasing wavelength of about 1.9 μm was passively modelocked, producing 190-fs pulses, using a butt-coupled InGaAs SESAM that also served as the output coupler (Sharp et al., 1996). The saturable absorber layer was proton implanted to reduce the carrier lifetime. We can expect continued efforts using different SESAM designs to passively modelock fiber lasers. For a more recent review of passively modelocked fiber lasers, the reader is referred to Fermann et al. (1997) and Nelson et al. (1997).

II. Semiconductor Saturable Absorber Mirrors (SESAMs)

1. MACROSCOPIC PROPERTIES OF SESAMs

The macroscopic properties of a saturable absorber are the modulation depth ΔR, the nonsaturable loss ΔR_{ns}, the saturation fluence F_{sat}, the saturation intensity I_{sat}, and the impulse response or recovery time. These parameters determine the operation of a passively modelocked or Q-switched laser. The required characteristics for stable laser operation will be discussed in Sections II.4 and II.5. The (*maximum*) *modulation depth* ΔR (Fig. 3) is the maximum amount of saturable losses of the absorber, which

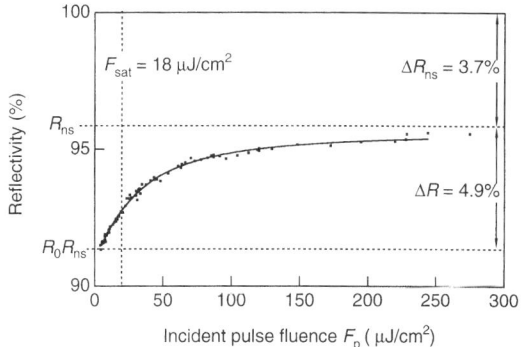

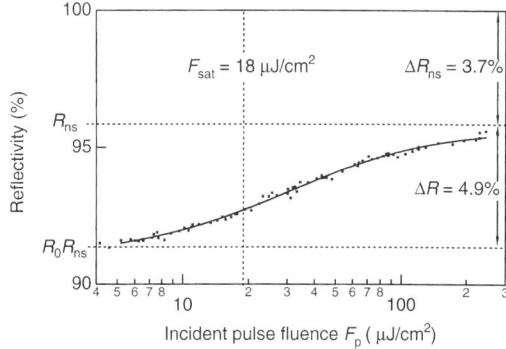

FIG. 3. Measured nonlinear reflectivity as a function of the incident pulse fluence F_p (i.e., pulse energy per unit area) on a SESAM using (a) linear and (b) log scales. The measurement was made with a SESAM supporting 34-fs pulses in a Ti:sapphire laser (Brovelli *et al.*, 1992). The measured data are fitted with Eqs. (5) and (6), which determines the saturation fluence F_{sat}, modulation depth ΔR, and nonsaturable loss $\Delta R_{ns} = 1 - R_{ns}$.

can be bleached by an incident *pulse fluence* (i.e., pulse energy per unit area) F_p that is much larger than the saturation fluence of the absorber. The maximum absorber amplitude loss coefficient q_0 is then given by

$$\Delta R = 1 - e^{-2q_0} \approx 2q_0, \quad q_0 \ll 1 \tag{1}$$

The nonsaturable loss ΔR_{ns} of the SESAM are the residual losses for an incident pulse energy density much larger than the saturation fluence (Fig. 3). Included in ΔR_{ns} are the reflectivity of the bottom mirror R_b, scattering losses resulting from impurities at the surface of the sample, residual absorption from defect states, and losses introduced by two-photon and free carrier absorption. The *saturation fluence* $F_{sat,A}$ is given by

$$F_{sat,A} = \frac{hv}{2\sigma_A} \tag{2}$$

where hv is the photon energy, σ_A is the absorber cross section, and the factor of two is for standing wave cavities. The *absorption coefficient* α of the material is then given by

$$\alpha = \sigma_A N_D \tag{3}$$

where N_D is the density of absorber atoms — the density of states in semiconductors, for example. The *saturation intensity* $I_{sat,A}$ is given by

$$I_{sat,A} = \frac{hv}{2\sigma_A \tau_A} = \frac{F_{sat,A}}{\tau_A} \tag{4}$$

where τ_A is the *absorber recovery time*.

Both the saturation fluence $F_{sat,A}$ and the absorber recovery time τ_A are determined experimentally as for example in Fig. 3 and Fig. 6. These experimentally determined values for $F_{sat,A}$ and τ_A are used throughout this chapter. The saturation intensity follows, then from Eq. (4). Standard pump-probe techniques determine the impulse response and therefore τ_A. In the femtosecond pulse regime, we have to consider more than one absorber recovery time, which will be discussed in more detail later in this chapter. The saturation fluence $F_{sat,A}$ is determined and defined by measurement of the nonlinear change in reflectivity of a pulse as a function of increased incident pulse energy (Fig. 3). The incident beam area on the device under test is measured with a charge coupled device (CCD) camera and a frame grabber to obtain high accuracy and to minimize systematic errors in the measurements of the saturation fluence. Calibration of the length scale on the CCD camera is obtained with lithographically defined circuit structures placed in the focus of the incident beam at the same position as the SESAM device under test. Figure 3 shows a typical measurement of a SESAM device that supported 34-fs pulses in a Ti:sapphire laser, as discussed in more detail in Sections III and IV. The change in reflectivity induced by a pulse whose duration is short compared with the absorption recovery time (i.e., slow absorber approximation) can be calculated from the common traveling waverate equations (Siegman, 1986; Agrawal and Olsson, 1989). If $R_0 R_{ns}$ is the unsaturated low-intensity reflectivity of the sample, the final reflectivity R_f after the passing of a pulse with a pulse fluence F_p is given by

$$R_f = \frac{R_0}{R_0 - (R_0 - 1)\exp(-F_p/F_{sat,A})} \tag{5}$$

and the reflectivity the pulse itself experiences is

$$R(F_p) = \frac{F_{out}}{F_p} = R_{ns} \frac{\log\left(\frac{R_0 - 1}{R_f - 1}\right)}{\log\left(\frac{R_0 - 1}{R_f - 1}\right) - \log\left(\frac{R_0}{R_f}\right)} \tag{6}$$

where $R_{ns} = R(F_p \to \infty)$. In Eq. (6), the transverse mode profile is approximated by a constant value, and the nonsaturable losses are not considered to be distributed along the saturable absorber. A nonsaturable loss distributed along the absorber could not be solved analytically anymore. From a phenomenological point of view, however, we used Eqs. (5) and (6) to fit the experimentally measured $R(F_p)$ (Fig. 3) in order to obtain $F_{sat,A}$, ΔR, and ΔR_{ns} for the SESAM. As we have demonstrated many times, Eq. (6) gives an excellent fit to the measured data, thus justifying our simple model. One has to be aware of the fact that the saturation fluence measured in reflection is smaller than that measured in transmission owing to standing wave effects. The saturation fluence measured in reflection, however, is relevant for the laser.

2. Requirements for Passive Modelocking

a. Passive Modelocking Mechanisms

Passive modelocking mechanisms are well explained by three fundamental models: slow saturable absorber modelocking with dynamic gain satura-

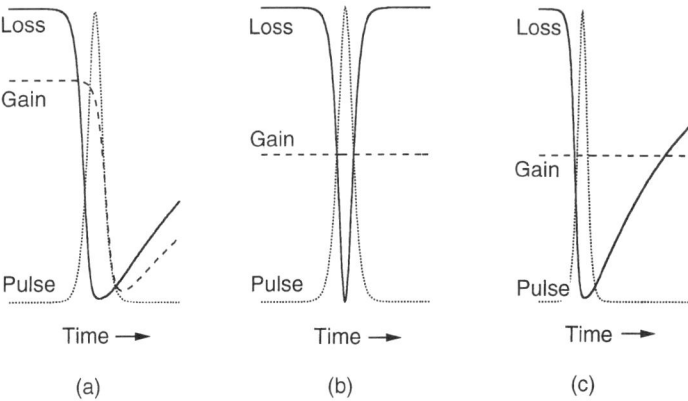

FIG. 4. Passive modelocking mechanisms explained by three fundamental models: (a) slow saturable absorber modelocking with dynamic gain saturation (New, 1974; Haus, 1975c), (b) fast saturable absorber modelocking (Haus, 1995b; Haus et al., 1991b), and (c) soliton modelocking (Kärtner and Keller, 1995b; Kärtner et al., 1996; Jung et al., 1995b).

tion (New, 1974; Haus, 1995c) (Fig. 4a), fast saturable absorber modelocking (Haus, 1975b; Haus et al., 1991b), and soliton modelocking (Kärtner and Keller, 1995b; Kärtner et al., 1996; Jung et al., 1995b) (Fig. 4c). In the first two cases, a short net-gain window forms and stabilizes an ultrashort pulse. This net-gain window also forms the minimal stability requirement—i.e., the net loss immediately before and after the pulse approximately defines its extent. For solid state lasers, we cannot apply slow saturable absorber modelocking as shown in Fig. 4(a), because no significant dynamic gain saturation is taking place as a result of the long upper state lifetime of the laser. Dynamic gain saturation means that the gain undergoes a fast pulse-induced saturation that then recovers again between consecutive pulses (Fig. 4a). The upper state lifetimes of solid state lasers are typically in the microsecond to millisecond regime—much longer than the pulse repetition period, which is typically in the nanosecond regime. We therefore observe no significant dynamic gain saturation, and the gain is saturated only to a constant level by the average intracavity intensity. In Fig. 4(a), an ultrashort net-gain window can be formed by the combined saturation of absorber and gain for which the absorber has to saturate and recover faster than the gain, whereas the recovery time of the saturable absorber can be much longer than the pulse duration.

This is different in the fast saturable absorber model, where no dynamic gain saturation occurs and the short net-gain window is formed by a fast recovering saturable absorber alone (Fig. 4b). This was initially believed to be the only stable approach to passive modelocking of a solid state laser. Additive pulse modelocking (APM) (Blow and Nelson, 1988; Kean et al., 1989; Ippen et al., 1989) was the first fast saturable absorber modelocking technique for solid state lasers. However, APM required interferometric cavity length stabilization because self-phase modulation in a fiber inside the coupled cavity generated a nonlinear phase shift that added constructively at the peak of the pulse in the main cavity and destructively in the wings, thus shortening the pulse duration inside the main cavity. Kerr lens modelocking (KLM) (Spence et al., 1991) was the first useful demonstration of an intracavity fast saturable absorber modelocking technique for a solid state laser and, because of its simplicity, replaced coupled cavity modelocking techniques. KLM is based on a very close to ideal fast saturable absorber for which the modulation depth is produced by either a decrease in losses because of self-focusing through a hard aperture (Keller et al., 1991a; Negus et al., 1991; Salin et al., 1991) or an increase in gain in the laser as a result of increased overlap of the laser mode with the pump mode in the laser crystal (Salin and Squier, 1992; Piché and Salin, 1993). Only in the ultrashort pulse regime, more complicated space-time coupling occurs and wavelength-dependent effects start to limit further pulse reduction (Christov et al., 1995; Cerullo et al., 1996; Cundiff et al., 1996).

Besides the tremendous success of KLM, there are some significant limitations for practical or "real-world" ultrafast lasers. First, the cavity is typically operated near one end of its stability range, where the Kerr-lens-induced change of the beam diameter is large enough to sustain modelocking. This results in a requirement for critical cavity alignment whereby mirrors and laser crystal have to be positioned to a typical accuracy of several hundred microns. Once the cavity is correctly aligned, KLM can be very stable and under certain conditions even self-starting (Cerullo et al., 1994a, 1994b). However, self-starting KLM lasers in the sub-50-fs regime have not yet been demonstrated without additional starting mechanisms (such as, for example, a SESAM). This is not surprising, because in a 10-fs Ti:sapphire laser with a 100-MHz repetition rate, the peak power changes by six orders of magnitude when the laser switches from cw to pulsed operation. Therefore, nonlinear effects that are still effective in the sub-10-fs regime are typically too small to initiate modelocking in the cw operation regime. In contrast, if self-starting is optimized, KLM tends to saturate in the ultrashort pulse regime or the large self-phase modulation (SPM) will drive the laser unstable. Soliton modelocking addresses this issue by decoupling the saturable absorber from the cavity.

In soliton modelocking, the pulse shaping is done solely by soliton formation—i.e., the balance of group delay dispersion (GDD) and self-phase modulation (SPM) at steady state—with no additional requirements on the cavity stability regime. An additional loss mechanism, such as a saturable absorber (Kärtner and Keller, 1995b; Kärtner et al., 1996) or an acousto-optic modelocker (Kopf et al., 1994a; Kärtner et al., 1995c), is necessary to start the modelocking process and to stabilize the soliton. In soliton modelocking, we have shown that the net-gain window (Fig. 4c) can remain open more than 10 times longer than the ultrashort pulse, depending on the specific laser parameters (Jung et al., 1995b; Kärtner et al., 1998). With the same modulation depth, one can obtain almost the same minimal pulse duration as with a fast saturable absorber, as long as the absorber recovery time is roughly less than 10 times longer than the final pulse width. In addition, high dynamic range autocorrelation measurements showed excellent pulse pedestal suppression over more than seven orders of magnitude in 130-fs pulses of an Nd:glass laser (Kopf et al., 1995b). Such a pedestal suppression is very similar to or even better than KLM pulses in the 10-fs pulse width regime (Jung et al., 1997a). Even better performance can be expected if the saturable absorber also shows a negative refractive index change coupled with the absorption change, as is the case for semiconductor materials (Kärtner et al., 1996).

Soliton modelocking can be explained as follows (Fig. 5). The soliton loses energy as a result of gain dispersion and losses in the cavity. Gain dispersion and losses can be treated as perturbations to the nonlinear Schrödinger equation for which a soliton is a stable solution (Kärtner et al., 1995c). This lossed energy, called "continuum" in soliton perturbation theory (Kaup, 1990), is initially contained in a low-intensity background pulse, which undergoes negligible bandwidth broadening as a result of SPM, but spreads in time in response to GDD (Fig. 5a). This continuum exhibits a higher gain than does the soliton, because it exhibits only the gain at line center (while the soliton exhibits an effectively lower average gain owing to its larger bandwidth) (Fig. 5b). After a sufficient buildup time, the continuum would actually grow until it reached lasing threshold, destabilizing the soliton. However, we can stabilize the soliton by introducing a "slow" saturable absorber into the cavity. This "slow" absorber is fast enough to add sufficient additional loss for the growing continuum that spreads in time so that it no longer reaches lasing threshold.

Soliton formation is generally very important in femtosecond lasers, which has already been recognized in colliding pulse modelocked dye lasers (Kaup, 1990; Salin et al., 1986, 1988) and fast saturable absorber modelocked lasers in general (Haus et al., 1991b) and, more specifically, in KLM Ti:sapphire lasers (Brabec et al., 1991, 1992; Spielman et al., 1994). However, no analytic solution has been presented for the soliton pulse shortening, and it has always been assumed that for a stable solution the modelocking mechanism without soliton effects has to generate a net-gain

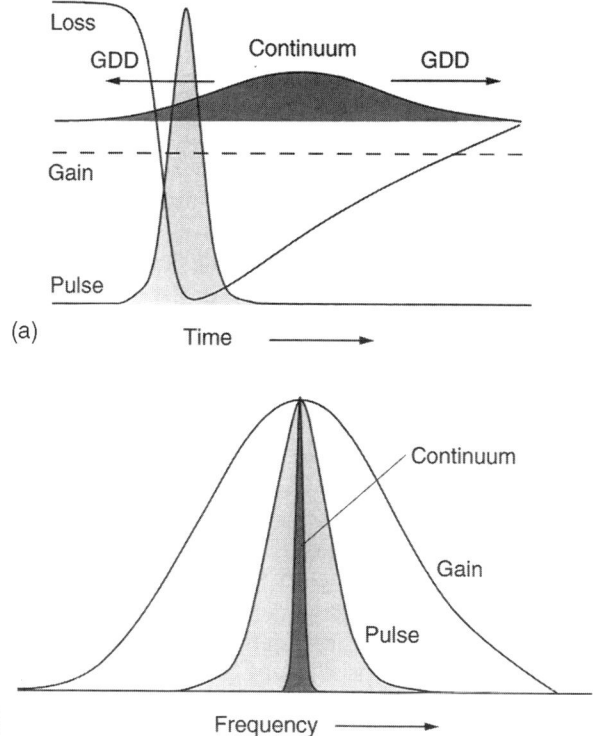

FIG. 5. Soliton modelocking in (a) time and (b) frequency domain. The continuum spreads in time owing to group delay dispersion (GDD) and thus undergoes more loss in the "slow" absorber, which is saturated by the shorter soliton pulse. However, the longer continuum pulse has a narrower spectrum and thus undergoes more gain than the spectrally broader soliton pulse.

window as short as the pulse (Fig. 4a and b). In contrast to these cases, in soliton modelocking we present an analytic solution based on soliton perturbation theory, where soliton pulse shaping is clearly assumed to be the dominant pulse formation process and the saturable absorber required for a stable solution is treated as a perturbation. Then, the net-gain window can be much longer than the pulse (Fig. 4c). Stability of the soliton against the continuum then determines the shortest possible pulse duration (see Eq. 15). This is a modelocking model fundamentally different from any model previously described. We therefore refer to it as soliton modelocking, emphasizing the fact that soliton pulse shaping has been assumed to be dominant.

b. Recovery Time τ_A of the Saturable Absorber

In the picosecond modelocking regime, we use SESAMs as fast saturable absorbers (Haus, 1995b; Haus et al., 1992) (Fig. 4b). Semiconductor absorbers have an intrinsic bitemporal impulse response—intraband carrier-carrier scattering and thermalization processes on the order of 10 to 100 fs as well as interband trapping and recombination processes that can be on the order of picoseconds to nanoseconds depending on the growth parameters (Kaminska et al., 1989; Gupta et al., 1992; Siegner et al., 1996). The fast relaxation process can typically be neglected in the picosecond regime, because it produces only a negligibly small modulation depth (Fig. 6). For a long 4-ps excitation pulse, the photogenerated carriers thermalize so fast that the unthermalized initial energy distribution does not contribute significantly to absorption bleaching. The typical bitemporal impulse response is visible only with shorter excitation pulses. Normally grown semiconductor materials have carrier recombination times in the nanosecond regime, which do not produce fast saturable absorbers and, because of the small saturation intensity, tend to drive many solid state lasers into Q-switching instabilities (see Fig. 9a and Eq. 23). To obtain a fast saturable absorber with a larger modulation depth, we artificially reduce the recovery time of the semiconductor saturable absorber into the picosecond regime by low-temperature MBE growth (Siegner et al., 1996; Haiml et al., 1998b), ion bombardment (Lederer et al., 1997), high-level Be-doping (Haiml et al.,

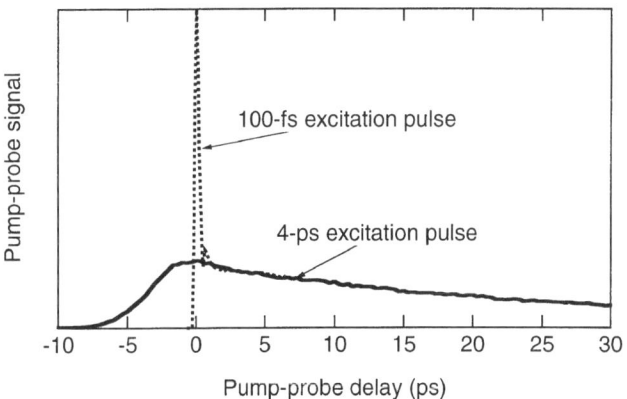

FIG. 6. Typical measured impulse response of a SESAM with different excitation pulse durations. The saturable absorber was grown at low temperature, which reduced the recovery time of the saturable absorber to about 20 ps. The short intraband thermalization recovery time is no longer visible with a 4-ps excitation pulse.

1998a; Prasad *et al.*, 1998), or some other technique. There are different trade-offs and advantages of each of these techniques. The goal is that the technique used to reduce the recovery time does not significantly decrease the strength of the optical nonlinearity and also does not introduce more nonsaturable losses. We typically use low-temperature-grown III–V semiconductors, which exhibit fast carrier trapping into point defects formed by the excess group-V atoms incorporated during the low-temperature (LT) growth (Kaminska *et al.*, 1989; Gupta *et al.*, 1992; Siegner *et al.*, 1996). We have demonstrated that annealing (Haiml *et al.*, 1995b) as well as Be doping (Haiml *et al.*, 1998a) of LT GaAs decreases the nonsaturable losses and increases the nonlinearity by up to one order of magnitude to the value of GaAs grown at normal temperatures. Importantly, the time response does not degrade on annealing or Be doping, but remains fast or gets even faster.

In the femtosecond modelocking regime, we could still use the SESAM as a fast saturable absorber, where the recovery time is determined by intraband nonlinearities (Fig. 6). The longer trapping and/or recombination time provides efficient self-starting of the modelocking process. Intraband thermalization and relaxation processes occur on different timescales depending on the excitation energy and density (Shah, 1996). A faster initial recovery time and larger modulation depth is obtained for excitation high into the conduction band in quantum wells. However, this transition is more difficult to saturate, and intervalley scattering limits the bandwidth. Excitonic nonlinearities are fast and have low saturation intensity, but are narrowband nonlinearities. Therefore, the fast intraband semiconductor nonlinearities would limit the performance of the modelocked laser and could become a critical design parameter. This will be discussed in more detail for the Ti:sapphire in Section IV. In the femtosecond regime, we typically use the intracavity SESAM either to start KLM or to start and stabilize soliton modelocking, which reduces the critical design issues of SESAMs in the femtosecond regime. The faster saturable absorption in the SESAM plays an important role in stabilizing femtosecond pulses in soliton modelocking, while the slower response is important for starting the pulse formation process. The upper limit of the slow response duration is set by the onset of Q-switching instabilities, which occur much earlier for lasers with long upper state lifetimes. For example, a femtosecond Yb:glass laser (Hönninger *et al.*, 1998a) with an upper state lifetime of about 1 ms has a much stronger tendency for Q-switching instabilities than a femtosecond Ti:sapphire laser with an upper state lifetime of about 2.5 μs (Brovelli *et al.*, 1995a). Therefore, especially in the picosecond regime, a more careful design of the slow response time of the SESAM is required for stable cw modelocking (Hönninger *et al.*, 1998b). In the femtosecond regime, the tendency for self-Q-switching instabilities is reduced, which we believe is most likely attributable to soliton formation. We have shown that with soliton modelocking the faster recovery time of the SESAM can be about 10 to 30 times slower than the final pulse duration (Jung et al., 1995b). This strongly relaxes the requirements on the ultrafast dynamics of the semiconductor absorber. In this case, no quantum well nonlinearities are absolutely necessary, and therefore bulk absorber layers are in many cases sufficient. The reduced requirements on the absorber dynamics also allowed us to demonstrate 50-nm tunability of a diode-pumped, soliton-modelocked Cr:LiSAF laser with a one-quantum-well low-finesse A-FPSA (Kopf et al., 1996a, 1997b) and 300-nm tunability with sub-30-fs pulses in a Ti:sapphire laser (Sutter et al., 1998b). In addition, pulses as short as 6.5 fs (Jung et al., 1997c) have been generated with a Ti:sapphire laser using a broadband low-finesse A-FPSA with a single 15-nm-thick quantum well absorber layer with a bandgap at approximately 860 nm (Fluck et al., 1996a; Jung et al., 1997b). The spectrum of the 6.5 fs pulse extends approximately from 680 to 900 nm. We would not obtain broad tunability and sub-10-fs pulse generation if, for example, the narrow-band excitonic nonlinearities in the SESAM provided the dominant pulse formation process.

c. Modulation Depth ΔR and Nonsaturable Losses ΔR_{ns}

According to all modelocking theories (Kärtner et al., 1995a; Haus, 1975b; Haus et al., 1991b), we find that the pulse width τ_p is typically inversely proportional to the modulation depth $\Delta R \approx 2q_0$ (see Eq. 1) of the saturable absorber:

$$\tau_p \propto \frac{1}{q_0^\beta}, \qquad \beta > 0 \tag{7}$$

The exponential factor β varies in the different modelocking theories. A high modulation depth always helps to achieve shorter pulses (Eq. 7) and relaxes the requirements for self-starting modelocking (see Eq. 22). However, an upper limit of the modulation depth is given by the onset of Q-switched modelocking (see Eq. 25). In addition, every saturable absorber introduces undesired nonsaturable losses (see Fig. 3), which degrade the performance of the laser. Thus, to keep the nonsaturable absorption at a minimum, it is advantageous to use as much of a given maximum modulation depth as possible for the generation of short pulses. However, the onset of multiple pulsing instabilities or damage sets an upper limit (see Section II.4.d).

For an ideal fast saturable absorber, Haus et al. (1991) developed an analytic solution for the pulse width assuming a weakly saturated fast saturable absorber:

$$q(t) = \frac{q_0}{1 + I(t)/I_{\text{sat,A}}} \approx q_0 - \gamma I(t), \quad I(t) \ll I_{\text{sat,A}} \tag{8}$$

with γ given by

$$\gamma \equiv \frac{q_0}{I_{\text{sat,A}}} \tag{9}$$

He then predicts an unchirped sech² pulse shape

$$I(t) = I_0 \operatorname{sech}^2\left(\frac{t}{\tau}\right) \tag{10}$$

with a full width at half maximum (FWHM) pulse width of

$$\tau_p = 1.7627\tau = 1.7627\frac{4D_g}{\gamma F_p} \tag{11}$$

(Haus, 1975b; Haus et al., 1992), where I_0 is the peak intensity of the pulse, F_p is the pulse energy fluence incident on the fast saturable absorber, and D_g is the gain dispersion of the laser medium given by

$$D_g = \frac{g}{\Omega_g^2} = \frac{1}{\pi^2}\frac{g}{(\Delta v_g)^2} \tag{12}$$

where g is the saturated amplitude gain coefficient per cavity round trip, $\Omega_g = \Delta\omega_g/2$ is the half width at half maximum (HWHM) gain bandwidth in radians (i.e., angular frequency), and $\Delta\omega_g = 2\pi\Delta v_g$, where Δv_g is the FWHM gain bandwidth of the laser medium.

The shortest possible pulses can be obtained when we use the full modulation depth of the fast saturable absorber. We obtain an analytic solution only if we assume an ideal fast saturable absorber that saturates linearly with the pulse intensity [i.e., $q_{\text{ideal}}(t) = q_0 - \gamma I(t)$] over the full modulation depth (Fig. 7). For a maximum modulation depth, we then can assume that $\gamma I_0 = q_0$. For a soliton-shaped pulse (see Eq. 10), the pulse energy fluence is given by $F_p = \int I(t)\,dt = 2I_0\tau$. The minimal pulse width for a fully saturated ideal fast saturable absorber (Kärtner et al., 1998) then follows from Eq. (11):

$$\tau_{p,\text{min}} = \frac{1.7627}{\Omega_g}\sqrt{\frac{2g}{q_0}} \tag{13}$$

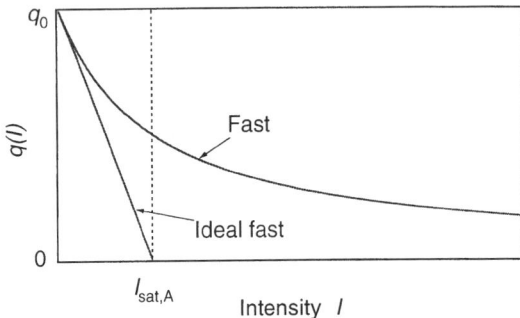

FIG. 7. Fast saturable absorber given by Eq. (8) and ideal fast saturable absorber given by $q_{\text{ideal}}(I) = q_0(1 - I/I_{\text{sat,A}})$.

This occurs right at the stability limit when the filter loss owing to gain dispersion is equal to the residual loss a soliton pulse undergoes in an ideal fast saturable absorber:

$$\text{Filter loss} = \frac{D_g}{3\tau^2} = \frac{q_0}{6} = q_s = \text{residual saturable absorber loss} \quad (14)$$

(Kärtner et al., 1998). The residual saturable absorber loss q_s results from the fact that the soliton pulse initially undergoes loss to saturate fully the fast saturable absorber. This residual loss is exactly $q_0/6$ for a sech^2 pulse shape. This condition results in a minimal FWHM pulse duration given by Eq. (13). Including soliton formation in the fast saturable absorber model, an additional pulse shortening of a factor of 2 was predicted (Haus et al., 1991b).

For soliton modelocking, we developed an analytic solution using soliton perturbation theory (Kärtner and Keller, 1995b; Kärtner et al., 1996) that predicts a minimum pulse duration of a soliton-like pulse:

$$\tau_{p,\text{min}} = 1.7627 \left(\frac{1}{\sqrt{6}\,\Omega_g}\right)^{3/4} \Phi_0^{-1/8} \left(\frac{\tau_A g^{3/2}}{q_0}\right)^{1/4} \quad (15)$$

where Φ_0 is the phase shift of the soliton per cavity round trip (Eq. 16). Here, we assume that the saturable absorber is fully saturated (i.e., $F_p \gg F_{\text{sat,A}}$), a linear approximation for the exponential decay of the slow saturable absorber, and that the absorber recovers on a timescale τ_A that is longer than the soliton pulse duration. The minimum pulse duration is achieved right at the stability limit, where the continuum starts to exhibit less loss than the soliton.

In soliton modelocking, the dominant pulse formation process is assumed to be soliton formation. Therefore, the pulse has to be a soliton for which the negative group delay dispersion (GDD) is balanced with the self-phase modulation (SPM) inside the laser cavity:

$$\Phi_0 = \frac{\delta E_p}{4\tau} = \frac{|D|}{\tau^2} \qquad (16)$$

We continue to follow Haus's notation, where E_p is the intracavity pulse energy and δ is the SPM coefficient given by $\delta = kn_2L/A_{\text{eff}}$, with $k = 2\pi/\lambda$ the vacuum wave number and A_{eff} the effective mode area. The effective mode area is equal to the mode area of the laser beam inside the gain medium if SPM occurs only in the gain medium and the laser beam size is approximately constant in the gain medium. n_2 is the nonlinear refractive index coefficient that determines the Kerr induced change in the refractive index $\Delta n = n_2 I$, and L is the length of the medium during which the pulse undergoes SPM during a cavity round trip. D is half of the total negative GDD inside the laser cavity given by $D = \frac{1}{2}(dT_g/d\omega)$, where T_g is the total group delay inside the laser cavity per round trip. The pulse duration is then given by the simple soliton solution

$$\tau_p = 1.7627 \frac{4|D|}{\delta \cdot E_p} \qquad (17)$$

This means that the pulse duration scales linearly with the negative group delay dispersion inside the laser cavity. In the case of an ideal fast saturable absorber, an unchirped soliton pulse is obtained only at a very specific dispersion setting for which $|D|/\delta = AD_g/\gamma$ (i.e., Eq. 17 equal to Eq. 11 with A the laser beam area incident on the SESAM), whereas for soliton modelocking an unchirped transform-limited soliton is obtained for all dispersion levels as long as the stability requirement against the continuum is fulfilled. This fact has also been used to confirm experimentally that soliton modelocking is the dominant pulse formation process and not a fast saturable absorber such as KLM (Aus der Au et al., 1997; Kärtner et al., 1998; Jung et al., 1995b). For example, this is shown in Fig. 8(a) for a soliton modelocked diode-pumped Nd:glass laser. The breakup into two or even more pulses will be discussed below. KLM has the additional disadvantage that stable modelocking is achieved only close to the cavity stability limit. Higher-order dispersion only increases the pulse duration, and therefore is undesirable and is assumed to be compensated.

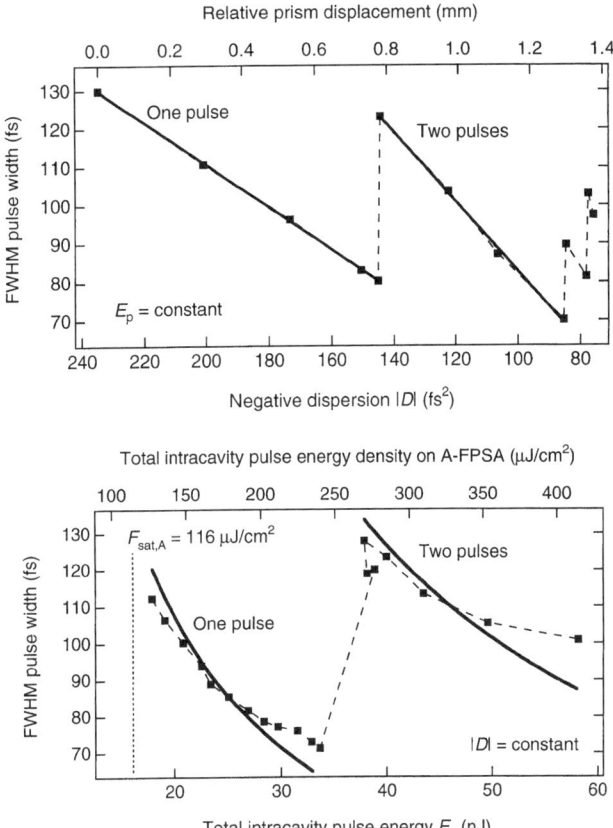

FIG. 8. Soliton modelocked diode-pumped Nd:glass laser using an intracavity SESAM (Aus der Au et al., 1997): (a) pulsewidth as a function of negative group delay dispersion (GDD) at constant pulse energy (i.e., 2.8 times the saturation fluence F_{sat}) and (b) pulsewidth as a function of pulse energy at constant GDD. The solid lines are the theoretical predictions of ideal soliton pulses. Breakups into two or more pulses per cavity round trip are observed at higher pulse energies and lower GDD.

So far we have neglected phase nonlinearities inside the SESAM device. Theory predicts that they would increase the stability range for stable soliton modelocking and even shorter pulse durations should be possible (Kärtner et al., 1996). For ultrashort pulse generation, a large modulation depth over the full bandwidth of the pulse spectrum is desirable. This is not a trivial task, because the saturable absorption of semiconductors is wavelength-dependent, as will be discussed later in Section IV.2.

d. Saturation Fluence $F_{sat,A}$ and the Onset of Multiple Pulsing Instabilities

We typically operate the saturable absorber with an incident pulse energy density F_p a few times greater than the saturation fluence $F_{sat,A}$ (see Fig. 3) nearly to obtain close to the maximum modulation depth ΔR. However, there is also an upper limit to F_p, determined by the onset of multiple pulsing (Aus der Au et al., 1997; Kärtner et al., 1998). Given a pulse fluence many times the saturation fluence $F_{sat,A}$, we can see from Fig. 3 that the reflectivity of a single pulse is strongly saturated and is no longer a strong function of the pulse energy. In addition, shorter pulses undergo a reduced average gain, owing to the limited gain bandwidth of the laser, which introduces a filter loss. Beyond a certain pulse energy, two soliton pulses with lower power, longer duration, and narrower spectrum are preferred, since their filter loss (see Eqs. 12 and 14) introduced by the limited gain bandwidth decreases so much that the slightly increased residual loss of the less saturated saturable absorber cannot compensate for it. This results in a lower total round trip loss and thus a reduced saturated or average gain for two pulses compared with one pulse. The threshold for multiple pulsing is lower for shorter pulses — i.e., with spectra that are broad in comparison with the gain bandwidth of the laser. Our experimentally determined rule of thumb for the pulse fluence on the saturable absorber is three to five times the saturation fluence. A more detailed description of multiple pulsing is given elsewhere (Kärtner et al., 1998). In general, the incident pulse fluence on the saturable absorber can be adjusted by the mode area — i.e., by how strongly the cavity mode is focused on the saturable absorber.

If we assume soliton modelocking, where the pulse is completely shaped by the soliton-like pulse shaping process, the pulse width is given by an ideal soliton (see Eq. 10). If the pulse breaks up into two pulses, while the pulse train maintains the same average power, two things happen: the pulse energy drops by a factor of 2, and the pulsewidth increases by a factor of 2 (see Eq. 17). This agrees well with the experimental observation in an Nd:glass laser (see Fig. 8) (Aus der Au et al., 1997). The pulse duration scales linearly with GDD, as expected from a soliton (Eq. 17), until the pulses become too short and each one breaks up into two pulses (see Fig. 8a), because the round trip loss (including filter loss and residual saturable absorber loss) undergone by the two pulses is lower than the loss undergone by the single pulse. This condition can also be obtained by increasing the pulse energy at constant GDD (see Fig. 8b). In this case, the absorber becomes more strongly saturated until, again, the only slightly increased residual saturable absorber loss cannot compensate for the lower filter loss for two pulses compared with the loss for one pulse. Thus, soliton modelocking gives a full explanation of the multiple pulse breakup, and no new modelocking mechanism is required to explain this experimental result, as suggested recently (Collings *et al.*, 1997).

e. Modelocking Buildup Time

The difference in absorption bleaching from cw to pulsed operation is the driving force for the modelocking buildup phase. Simple arguments discussed below show that when the saturable absorber is only weakly saturated at cw power we can obtain lower loss when the laser is cw modelocked and produces pulses that are shorter than or similar to the absorber recovery time. Initially, during the modelocking buildup phase, small noise fluctuations undergo lower loss because they can start to saturate the absorber. Therefore, the modelocking buildup time scales inversely with the slope dR/dI for $I \ll I_{\text{sat,A}}$ or $I \approx 0$ (i.e., weakly saturated for cw light):

$$T_{\text{build-up}} \propto \frac{1}{(dR/dI)|_{I \approx 0} I} \tag{18}$$

From Fig. 9, it becomes clear that small intensity fluctuations initially introduce larger reflectivity changes of the saturable absorber if the slope is larger. Therefore, the modelocking buildup time decreases with smaller saturation intensities $I_{\text{sat,A}}$. However, the onset of Q-switching sets a lower limit for $I_{\text{sat,A}}$ (see Eq. 23).

Absorption bleaching for a fast saturable absorber can be modeled with a nonlinear absorption coefficient $q = q_0/(1 + x)$ (see Eq. 8), where x determines the absorption bleaching. To keep the arguments simple, we assume that the average power does not change between cw and modelocked operation. This is a good approximation because the SESAM modulation depth cannot be too large for stable cw modelocking without Q-switching (see Eq. 25). At cw lasing, x is given by $x_{\text{cw}} = I/I_{\text{sat,A}}$. Under modelocked operation, the absorption bleaching and hence the reflectivity are increased as a result of the increased numbers of carriers generated within one laser pulse duration. For a fast saturable absorber—i.e., $\tau_A \ll \tau_p$—the absorption bleaching x in the modelocked pulse regime is given by

$$x_p = \frac{I(t)}{I_{\text{sat,A}}} \approx \frac{I}{I_{\text{sat,A}}} \frac{T_R}{\tau_p} \gg x_{\text{cw}} \tag{19}$$

For modelocked lasers, τ_p is typically much shorter than T_R, and therefore $x_p \gg x_{\text{cw}}$. Equation (19) also shows that the nonlinearity (i.e., the difference between cw and pulsed reflectivity) decreases with increasing pulse repetition rate and increases with shorter pulses. This also gives a simple explanation of why passive modelocking becomes more difficult at higher pulse repetition rates (Keller et al., , 1992d).

The growth rates of adjacent axial modes can be used to give more accurate estimates of the threshold for self-starting modelocking and the modelocking buildup time (Haus, 1976). In an ideal homogeneously broadened laser without any spatial hole burning and without a saturable absorber, only one longitudinal cavity mode at the center of the gain is running, and it saturates the gain, so that all neighboring modes are below threshold. A saturable absorber produces self-amplitude modulation (SAM), which couples the cavity mode at line center and the cavity mode m times the mode spacing $2\pi/T_R$ away from the line center. This two-mode coupling provides additional gain for the adjacent modes to reach threshold. The growth rates of these modes can be used as estimates for the modelocking buildup time T_{MBT}. Note that we assume that the growing modes become phaselocked automatically and form a pulse. This may not always be the case and has been considered previously (Krausz et al., 1991). For a short absorber recovery time τ_A compared with the round trip time T_R and a long upper state lifetime τ_L of the gain much greater than T_R, Eq. (8) in Kärtner et al. (1995a) can be simplified to the approximate condition for $\tau_A \ll T_R$ and $\tau_L \gg T_R$:

$$\frac{1}{T_{MBT}} \approx \left(\frac{2q_0}{AF_{sat,A}} \cdot \tau_A - \frac{2g_0 T_R^2}{(2m\cdot\pi)^2 \tau_L^2 A_L I_{sat,L}}\right) \cdot P \approx \frac{2q_0}{F_{sat,A}} \tau_A \frac{P}{A} \qquad (20)$$

(Jung et al., 1997b), where P is the intracavity laser power, A is the laser mode area incident on the saturable absorber, g_0 is the small-signal amplitude gain coefficient (Weingarten et al., 1994), A_L is the mode area inside the laser gain material, $I_{sat,L} = h\nu/(2\sigma_L \tau_L)$ is the saturation intensity of the laser, and σ_L is the gain cross section. The factor of 2 in $I_{sat,L}$ occurs in a standing wave cavity. The second term in Eq. (20) determines the threshold for self-starting cw modelocking, because the modelocking build-up time has to be $T_{MBT} \geqslant 0$. Far above this threshold, we can neglect this term and the modelocking buildup time is shorter at larger intracavity power and therefore at larger small-signal gain. This is expected because the small noise pulse initially grows with the small-signal gain. This result also agrees with the simple arguments given above, because the first term in Eq. (20) corresponds to Eq. (18), using Eqs. (1) and (8):

$$\left|\frac{dR}{dI}\right| I \approx 2\left|\frac{dq}{dI}\right| I \approx 2\gamma I = \frac{2q_0}{I_{sat,A}} = \frac{2q_0}{F_{sat,A}} \tau_A \frac{P}{A}, \qquad I \ll I_{sat,A} \qquad (21)$$

Therefore, the fast response of the artificial absorption owing to KLM is detrimental for self-starting. Self-starting KLM lasers in the 10-fs pulse regime have not yet been demonstrated without an additional starting mechanism such as feedback-initiated starting (Kasper and Witte, 1996; Xu et al., 1996) or a SESAM (Fluck et al., 1996a; Jung et al., 1997b). A clear advantage of using SESAMs is simple and reliable starting of the modelocking process.

The condition that $T_{MBT} \geqslant 0$ in Eq. (20) determines the threshold for self-starting modelocking. Thus, for a given laser design, the saturable absorber has to be adjusted to fulfill the following condition to obtain self-starting modelocking (Eqs. 20 and 21 with $T_{MBT} \geqslant 0$):

$$q_0 \gg \frac{g_0}{(2m \cdot \pi)^2} \frac{T_R^2}{\tau_A \tau_L} \frac{\sigma_L}{\sigma_A} \frac{A}{A_L} \qquad (22)$$

We want to operate the laser well above this threshold condition, because other perturbations of the laser can hinder the growth of initial intensity fluctuations formed by mode beating of two cavity modes. Spurious reflections inside the laser cavity can compete with the original noise pulse and substantially increase the threshold for self-starting modelocking (Haus and Ippen, 1991). The condition for self-starting modelocking provides an upper limit for the magnitude of the spurious reflections. The problems with spurious reflections can be strongly reduced with unidirectional ring lasers compared to standing wave cavities. This has been experimentally confirmed with both a self-starting passively modelocked Ti:sapphire and an erbium fiber laser for which an unidirectional ring resonator strongly reduces backscatter coupling (Tamura et al., 1993). Generally, in linear cavities it is common practice to start KLM by shaking one end mirror or a prism used for dispersion compensation. This fast-changing phase shift could destroy temporarily the interferometric effects of the internal reflections. Within this time, the noise pulses are given a chance to grow so short that they become self-sustaining. A more phenomenological criterion for the threshold for self-starting passive modelocking postulates a coherence time for the initial noise pulse (Krausz et al., 1992a). Significant pulse shortening of this pulse must then occur in a time that is short compared with this coherence time. The coherence time can be experimentally determined by the 3-dB full width of the first beat note of two adjacent axial modes in the free-running laser using a microwave spectrum analyzer. However, no model for predicting this coherence time has been presented. Normally we can minimize spurious intracavity reflections with Brewster cut surfaces and wedged output couplers. With these precautions we have successfully relied on our experimental guideline to keep the modelocking buildup time smaller than about 1 ms by choosing the laser and saturable absorber parameters accordingly (see Eq. 20).

In Eqs. (20) and (22) we use only parameters that can be determined experimentally. However, not all parameters can be chosen freely. Obviously, a large modulation depth helps both to generate shorter pulses (see Eq. 7) and to reach threshold for self-starting modelocking (Eq. 22), but an upper limit is set by Q-switching instabilities (see Eq. 25). Using the SESAM as a fast saturable absorber in the picosecond regime will set an upper limit for $\tau_A \approx 1$ ps. In the femtosecond regime, we can benefit from the bitemporal response of the SESAM, which allows for a longer τ_A. An upper limit is then set by the onset of Q-switching instabilities that occur at low saturation intensities $I_{sat,A}$, because the slope dR/dI becomes too large (see Eq. 23). Low cavity losses l and a large small-signal gain g_0 (thus, a small laser mode inside the gain medium) are advantageous for suppressing Q-switching instabilities (see Eqs. 23 and 24) and finally obtaining a short modelocking buildup time when the threshold is reached (see Eq. 20). A large small-signal gain will not be a problem for self-starting modelocking, because typically $\sigma_L/\sigma_A \ll 1$ and $T_R^2/(\tau_A \tau_L) \leqslant 1$, assuming $T_R = 1$ ns, $\tau_A = 1$ ps, and $\tau_L \geqslant 1$ μs (see Eq. 22). A semiconductor absorber cross section σ_A is typically about 10^{-14} cm^2, which is significantly higher than the gain cross section σ_L of a solid state laser, which is in the range from 10^{-19} to 10^{-22} cm^2, resulting in σ_L/σ_A of 10^{-5} to 10^{-8}.

f. Onset of Q-Switching Instabilities

A long absorber recovery time τ_A and a large absorber cross section σ_A result in a small saturation intensity (see Eq. 4) and are desirable for self-starting modelocking (see Eqs. 20 and 22). However, there is a trade-off: if the saturation intensity of the absorber is too small, the laser will start to Q-switch. The condition for no Q-switching, derived in Haus (1976); Kärtner et al. (1995a); Hönninger et al. (1998b), is

$$\left|\frac{dR}{dI}\right| I < \frac{g_0}{l} \frac{T_R}{\tau_L} \approx \frac{T_R}{\tau_{stim}} \qquad (23)$$

for the approximation $\Delta R \approx 2q_0$ (according to Eq. 1) — i.e., a small modulation depth. This approximation is very well fulfilled for passively cw modelocked solid state lasers, because a large ΔR tends to drive the laser into Q-switched modelocking (see Eq. 25). g_0 is the small-signal gain coefficient of the laser given by $g_0 = rl$, where l is the total amplitude loss coefficient of the laser cavity and r is the pump parameter that determines how many times the laser is pumped above threshold; T_R is the cavity round trip time; and τ_L is the upper state lifetime of the laser. The stimulated lifetime τ_{stim} of the upper laser level is given by $\tau_{stim} = \tau_L/(r - 1) \approx \tau_L/r$ for $r \gg 1$. For solid state lasers with high small-signal gains it is more accurate to measure g_0 with the relaxation oscillation frequency (Weingarten et al., 1994). From Eq. (23), it then follows that Q-switching can be more easily suppressed for a small slope dR/dI (i.e., a large saturation intensity), a large small-signal gain g_0, a small cavity loss l (i.e., a large pump parameter r), and a large cavity round trip period T_R (for example, a low modelocked pulse repetition rate). Equation (23) also indicates that solid state lasers with large upper state lifetimes τ_L have an increased tendency for self-Q-switching instabilities.

The physical interpretation of the Q-switching threshold (Eq. 23) is as follows. The left side of Eq. (23) determines the reduction in loss per cavity round trip owing to the bleaching of the saturable absorber. This loss reduction increases the intensity inside the laser cavity. The right side of Eq. (23) determines how much the gain per round trip saturates, compensating for the reduced loss and keeping the intensity inside the laser cavity constant. If the gain cannot respond fast enough, the intensity continues to increase as the absorber is bleached, leading to self-Q-switching instabilities or stable Q-switching.

Equations (22) and (23) give upper and lower bounds for the saturation intensity that results in self-starting cw modelocking without self-Q-switching. Of course, we can also optimize a saturable absorber for Q-switching by selecting a small saturation intensity and a short cavity length i.e., a short T_R. This will be discussed in more detail in Section II.3.

If we use a fast saturable absorber with recovery time much shorter than the cavity round trip time ($\tau_A \ll T_R$), the condition given by Eq. (23) is typically fulfilled, and much shorter pulses can be formed. But now an additional stability requirement has to be fulfilled to prevent Q-switched modelocking (see Fig. 1). For this further discussion, we assume a slow saturable absorber—i.e., that the steady state pulse duration τ_p is shorter than the recovery time τ_A of the saturable absorber, $\tau_p < \tau_A$. In this case, the saturation (see Fig. 9b) is determined by the saturation fluence $F_{sat,A}$ and the incident pulse fluence F_p on the saturable absorber. The loss reduction per round trip is now a result of bleaching of the saturable absorber by the short pulses, not the cw intensity. This is a much larger effect when $\tau_A \ll T_R$. Therefore, in analogy to Eq. (23), we can show that the condition for preventing Q-switched modelocking, given by (Kärtner et al., 1995a;

(a)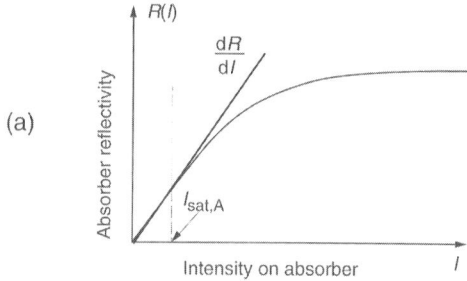

(b)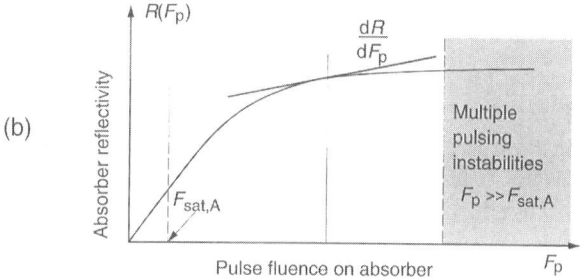

FIG. 9. Nonlinear reflectivity change of a saturable absorber mirror resulting from absorption bleaching with (a) cw intensity and (b) short pulses.

Hönninger et al., 1998b), is

$$\left|\frac{dR}{dF_p}\right| F_p < \frac{g_0}{l} \frac{T_R}{\tau_2} \approx \frac{T_R}{\tau_{\text{stim}}} \quad (24)$$

again for the approximation $\Delta R \approx 2q_0$ (according to Eq. 1)—i.e., a small modulation depth. We can easily fulfill this condition by choosing $F_p \gg F_{\text{sat},A}$ (see Fig. 9b). This also optimizes the modulation depth, resulting in reduced pulse duration. In general, the pulse energy density incident on the saturable absorber can be adjusted by the incident mode area—i.e., by how strongly the cavity mode is focused on the saturable absorber. However, there is also an upper limit on F_p, determined by the onset of multiple pulsing or damage, as discussed in Section II.4.b. Our experimentally determined rule of thumb for the pulse fluence on the saturable absorber is three to five times the saturation fluence. For solid state lasers with long upper state lifetimes (i.e., $> \mu$s), we can show that we can reduce the condition of Eq. (24) to a much simpler expression:

$$F_p^2 > F_{sat,L} F_{sat,A} \Delta R \frac{A_L}{A} \qquad (25)$$

(Hönninger et al., 1998b), where F_p is the pulse fluence on the saturable absorber (i.e., SESAM), $F_{sat,L}$ and $F_{sat,A}$ are the saturation fluences of the laser and absorber, respectively, A_L is the effective mode area in the laser media, and A is the effective mode area in the saturable absorber. This equation has the advantage that the condition for no Q-switched modelocking is described by directly accessible experimental parameters. This equation clearly demonstrates that ΔR cannot be too large for stable cw modelocking even though a larger ΔR results in shorter pulses (see Section II.4.c). Experimentally, we have some free adjustable parameters such as $F_{sat,A}$, A_L, and A. Normally, we keep A_L as small as possible for a low laser threshold, which agrees well with the stability requirement of Eq. (25). The upper limit of the pulse fluence on the saturable absorber is determined by damage and the onset of multiple pulsing, where the latter limit normally occurs first.

In the picosecond regime, we experimentally confirmed the threshold condition for the onset of Q-switched modelocking as described in Eqs. (24) and (25) (Hönninger et al., 1998b). However, it is interesting to note that for femtosecond pulse generation we observed a significant reduction of the tendency for self-Q-switching. In soliton modelocked lasers we can explain this by an extended theory which takes into account soliton shaping effects and gain filtering (Hönninger et al., 1998b).

g. Laser Mode Size in Gain and Absorber

We keep the laser mode inside the gain as small as possible to obtain a low laser threshold and a large small-signal gain. A lower limit is set by damage or by the beam quality of the pump laser (as discussed in Section IV). Normally, we can keep the mode area incident on the SESAM as a freely adjustable parameter to set the pulse energy density incident on the saturable absorber to a few times the saturation fluence to prevent Q-switch modelocking (Eq. 24, Fig. 9b) and the multiple pulsing instabilities (i.e., typically $F_p \approx 5 \cdot F_{sat,A}$). In this case we can use Eq. (2) to reduce Eq. (25) to

$$F_p \approx 5 \cdot F_{sat,A} \Rightarrow 25 > \frac{\sigma_A}{\sigma_L} \Delta R \frac{A_L}{A} \qquad (26)$$

A GaAs saturable absorber that is not embedded inside a device structure typically has an absorber cross section σ_A of about 10^{-14} cm^2, which results in a saturation fluence of approximately 10 μJ/cm^2 at 800 nm, for example. Most solid state lasers have gain cross sections σ_L ranging from about 10^{-18} to 10^{-22} cm^2, and thus the ratio $\sigma_A/\sigma_L \gg 1$ tends to be rather large. For cases in which the cavity design is more restricted and the incident mode area on the saturable absorber is not freely adjustable, integrating the semiconductor saturable absorber inside a device structure offers an interesting solution for adjusting the effective absorber cross section σ_A^{eff} as observed from outside the device. For example, we can continuously decrease σ_A^{eff} of a semiconductor saturable absorber by integrating the absorber layer inside a Fabry-Perot saturable absorber operated at antiresonance (A-FPSA). Depending on the top reflector, we reduce the intensity inside the saturable absorber, thereby increasing the effective saturation fluence $F_{\text{sat,A}}^{\text{eff}}$ and decreasing σ_A^{eff} as observed outside the A-FPSA. In case of a thin absorber layer, we can adjust $F_{\text{sat,A}}^{\text{eff}}$ by shifting the thin absorber layer with respect to the standing wave intensity profile of the reflected laser light inside the SESAM. We will discuss the different SESAM designs in Section III.

It is interesting to realize that with this additional device design we can adjust the saturation fluence such that we have sufficient variable parameters to obtain stable cw modelocking even with the same mode size in the gain and absorber (i.e., $A = A_L$). This is particularly interesting if we want to design more compact ultrafast lasers in which the SESAM is directly attached to one end of the gain crystal and forms one reflecting mirror in the laser cavity. Another interesting application is a passively modelocked fiber laser, in which we attach the SESAM directly to the end of the fiber. An upper limit of σ_A^{eff} will be given by the condition for self-starting modelocking (see Fig. 22).

3. REQUIREMENTS FOR PASSIVE Q-SWITCHING

A solid state laser with short cavity length and a saturable absorber with a small saturation intensity tends to fulfill the condition for passive Q-switching (see Eq. 23). In this case, the laser intensity strongly bleaches the saturable absorber and reduces the cavity loss such that the intensity inside the laser grows too fast for the gain saturation to follow (Fig. 10). Thus, the intensity continues to increase until finally the gain is saturated to the cavity loss level. However, the intensity is now so high that the gain saturation continues during the pulse decay time. The decay is not dominated by the cavity decay rate for $q_0 < l$. After the Q-switched pulse, the gain is depleted

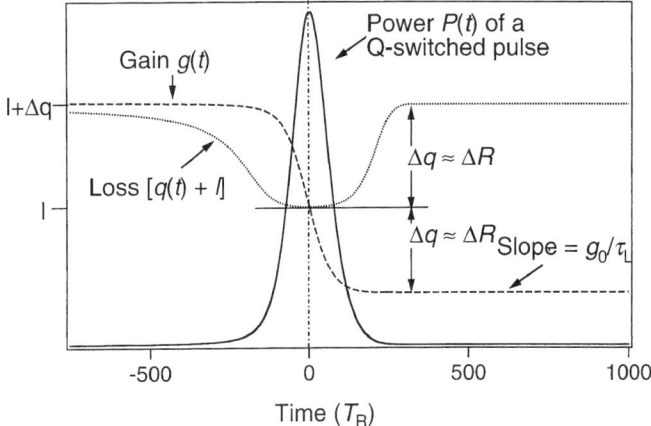

FIG. 10. Numerical simulation of the Q-switching dynamics of a passively Q-switched microchip laser with $\Delta R \leq L$, where L is the total cavity loss, and $F_p \gg F_{sat,A}$.

far below threshold and has to be increased again by the pump power. The growth rate between the pulse, is then given by g_0/τ_L, as long as we can neglect spontaneous emission.

In the Q-switching regime, a fast saturable absorber is not required. For short Q-switched pulses, we need a short cavity length, which also lowers the threshold for Q-switching (see Eq. 23). For optimum performance we design the SESAM to be fully saturated by the Q-switched pulses and to be fully recovered between consecutive pulses (Spühler *et al.*, 1998; Braun *et al.*, 1997). Therefore, it is advantageous to use a SESAM with a long carrier lifetime that reduces the saturation intensity such that the SESAM can be easily saturated. The long carrier lifetime has the additional advantage that the semiconductor absorber has fewer defects, resulting in minimal nonsaturable loss. Nonsaturable losses are undesirable inside a laser cavity because they reduce the small-signal gain and the efficiency of the laser.

Numerical simulations of the Q-switching dynamics show that we obtain the largest extracted pulse energy under given pumping conditions if the modulation depth of the absorber is as large as the output coupling and the absorber is fully bleached by the pulse energy (Spühler *et al.*, 1998). For this case we find that the center portion of the pulse is proportional to $\mathrm{sech}^2(t/\tau)$ with an FWHM pulse width of

$$\tau_p = 1.76\tau \approx 1.76 \frac{2T_R}{\Delta R} \qquad (27)$$

for $\Delta R \approx 2q_0$ (Spühler et al., 1998; Braun et al., 1997). Therefore, for a fully saturated absorber, the pulse duration is constant with increased pump power. This equation is identical to the expression derived for rapid Q-switching (Zayhowski et al., 1994). The pulse energy is then given for $\Delta R \approx 2q_0$ by

$$E_p \cong \frac{hv}{\sigma_L} A_L \Delta R \frac{l_{out}}{l} \qquad (28)$$

(Spühler et al., 1998), where l_{out} is the output coupling loss coefficient and we assume small cavity loss with $q_0 \leq l$ and a fully saturated absorber — i.e., $F_p \gg F_{sat,A}$. As long as the pulse repetition rate f_{rep} is much greater than the inverse of the upper state lifetime τ_L of the laser (i.e., $f_{rep} \geqslant 2/\tau_L$), it can be shown that

$$f_{rep} \cong \frac{g_0}{2\Delta R \tau_L} \qquad (29)$$

(Spühler et al., 1998; Braun et al., 1997), where g_0 is the small-signal gain coefficient. Since f_{rep} is a linear function of g_0, the pulse repetition rate is proportional to the pump power, and the pulse energy is expected to be approximately constant as a function of the pump power as long as the SESAM is fully saturated.

In contrast to cw modelocking, there is no upper limit on the modulation depth ΔR. Thus, both Eqs. (27) and (28) suggest that for optimized pulse duration and pulse energy a large ΔR is desirable. The pump power can then be used to adjust for the desired pulse repetition rate. The pump power does not affect the pulse energy. More detailed design guidelines for passively Q-switched solid state lasers are given in Spühler et al. (1998).

4. Microscopic Properties

Nonstoichiometric semiconductor materials such as low temperature (LT) grown semiconductors (Smith et al., 1988) are markedly different materials than conventional semiconductors (Witt, 1993; Look, 1993; Melloch et al., 1995), exhibiting a four-order-of-magnitude increase in trap densities (Kaminska et al., 1989), a one-order-of-magnitude increase in breakdown fields (Smith et al., 1988; Yin et al., 1990), a 5- to 6-order-of-magnitude increase in resistivity, and photoelectron lifetimes in the conduction band that are reduced from nanoseconds to the sub-picosecond regime (Gupta et al., 1992). LT-grown III–V semiconductors have found various applications in optoelectronics and laser science as a result of the short carrier trapping times of these materials. A short carrier trapping time is essential for fast photoconductive switches with picosecond and sub-picosecond response times (Gupta et al., 1992; Whitaker, 1993). The carrier trapping times in LT semiconductors determine the recovery times of fast semiconductor saturable absorbers, which are used to generate ultrashort pulses with solid state lasers. While the response time of a device after a single optical excitation depends on the timescale of the initial carrier trapping, the maximum repetition rate, at which photoconductive switches or saturable absorbers can be used, is limited by the time it takes for the LT semiconductor to relax to the ground state. This time is determined by the recombination time of the optically excited electron-hole pairs.

The fast carrier trapping in LT semiconductors is attributable to the incorporation of excess group V atoms during LT growth. The excess atoms in LT GaAs form As antisite (As_{Ga}) point defects (Kaminska et al., 1989) with energies close to the center of the bandgap, which act as trap states (Gupta et al., 1992). The As_{Ga} point defect is a double donor, which is generally found in the neutral charge state As_{Ga}^0 and in the single positive charge state As_{Ga}^+. Typical concentrations are $[As_{Ga}^0] \approx 10^{20} \text{cm}^{-3}$ and $[As_{Ga}^+] \approx 10^{19} \text{cm}^{-3}$ at the lowest growth temperatures at which crystalline LT GaAs can be grown (about 200 °C). These concentrations decrease with increasing growth temperature (Look et al., 1992; Liu et al., 1994). The positively charged As_{Ga}^+ defects act as electron traps very similar to EL2 defects in semi-insulating GaAs (Look et al., 1992; Manasreh et al., 1990), which are well-known electron traps (Kaminska and Weber, 1993). The positive As_{Ga}^+ defects are compensated by native acceptors. Recently, it has been shown that Ga vacancies (V_{Ga}) are the native acceptors in LT GaAs (Liu et al., 1995; Hautojärvi et al., 1993). The V_{Ga} states are located in the lower half of the bandgap with a concentration equal to about one-third of the As_{Ga}^+ concentration (Luysberg et al., 1997).

For our time-resolved experimental investigation, we studied 1-μm GaAs thin films grown by molecular beam epitaxy on GaAs substrates at growth temperatures of 250, 300, and 350 °C, with and without subsequent annealing. The As/Ga flux ratio was 30. X-ray diffraction measurements showed that all samples were crystalline, as expected. To prepare the samples for transmission experiments, we etched off the substrates and deposited anti-reflection coatings. Differential transmission experiments were carried out at room temperature with 100-fs pulses from a modelocked Ti:sapphire laser. In these experiments, we excited and probed above the bandgap at 830 nm. The pump and probe pulses were cross-polarized, with the probe intensity one order of magnitude lower than the pump intensity. We showed that carrier trapping becomes faster while carrier recombination becomes slower

4 SEMICONDUCTOR NONLINEARITIES FOR SOLID-STATE LASER   247

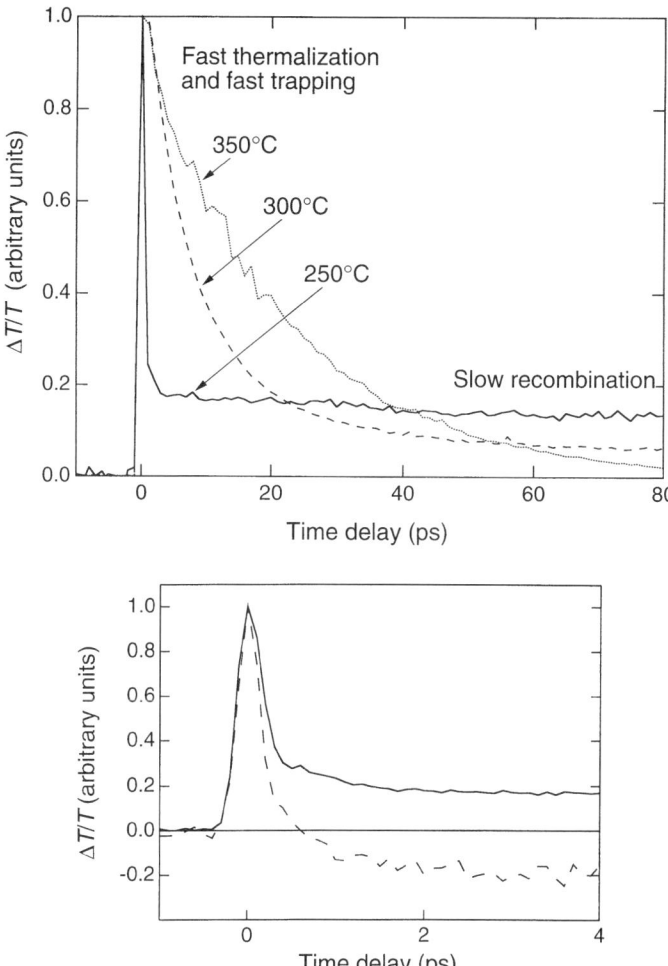

FIG. 11. Room-temperature differential transmission from (a) LT GaAs grown at 250 °C (solid line, 300 °C (dashed line), and 350 °C (dotted line) at a carrier density $N \approx 5 \times 10^{19}$ cm^{-3}; and (b) the 250 °C sample at $N \approx 5 \times 10^{18}$ cm^{-3} (dashed line) and at $N \approx 5 \times 10^{19}$ cm^{-3} (solid line). All curves have been individually normalized.

as the growth temperature of LT GaAs is decreased (Fig. 11a) (Siegner *et al.*, 1996). Our experimental results indicate that electrons and holes are trapped in different point defects. Trapping of electrons and holes in different, spatially separated defects is also consistent with the relatively large recombination time of several 100 ps. Since electrons are trapped by

As_{Ga}^{+}, the most likely candidate for the hole trap is the V_{Ga} defect. At high excitation levels, this slow recombination time forms a bottleneck for fast trapping (Siegner et al., 1996). This is a serious limitation for our saturable absorber application because it results in a reduced modulation depth of the fast picosecond recovery time (i.e., as a result of trapping) and an increased modulation depth of the much slower recovery time (i.e., as a result of recombination of trapped carriers) (Fig. 11b). The induced absorption signal (i.e., $\Delta T/T < 0$) at a lower excitation level in Fig. 11(b) is based on defect-to-band transition of the trapped electrons and becomes more pronounced at higher defect concentrations—i.e., at lower LT growth temperatures. The induced absorption signal decays slowly with the recombination rate of trapped carriers. Further investigations with the goal of reducing the slow recombination times of trapped carriers are in progress.

The growth temperature dependence of the saturation fluence, modulation depth, and nonsaturable losses of 0.5-μm-thick LT-grown GaAs grown on an AlGaAs/AlAs Bragg mirror is summarized in Fig. 12. For growth temperatures does to about 300 °C, we observe no significant degradation. However, at lower growth temperatures we obtain a larger saturation fluence, a much smaller modulation depth, and significantly higher nonsaturable losses—a high price to pay for a fast saturable absorber. However, at 300 °C we typically obtain a short trapping time of a few picoseconds. Therefore, we typically use growth temperatures of 300 °C or higher for picosecond pulse generation. The increase in nonsaturable losses is most likely a result of defect-to-band absorption, which excites electrons high into the conduction band and cannot be saturated at these lower pulse energies. The change in the saturation fluence is not as important, because the SESAM device design can adjust the saturation fluence.

We have demonstrated that annealing (Haiml et al., 1998b) as well as Be

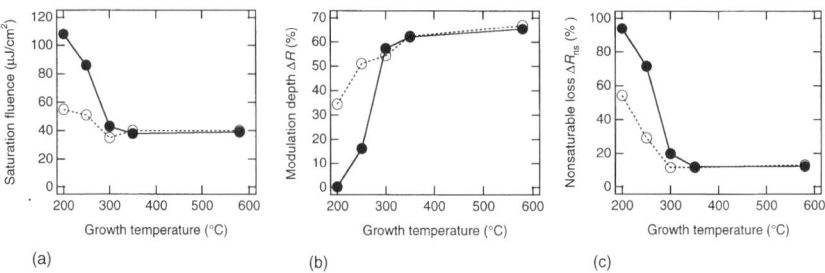

FIG. 12. (a) Saturation fluence, (b) modulation depth, and (c) nonsaturable loss of 0.5-μm-thick LT-grown GaAs layer grown on an AlGaAs/AlAs Bragg reflector as a function of MBE growth temperatures. Solid circle, as-grown; open circle, annealed. These parameters are determined from the measured nonlinear reflectivity (Eqs. 5 and 6).

doping (Haiml et al., 1998a) of LT GaAs decreases the nonsaturable losses and increases the nonlinearity by up to one order of magnitude to the value of GaAs grown at normal temperatures (Fig. 12). Importantly, the time response does not degrade on annealing or Be doping, but remains fast or gets even faster. In particular, Be doping can reduce the trapping time to only 100 fs without degrading the absorption modulation. Our data confirm that the high concentration of As antisites in as-grown undoped LT GaAs decreases the initial response time. However, a fast response is accompanied by only small absorption modulation and large nonsaturable losses. We attribute the nonsaturable losses to the transition between neutral As antisites and band states 0.7 eV above the bottom of the conduction band. This transition can be saturated only at very high fluences because of the large neutral antisite concentration and the large density of final states. Both annealing and Be doping considerably reduce the neutral antisite concentration by precipitate formation and ionization of neutral antisites, respectively. As a consequence, the neutral antisite–conduction band transition can at least be partially saturated, which decreases the nonsaturable losses and, in turn, increases the absorption modulation. The improvement of the nonlinearity resulting from annealing or Be doping is most pronounced for LT GaAs grown between 200 and 300 °C. From the picosecond or subpicosecond initial decay of the pump-probe response, we conclude that precipitates in annealed LT GaAs as well as charged antisites or antisite-Be complexes (Talwar et al., 1993) in doped material are highly efficient carrier traps.

III. SESAM Designs

1. ANTIRESONANT FABRY-PEROT SATURABLE ABSORBER (A-FPSA)

SESAMs offer a distinct range of operating parameters not available with other approaches. We can modify the absorber cross section of the saturable absorber if we integrate the absorber layer into a device structure. This allows us to modify the effective absorber cross section beyond its material value. We use various designs of SESAMs to achieve many of the desired properties. Figure 2 shows the different SESAM designs in historical order. The first intracavity SESAM device was the antiresonant Fabry-Perot saturable absorber (A-FPSA) (Keller et al., 1992c), initially used in a design regime with a rather high top reflector, which we now call more specifically the high-finesse A-FPSA. The A-FPSA is typically formed by the lower semiconductor Bragg mirror and a dielectric top mirror, with a saturable absorber and possibly transparent spacer layers between them. The thickness of the total absorber and spacer layers, the Fabry-Perot thickness d, is adjusted such that the A-FPSA is operated at antiresonance (Fig. 13). The Fabry-Perot thickness d (which can include nonabsorbing transparent material to reduce the total absorption) has to fulfill the condition for antiresonance:

$$\varphi_{\text{rt,a}} = \varphi_b + \varphi_t + 2k\bar{n}d = (2m+1)\pi \tag{30}$$

where φ_{rt} is the round trip phase inside the A-FPSA, $\varphi_{\text{rt,a}}$ is φ_{rt} at antiresonance (i.e., destructive interference), φ_b and φ_t are the phase shifts of the reflectivity at the bottom and top mirrors, $\bar{n}$ is the average refractive index of the Fabry-Perot thickness layer, $k = 2\pi/\lambda$ is the wavevector, λ is the wavelength in vacuum, and m is an integer number. At antiresonance, the interference of the partially reflected waves at the Fabry-Perot mirrors is destructive. Therefore, operation at antiresonance results in a device that is broadband and has minimal group velocity dispersion (Fig. 13). The bandwidth of the A-FPSA is limited by either the free spectral range of the Fabry-Perot or the bandwidth of the mirrors.

The top reflector of the A-FPSA provides an adjustable parameter that determines the intensity entering the semiconductor saturable absorber and therefore the effective saturation fluence $F_{\text{sat,A}}^{\text{eff}}$ of the device:

$$F_{\text{sat}}^{\text{eff}} = \frac{1}{\xi} F_{\text{sat,A}}$$

with

$$\xi = \frac{1 - R_t}{[1 + \sqrt{R_t R_b} \exp(-2\alpha d_A)]^2 - 4\sqrt{R_t R_b}\exp(-2\alpha d_A)\cos^2(\varphi_{\text{rt}}/2)} \tag{31}$$

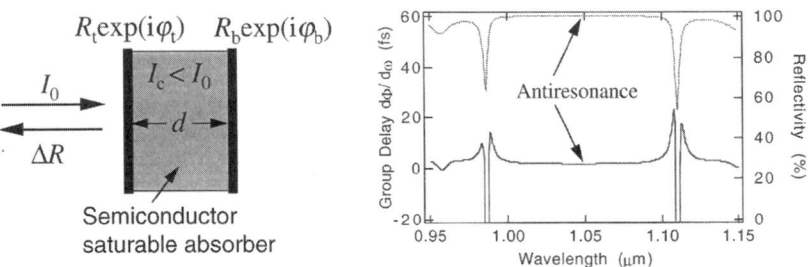

FIG. 13. Basic principle of the A-FPSA concept. With the top reflector we can control the intensity incident on the saturable absorber section. The thickness of this absorber section is adjusted for antiresonance. The typical reflectivity (dotted line) and group delay (solid line) are shown as functions of wavelength. At antiresonance, we have high, broadband reflection and minimal GDD.

(Brovelli et al., 1995c), where R_t is the intensity reflectivity of the top reflector, $F_{sat,A}$ is the saturation fluence of an AR-coated SESAM (i.e., $R_t = 0$), R_b is the intensity reflectivity of the bottom reflector, α is the amplitude absorption coefficient of the absorber layer, and d_A is the total thickness of all absorber layers. At antiresonance, $\varphi_{rt} = \varphi_{rt,a} = (2m + 1)\pi$ (Eq. 30), which results in $\cos^2(\varphi_{rt,a}/2) = 0$. A high top reflector results in a lower modulation depth for the same absorber thickness. However, with a thicker absorber layer, the modulation depth can be increased again. A low top reflector generally requires a thinner absorber layer to reduce the nonsaturable insertion loss of the device. The two design limits of the A-FPSA are the high-finesse A-FPSA (Fig. 2a) with a relatively high top reflector (i.e., $R_t \approx 96\%$) and the AR-coated SESAM (Fig. 2b) with no top reflection (i.e., $R_t \approx 0\%$) (Brovelli et al., 1995a). Using the incident laser mode area as an adjustable parameter, the incident pulse fluence F_p can be adapted to the saturation fluence $F_{sat,A}$ of both SESAMs for stable modelocking by choosing F_p a few times greater than $F_{sat,A}$ (see Section II) (Jung et al., 1995a).

a. High-Finesse A-FPSA

A specific design for a high-finesse A-FPSA is shown in Fig. 14(a) and (b) for a Ti:sapphire laser. For a center wavelength around 800 nm, we typically use an AlAs/AlGaAs Bragg mirror with an aluminum content large enough to introduce no significant absorption. A higher aluminum concentration in AlGaAs increases the bandgap energy and reduces the refractive index (Afromowitz, 1974). Thus, the AlAs/AlGaAs Bragg mirrors have less reflection bandwidth than the GaAs/AlAs Bragg mirrors because of the lower refractive index difference. However, we have demonstrated pulses as short as 19 fs from a Ti:sapphire laser (Jung et al., 1995a) with such a device (Fig. 15). In this case, the bandwidth of the modelocked pulse extends slightly beyond the bandwidth of the lower AlAs/AlGaAs mirror, because the much broader SiO_2/TiO_2 Bragg mirror on top reduces the bandwidth-limiting effects of the lower mirror. Reducing the top mirror reflectivity increases the minimum attainable pulse width as a result of the lower mirror bandwidth (Fig. 15).

The high-finesse A-FPSA in Fig. 14(a) consists of a 96% SiO_2/TiO_2 top reflector evaporated onto a 0.3-μm-thick low-temperature (375 °C) MBE-grown $Al_xGa_{1-x}As$ ($x = 0.09$) bulk saturable absorber layer on an $Al_{0.17}Ga_{0.83}As$/AlAs Bragg mirror. $Al_{0.09}Ga_{0.91}As$ has a bandgap at 809 nm and a refractive index of 3.65 at 800 nm (Afromowitz, 1974). At the wavelength of 800 nm, the lower Bragg reflector is formed by the accumulated reflection

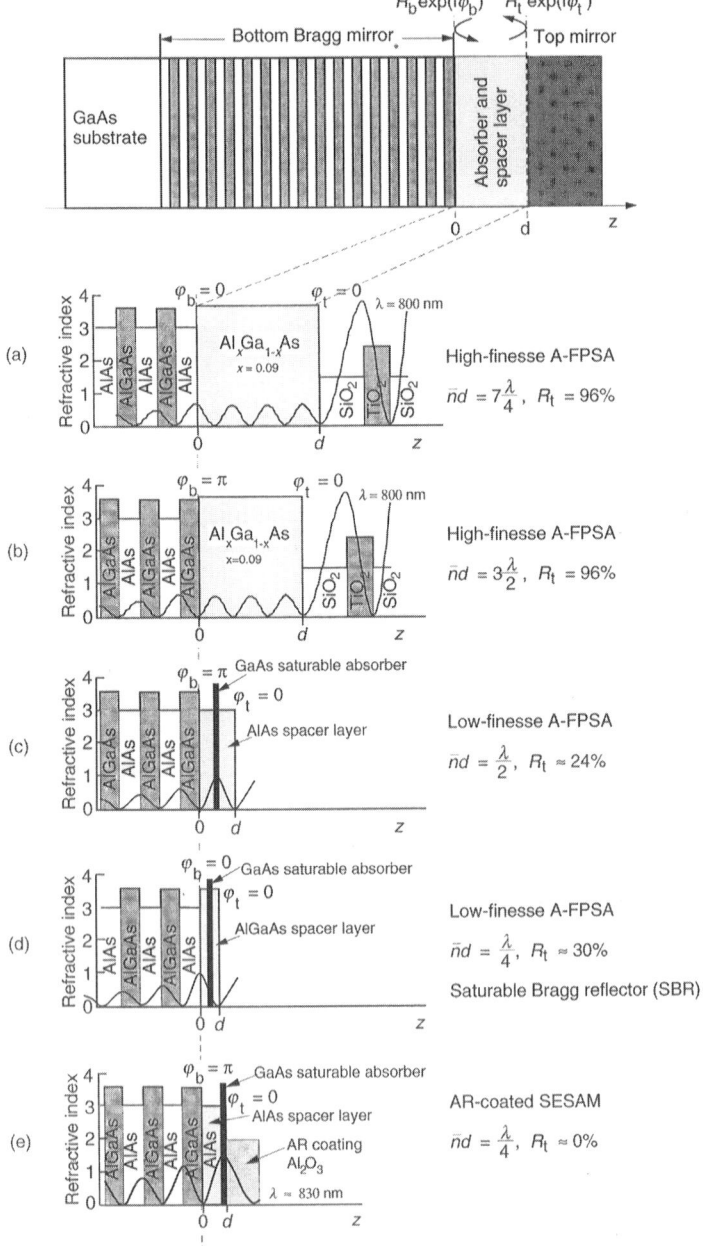

Fig. 14. Different SESAM designs with AlGaAs/AlAs Bragg mirrors for a Ti:sapphire laser.

at the $Al_{0.17}Ga_{0.83}As/AlAs$ interfaces with refractive indices of 3.56 and 3.01 and bandgap wavelengths of 757 and 422 nm, respectively. The Fabry-Perot thickness d is adjusted for antiresonance (see Eq. 30): the bottom AlAs/AlGaAs Bragg reflector produces an accumulated phase shift of $\varphi_b = 0$, and the top SiO_2/TiO_2 Bragg mirror produces a phase shift of $\varphi_t = 0$ (Brovelli et al., 1995b). The reflection from the first interface between the absorber layer and the first Bragg layer gives a phase shift of zero because of the lower refractive index of the Bragg mirror. All other interfaces in the Bragg mirrors add constructively to this first reflection and therefore maintain the phase shift of zero. For $m = 3$, the antiresonance condition is fulfilled, with an optical thickness of the absorber layer of $\bar{n}d = 7\lambda/4 = 3\lambda/2 + \lambda/4$ (see Eq. 30).

Figure 14(b) shows a similar design but with a lower Bragg reflector finishing with an $Al_{0.17}Ga_{0.83}As$ quarter-wave layer (Jung et al., 1995a). In this case, the accumulated phase shift from the lower Bragg mirror is π even though the first reflection is at zero phase shift because the refractive index of the saturable absorber layer is higher. However, this first partial reflection is negligible in comparison with the accumulated reflections from all other interfaces of the Bragg mirror. For simplicity, it would be easier to look at the A-FPSA design that has a Fabry-Perot thickness d consisting of a $3\lambda_n/2$-thick $Al_{0.09}Ga_{0.91}As$ absorber layer and a $\lambda_n/4$-thick transparent $Al_{0.17}Ga_{0.83}As$ layer (last quarter-wave laser from lower Bragg mirror). The partial reflection at this $Al_{0.09}Ga_{0.91}As/Al_{0.17}Ga_{0.83}As$ interface is negligible compared with the accumulated reflection from the bottom and top Bragg reflectors. Because of the thinner absorbing layer, this design has a smaller modulation depth and a lower insertion loss than the one in Fig. 14(a).

Why would we even consider finishing the lower Bragg reflector in Fig. 14(b) with a high-index $Al_{0.17}Ga_{0.83}As$ quarter-wave layer? The last layer does not add constructively to the Bragg reflection and therefore should be removed. The reason is based on our growth process. We typically grow our Bragg mirrors with MOCVD, which provides full wafers with very good uniformity. With MBE growth we would have to invest more time to maintain good thickness control because the growth calibration depends on the amount of material in the crucibles. However, only with MBE are we able to grow the saturable absorber at lower temperatures. This is particularly important for solid state lasers with much longer upper state lifetimes than that of Ti:sapphire. From a purely practical point of view, we typically use previously fabricated Bragg mirrors as a new substrate to grow the final SESAM structure. To minimize oxidation effects before regrowth, it is important to finish with a minimal aluminum content inside the last quarter-wave layer of the Bragg mirror.

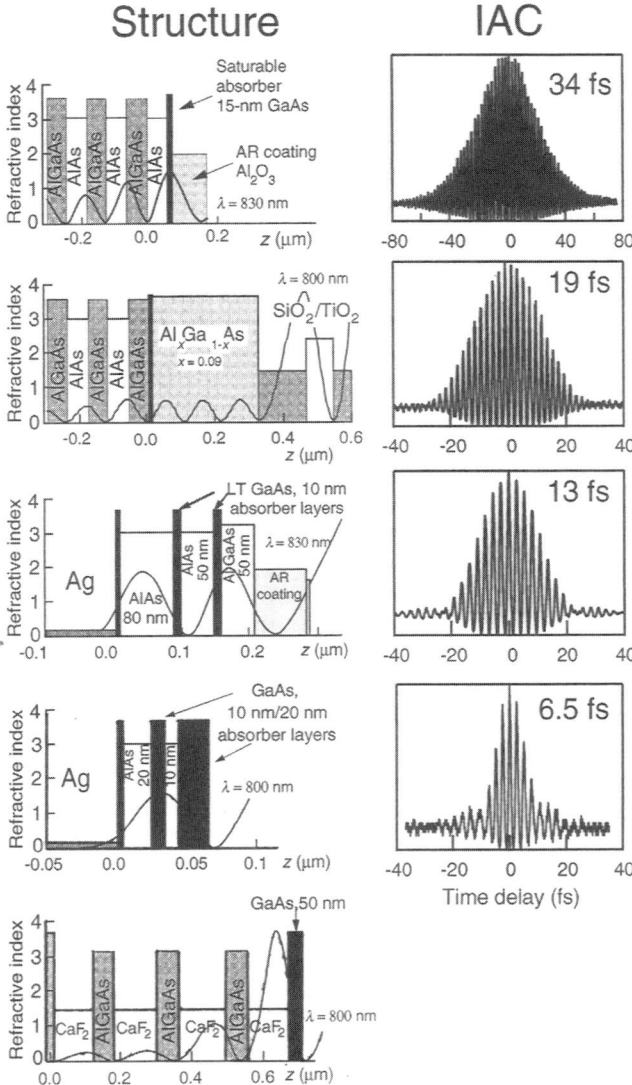

FIG. 15. Approaching limits in the SESAM designs for Ti:sapphire lasers. We show for each SESAM design the interferometric autocorrelation (IAC) of pulses obtained with a Ti:sapphire laser; the low-intensity SESAM reflectivity compared with pulse spectrum (dark area) and Ti:sapphire fluorescence (dashed curve); the measured impulse response using 13-fs excitation pulses, which determines the fast and slow recovery times of the SESAM; and the nonlinear reflectivity on a log scale, which determines saturation fluence $F_{sat,A}$, modulation depth ΔR, and nonsaturable loss ΔR_{ns} of the SESAM.

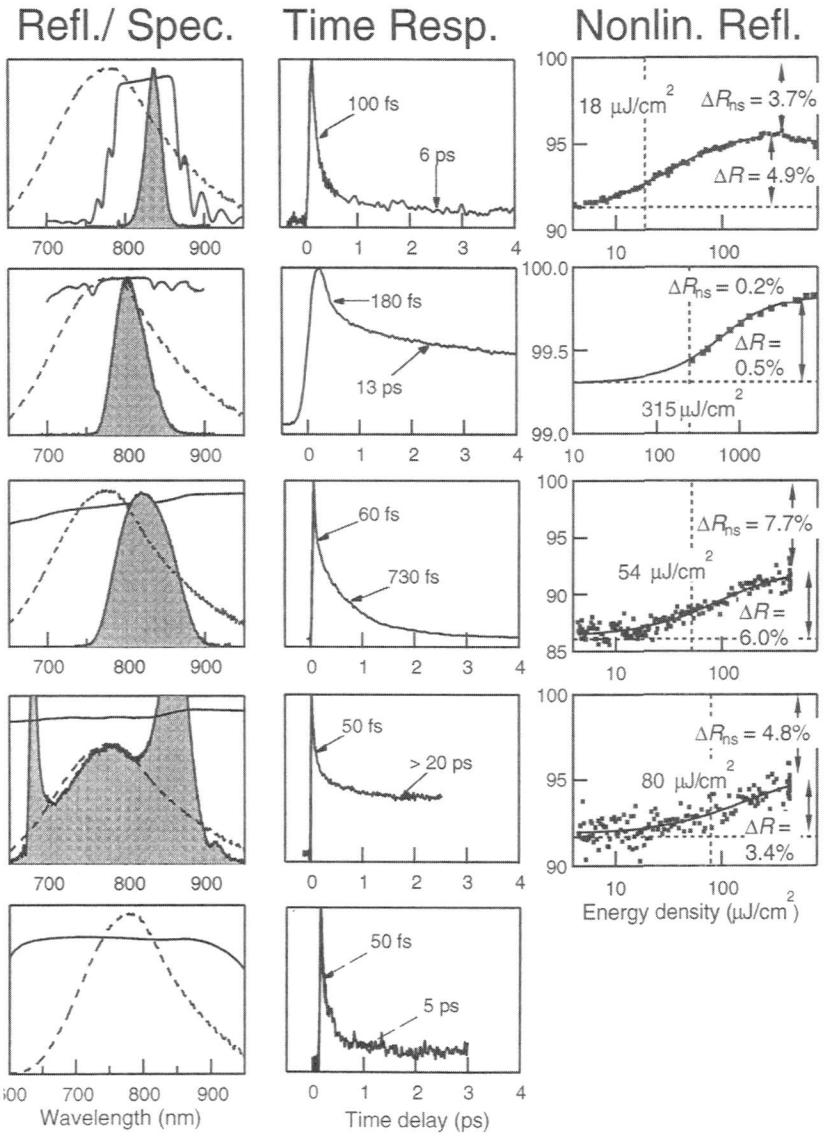

FIG. 15. (*Continued*).

b. Low-Finesse A-FPSA

A specific intermediate design is the low-finesse A-FPSA (Tsuda *et al.*, 1995b; Hönninger *et al.*, 1995; Jung *et al.*, 1995a), where the top reflector is formed by the approximately 30% Fresnel reflection of the semiconductor/air interface (see Fig. 2c). Reducing the top reflector typically requires a thinner saturable absorber and a higher bottom reflector to minimize nonsaturable insertion loss. Because the absorber layers are typically rather thin (i.e., less than 30 nm), the effective saturation fluence of the device can then be varied by changing the position of the buried absorber section within the Bragg reflector or simply within the last quarter-wave layer of the Bragg reflector, taking into account that an infinitely thin absorber layer at the node of a standing wave does not introduce any absorption. Different wavelengths have different positions for their nodes in the standing wave profile. This can be used to reduce the wavelength dependence of the absorber edge and obtain broadband performance (Kopf *et al.*, 1997b).

Figure 14(c) and (d) show two specific designs for a Ti:sapphire laser at a center wavelength of about 800 nm. The first design (Fig. 14c) uses the same lower $Al_{0.17}Ga_{0.83}As/AlAs$ Bragg mirror as in the high-finesse A-FPSA (Fig. 14b), a Fabry-Perot thickness of $\bar{n}d = \lambda/2$, and a thin GaAs absorber layer typically 15 to 30 nm thick embedded inside an AlAs spacer layer. The top reflector of the A-FPSA is formed by the simple Fresnel reflection of the AlAs/air interface, which is about 24%. The thickness of the Fabry-Perot is adjusted for antiresonance with $m = 1$ (Eq. 30), taking into account a π-phase shift from the bottom Bragg mirror and a 0-phase shift at the semiconductor/air interface. In Fig. 14(c) we chose to set the absorber layer at the maximum of the standing wave intensity pattern of the incident laser beam.

The second design (Fig. 14d) shows a low-finesse A-FPSA with a minimal optical Fabry-Perot thickness of a quarter wavelength (i.e., $\bar{n}d = \lambda/4$). The thickness of the Fabry-Perot is adjusted for antiresonance with $m = 0$ (Eq. 30), taking into account a 0-phase shift from the bottom Bragg mirror and a 0-phase shift at the semiconductor/air interface. This device has also been described as a saturable Bragg reflector (SBR) (Tsuda *et al.*, 1995b) because the 10-nm-thick GaAs absorber layer is embedded inside the last quarter-wave layer of an AlAs/AlGaAs Bragg reflector. The SBR is discussed in more detail in Section II.7.d.

c. Antireflection-Coated SESAM

The AR-coated SESAM device can be viewed as one design limit of the A-FPSA with an approximately 0% top reflector (Brovelli *et al.*, 1995a; Jung *et al.*, 1995a). An example of such a device design is shown in Fig. 14(e) for a Ti:sapphire laser. The thickness of the absorber layer has to be smaller than *d* to reduce the nonsaturable insertion losses of these intracavity saturable absorber devices. Such an AR-coated SESAM has started and stabilized a soliton modelocked Ti:sapphire laser, achieving pulses as short as 34 fs (Keller *et al.*, 1991a) with a modelocking buildup time of only about 3 μs (Fig. 15). Stable modelocking was achieved over the full stability regime of the laser cavity. The AR coating increases the modulation depth of this device compared with the low-finesse A-FPSA shown in Fig. 14(c) and acts as a passivation layer for the semiconductor surface, which may improve the long-term reliability of this SESAM device. The limitations of this device include the bandwidth of the lower AlAs/AlGaAs Bragg mirror and the potentially higher insertion loss compared with that of the high-finesse A-FPSA.

The specific design of the AR-coated SESAM shown in Fig. 14(e) consists of a simple AlAs/AlGaAs Bragg reflector with a single GaAs quantum well absorber in the last quarter-wavelength-thick AlAs layer. The additional AR coating is required to prevent Fabry-Perot resonance effects. The need for this additional AR coating may not be obvious but can be seen in low-intensity reflectivity measurements of this device with and without an AR coating (Fig. 16). The reflectivity dip in Fig. 16(b) at about 850 nm is a

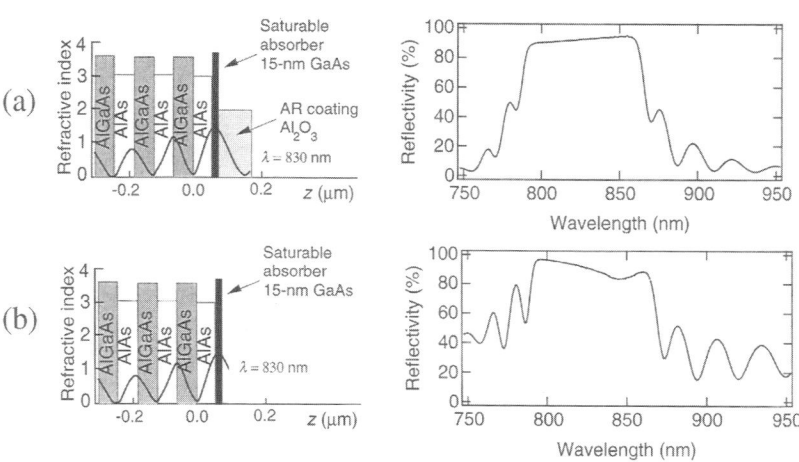

FIG. 16. Low-intensity reflectivity of (a) an AR-coated saturable Bragg reflector (SBR) and (b) an SBR, also referred to as a low-finesse A-FPSA. The resonance dip in the reflection of this SBR design can be explained with the design criteria of a low-finesse A-FPSA. In this case, an AR coating is required for femtosecond pulse generation.

result of the absorption in the GaAs quantum well and corresponds to a Fabry-Perot resonance. This strong wavelength-dependent reflectivity prevents short pulse generation and pushes the lasing wavelength of the Ti:sapphire laser to the high-reflectivity region of the device at shorter wavelengths at the edge of the Bragg mirror (Brovelli et al., 1995a). This resonance dip is formed by the lower part of the AlAs/AlGaAs Bragg reflector, the transparent AlAs layer with the GaAs absorber quantum well layer of total thickness $\bar{n}d = \lambda/4$, and the Fresnel reflection of the last semiconductor/air interface, when $\bar{n}$ is the average refractive index of the last AlAs/GaAs quarter-wave layer. This Fabry-Perot is at resonance because the round trip phase shift φ_{rt} is in accordance with Eq. (30): $\varphi_{rt} = \varphi_b + \varphi_t + 2k\bar{n}d = \pi + 0 + \pi = 2\pi$. A φ_{rt} of 2π allows for constructive interference and therefore fulfills the resonance condition of the Fabry-Perot. No AR coating would be required if the AlAs/AlGaAs Bragg reflector in Fig. 16(a) would end with the quarter-wavelength-thick AlGaAs layer that would then incorporate the GaAs quantum well. In this case, the phase shift of the lower part of the Bragg mirror is $\varphi_b = 0$ instead of π and therefore $\varphi_{rt} = \pi$, the condition for antiresonance (see Eq. 30). This design would then correspond to a specific low-finesse A-FPSA also referred to as the saturable Bragg reflector.

d. Saturable Bragg Reflector (SBR)

An earlier version of a nonlinear or saturable Bragg reflector design was introduced by Kim et al. (1989). In this case the nonlinear Bragg reflector operates on saturable absorption owing to band filling in all low-index quarter-wave layers with the narrower bandgap inside the Bragg reflector. This results in a distributed absorption over many layers. This device, however, would introduce too much loss inside a typical ion-doped solid state laser. Therefore, only one or a few thin absorbing sections inside the quarter-wave layers of the Bragg reflector are required.

A specific SBR with lower insertion loss (Fig. 14d) was introduced to cw modelock a Ti:sapphire and Cr:LiSAF laser (Tsuda et al., 1995b). This device is very similar to the previously introduced AR-coated SESAM device (Brovelli et al., 1995a) shown in Fig. 14(e). In this case, however, no AR coating is required on the AlAs/AlGaAs Bragg reflector as described above (Section III.7.c). With this device, pulses as short as 90 fs have been generated with a Ti:sapphire laser (Tsuda et al., 1995a), which are significantly longer than the 34-fs pulses obtained with the AR-coated SESAM device (Fig. 15). This is most likely attributable to the lower modulation depth of this specific SBR, which could be increased with a thicker absorber layer or more than one quantum well. It is important to realize that the Bragg reflector does not play a key role in the operation of this device. For example, the Bragg reflector can be replaced by a metal reflector (Fig. 15), as will be discussed in Section IV.

2. DISPERSIVE SESAMs

Recently, we combined both saturable absorption and dispersion compensation in a semiconductor Gires-Tournois-like structure called a dispersive saturable absorber mirror (D-SAM) (see Fig. 2d) (Kopf et al., 1996b). By replacing one end mirror of a diode-pumped Cr:LiSAF laser with this device, we achieved 160-fs pulses without further dispersion compensation or special cavity design. This is the first time that both saturable absorption and dispersion compensation have been combined within one integrated device. The D-SAM, in contrast to the A-FPSA, is operated close to the Fabry-Perot resonance, which tends to limit the available bandwidth of the device. In the future, chirped mirror designs that incorporate saturable absorber layers could also potentially provide both saturable absorption and negative dispersion, but with potentially more bandwidth.

3. ANTIRESONANT FABRY-PEROT p-i-n MODULATOR

We do not have to rely only on passive saturable absorption with semiconductors. Multiple quantum well (MQW) p-i-n modulators based on the quantum-confined Stark effect (Miller et al., 1984, 1985, 1896) are promising as active modulation devices for solid state lasers, sharing the advantages of passive SESAMs: they are compact, inexpensive, and fast, and can cover a wide wavelength range from the visible to the infrared. In addition, they require only a few volts of drive voltage or several hundred milliwatts of RF power. In general, however, semiconductor MQW modulators would normally introduce excessive insertion loss inside a solid state laser cavity and would also saturate at relatively low intensities (Fox et al., 1990; Sizer et al., 1994). We extended the antiresonant Fabry-Perot principle by integrating an active MQW modulator inside a Fabry-Perot structure, which we called an antiresonant Fabry-Perot Modulator (A-FPMod) (Brovelli et al., 1995d). We then actively modelocked a diode-pumped Nd:YLF laser.

One advantage of quantum well modulators compared with other modulators such as acousto-optic modulators or phase modulators is that they also can act as saturable absorbers leading to passive modelocking with much shorter pulses. Combining the effects of saturable absorption and absorption modulation within one single device, we have demonstrated the possibility of synchronizing passively modelocked pulses to an external RF signal (Brovelli et al., 1995d). At higher output powers, we were limited by the increased saturation of the active modulator.

4. GENERAL SESAM DESIGN

We can reduce the design problem of a SESAM to the analysis of multilayered interference filters with a certain mirror or filter response for both the amplitude and phase. The transfer matrix calculus is usually used to evaluate such general multilayer systems (Macleod, 1985). With a careful selection of the absorber positions (which also includes possibly different absorber materials) within the multilayer ensemble, an optimal performance for broad bandwidth, for example, can be found. However, this approach gives very little insight into how this SESAM design works. The previously discussed special cases of such SESAM design provide very simple design guidelines. For example, the antiresonant Fabry-Perot model provides very simple design guidelines, when the appropriate sets of layers or interfaces are collected together to form the discrete elements such as the bottom reflector, top reflector, and absorber/spacer layer (see Figs. 13 to 15). All the previously discussed designs are then very specific examples of multilayer filter designs and can always be used as starting points for further computer optimization algorithms if required.

IV. Passively Modelocked Solid State Lasers Using SESAMs

1. LASER DESIGN AND DIODE PUMPING

The discussion in Section II.4 shows that the large small-signal gain g_0 and low losses l reduce the tendency for self-Q-switching (Eqs. 23 to 26), reduce the modelocking buildup time (Eq. 20), and result in shorter pulses at steady state, where $g = l$ (see Eqs. 13 and 15). Small-signal gain can be maximized by minimizing pump volume while maintaining good spatial overlap of the pump laser and laser mode. This can be easily accomplished when a diffraction-limited pump laser is used, as, for example, an argon-ion-pumped Ti:sapphire laser. The lower limit of the pump volume is then set by diffraction, or possibly by pump-induced damage to the crystal. However, this is less simple when non-diffraction-limited pump lasers, such as diode laser arrays or bars, are used to pump solid state lasers.

The output of a laser diode array or broad-stripe diode differs significantly from many typical laser sources. In the so-called "fast" axis, perpendicular to the p-n-junction of the diode laser, the light diverges with large angle (typically 25 to 40°) from a narrow aperture of about 1 μm. In this direction the light is nearly diffraction-limited with $M_{\text{fast}}^2 \approx 1$. Note that the M^2 factor is the parameter that is used to characterize the beam quality. An M^2 factor of 1 means that the product of the beam waist diameter and the full-angle divergence is approximately equal to the wavelength (as is the case for an ideal Gaussian beam). This is not the case for non-diffraction-limited beams. A larger M^2 value indicates that the product of the beam waist and the divergence is larger — i.e., a given beam size diverges M^2 times faster than a diffraction-limited beam. Although the light in the fast axis is highly divergent, it can be efficiently collected with a "fast" high-numerical aperture lens and tightly focused owing to its diffraction-limited nature.

In the "slow" axis, parallel to the p-n junction of the diode laser, the beam typically has a divergence of about 10°. For single-stripe diodes, the emitting aperture is approximately 3 μm, resulting in a nearly diffraction-limited beam. For higher-power "arrays" of such apertures, the divergence is also about 10°, but the total aperture has increased to typically 50 μm to more than 200 μm, or, in the case of "arrays of arrays" (i.e., bars) to a width of approximately 1 cm. The diode laser light in the slow axis is therefore many times worse than diffraction-limited. High-brightness diode arrays with about 1 W of output power and stripe widths of approximately 100 μm typically have M_{slow}^2 values of about 10, whereas low-brightness bars with about 20 W of output power and stripe widths of approximately 1 cm have M_{slow}^2 values greater than 1000. The slow axis ultimately limits the spot size of the focused pump owing to the requirements of mode matching to the laser mode.

With such pump lasers, the lowest pump threshold can be achieved with the following optimized modematching (OMM) design guidelines applied to both fast and slow axes of the diode pump laser (Fan and Sanchez, 1990; Kopf et al., 1997a).

- Determine M^2 for the pump beam:

$$M^2 \equiv \frac{\theta}{\theta_G(2w_0 = D)} \tag{32}$$

where θ is the measured divergence angle of the pump source, D is the width of the pump source (i.e., the stripe width of a diode array or bar), and $\theta_G(2w_0 = D)$ is the theoretical divergence angle of an ideal Gaussian beam with a beam waist w_0 of $D/2$:

$$\theta_G(2w_0 = D) = \frac{\lambda}{\pi w_0} = \frac{2\lambda}{\pi D} \tag{33}$$

where λ is the wavelength of the pump source.
- Determine the "effective wavelength" λ_{eff}:

$$\lambda_{\text{eff}} \equiv M^2 \cdot \lambda_n \tag{34}$$

where λ_n is the wavelength of the pump source inside the laser medium. Standard ABCD matrix calculations can then be used to predict the propagation of a Gaussian beam with such an "effective wavelength." Physically, this means that propagation of a non-diffraction-limited beam is similar to propagation of an ideal diffraction-limited Gaussian beam, but with the new, longer "effective wavelength." A beam with a larger M^2 has a larger "effective wavelength and therefore a smaller confocal parameter for a given beam waist.
- Set confocal parameter $\approx$ absorption length:

$$b \approx L_a \tag{35}$$

- Calculate optimum beam waist radius $w_{0,\text{opt}}$ using the Rayleigh range formula for an ideal Gaussian beam with the "effective wavelength" given in Eq. (34) and a confocal parameter given in Eq. (35):

$$w_{0,\text{opt}} = \sqrt{\frac{\lambda_{\text{eff}} b}{2\pi}} = \sqrt{\frac{M^2 \lambda_n L_a}{2\pi}} \tag{36}$$

Equation (36) determines the smallest pump beam waist for which a good mode overlap over the absorption length of the pump and the cavity mode can be obtained. This is the minimum pump spot size in the gain medium and therefore determines the lowest pump threshold. From Eq. (36) it becomes clear that for a small spot size the absorption length L_a in the gain medium should be as short as possible. The absorption length, however, limits the maximum pump power at which thermal effects will start to degrade the laser's performance. This will be more severe for "thermally challenged" lasers that exhibit low thermal heat conductivity, such as ion-doped glasses, or upper state lifetime quenching, such as Cr:LiSAF — gain materials interesting for diode-pumped femtosecond lasers. Low thermal conductivity results in large thermal lenses and distortions that limit the maximum pump power. Upper state lifetime quenching, as observed in Cr:LiSAF, has the following results. As the temperature in the laser medium increases, the upper state lifetime of the laser drops, and the pump threshold increases. Beyond a critical temperature, the laser actually switches off. If the absorption length is too short for these materials, this critical temperature occurs at relatively low pump powers. There is an optimum doping level for best modematching to the available pump diodes and for minimizing pump-induced upper state lifetime quenching.

In standard diode pumping, we use high-brightness diode arrays (i.e., brightness as high as possible) and apply OMM (Eqs. 32 to 36) only in the slow axis of the diodes and weaker focusing in the fast axis. This results in an approximately circular pump beam that becomes slightly elliptical when the laser crystal is pumped at a Brewster angle. The standard diode pumping is explained in more detail by the example of a diode-pumped Cr:LiSAF laser (Kopf *et al.*, 1994b) (Fig. 17a). We used two high-brightness pump diodes emitting 400 to 500 mW from a 100-μm stripe width at a wavelength of about 670 nm. The pump beam was collected in its fast diverging direction with a round cylindrical fiber lens. The astigmatism introduced by the fiber lens was negligible, allowing the first round achromatic lens ($f = 200$ mm) to collimate both directions simultaneously. Then the beam was refocused through a cavity mirror by the second lens ($f = 120$ mm). Using a beam scan with a moving slit aperture, the minimum spot size at the crystal position in the slow axis was measured at 80 μm in diameter with a far-field full angle divergence of 8° (about 13 times diffraction-limited) and in the fast axis at 60 μm with a divergence of 2°. This provided modematching to the 100 μm diameter $\times$ 140 μm diameter of the laser mode over the approximately 1-mm absorption length of the Cr:LiSAF crystal. With this standard pumping approach, the average output power was limited by the mentioned thermal problems to 230 mW cw and 125 mW modelocked with 60-fs pulses (Kopf *et al.*, 1997b) (Fig. 18f). Standard diode pumping has also been successfully used with most other solid state lasers such as Nd:YAG. Such lasers are not "thermally challenged," and much higher average output power has been achieved with this approach.

Significantly more output power (Fig. 18e) can be obtained with a diode-pumped Cr:LiSAF laser for which OMM (Eqs. 32 to 36) is applied in both axes in combination with a long absorption length. Kopf *et al.* (1995a) applied OMM in both axes combined with efficient one-dimensional cooling (Fig. 19). Optimized modematching in both axes results in a highly elliptical laser mode in the crystal, because the pump beam can be focused to a much smaller beam radius in the diffraction-limited fast axis (Eq. 36) than in the slow axis. Additionally, we can extract the heat very efficiently using a thin

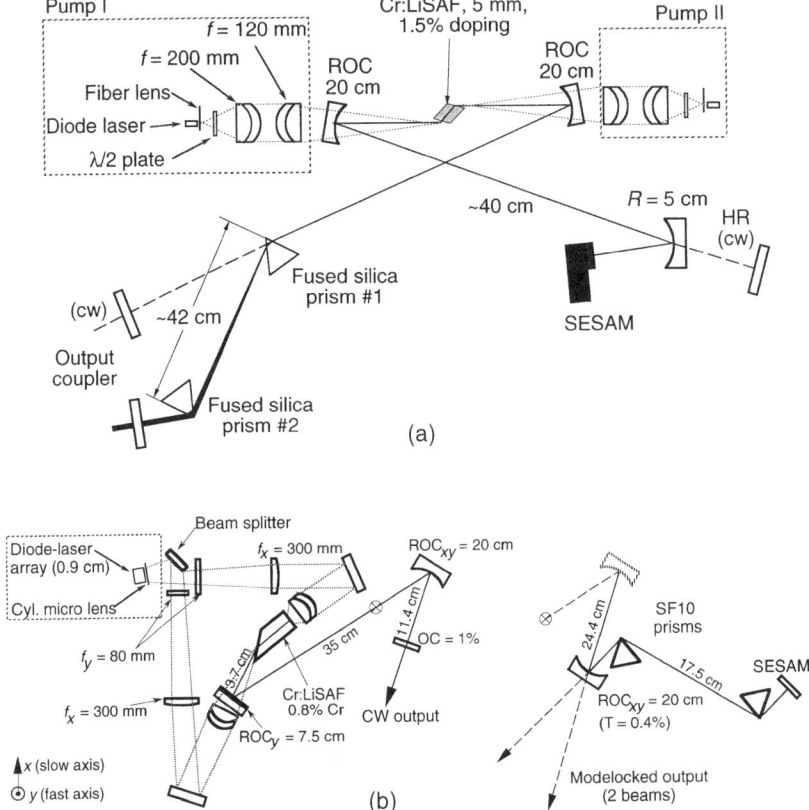

FIG. 17. Typical experimental setup of a diode-pumped solid state laser: (a) standard diode pumping in which optimized modematching (OMM) is applied only in the slow axis and (b) OMM applied in both the fast and slow axes combined with efficient one-dimensional heat removal. The fast axis (y-axis) lies in the saggital plane, and the slow axis (x-axis) lies in the tangential plane of the laser resonator.

crystal with a height of about 1 mm and obtain approximately one-dimensional heat flow (Fig. 19a). In a one-dimensional heat flow we can increase the total pump power by increasing the sheet width Δx to maintain constant pump intensity without increasing the temperature inside the laser medium (Fig. 19b). Using a 15-W, 0.9-cm-wide diode pump array with $M^2_{\text{slow}} = 1200$ and $M^2_{\text{fast}} = 1$ (Skidmore et al., 1995), the average output power of such a diode-pumped Cr:LiSAF laser (Fig. 17b) was 1.4 W cw and 500 mW modelocked with 110-fs pulses, and 340 mW modelocked with 50-fs pulses (Kopf

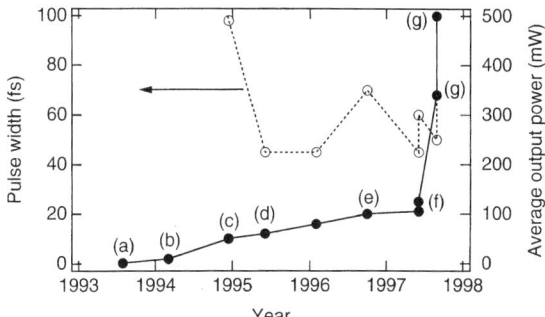

FIG. 18. Historical evolution of average output power of femtosecond diode-pumped Cr:LiSAF lasers using intracavity SESAMs. With KLM, the average output power was limited to less than about 50 mW. (a) First passive modelocking: RPM Cr:LiSAF, less than 1 mW, 300 ps (French *et al.*, 1993). (b) First intracavity SESAM, which was optimized for diode lasers and therefore too lossy: approximately 10 mW, 220 fs (Mellish *et al.*, 1994). (c) First sub-100-fs pulses with a SESAM optimized for solid state lasers: 50 mW, 98 fs (Kopf *et al.*, 1997a). (d) First sub-50-fs pulses with a SESAM (Kopf *et al.*, 1995c). (e) Diffraction-limited pump using an MOPA: 100 mW, 70 fs (Tsuda *et al.*, 1996a). (f) Best results to date with standard diode pumping using high-brightness, 100-µm-stripe-width pump diode arrays: 125 mW, 60 fs and 105 mW, 45 fs (Kopf *et al.*, 1997b). (g) Highest average output power using high-power 15-W low-brightness pump diodes with 0.9-cm stripe width, OMM in both axes, and efficient one-dimensional cooling: 500 mW, 110 fs and 340 mW, 50 fs (Kopf *et al.*, 1997c).

et al., 1997c) (Fig. 18e). In combination with a relatively long absorption length, we pumped a thin sheet-like volume with a width of $\Delta x \approx 1$ mm, a length of $L_a \approx 4$ mm, and a thickness of about 80 µm in the laser crystal. Note that there are two advantages of splitting the pump beam into two, as shown in Fig. 17(b): the M^2 of the pump is reduced by a factor of 2, and the pump power is distributed over about twice the volume, reducing the temperature rise by a factor of about 2. Recently, this approach has been applied to a diode-pumped Nd:glass laser, resulting in an average output power of more than 2 W cw and more than 1 W modelocked with pulses as short as 175 fs (Aus der Au *et al.*, 1998). This output power is approximately 10 times higher than for previously reported modelocked diode-pumped Nd:glass lasers.

To date, KLM diode-pumped Cr:LiSAF lasers have been demonstrated with only limited average output power. Such lasers generated pulses as short as 34 fs with an average power of 42 mW (Dymott and Ferguson, 1995). Shorter pulses of 18 fs were achieved at the expense of output power, because leakage through the high reflectors of the laser resonator was used as the output coupler to maintain the necessary intracavity intensity (Dymott and Ferguson, 1997). It is probably very difficult to achieve stable

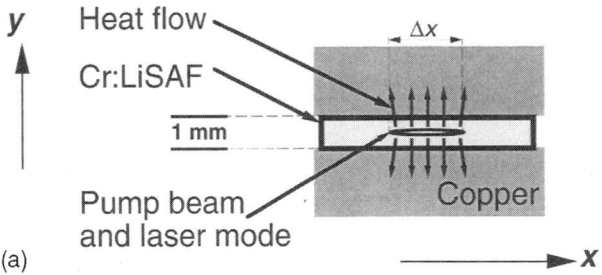

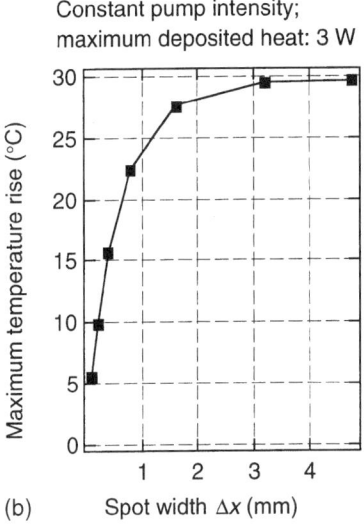

FIG. 19. (a) Schematic cross-sectional view of the laser crystal pumped by a strongly asymmetric diode laser beam. The crystal geometry causes approximately a one-dimensional heat flow to the copper heat sinks. (b) Calculated maximum temperature rise with increased pump spot width Δx at constant pump intensity—i.e., increased total pump power. An approximately one-dimensional heat flow results in constant peak temperatures in the crystal even when the pump power is increased while maintaining constant pump intensity with increased pump spot width Δx.

KLM in a laser cavity with a strongly asymmetric beam, as shown in Fig. 17(b). SESAMs and soliton modelocking allow us to optimize laser cavity design for good diode-pumped performance with femtosecond pulses.

4 SEMICONDUCTOR NONLINEARITIES FOR SOLID-STATE LASER

2. APPROACHING LIMITS: TI:SAPPHIRE LASERS

Until the end of the 1980s, ultrashort pulse generation was dominated by dye lasers (Fig. 20), where modelocking was based on a balanced saturation of both gain and loss, opening a steady state net-gain window as short as the pulse duration (see Fig. 4a). Pulses as short as 27 fs with an average power of about 10 mW were generated with a CPM (colliding pulse modelocked) Rhodamine 6G/DODCI dye ring laser (Valdmanis and Fork, 1986). Shorter pulse durations—as short as 6 fs—were achieved only through additional amplification and fiber-grating pulse compression at much lower repetition rates (Fork et al., 1987). The situation changed with the development and commercialization of the Ti:sapphire laser (Moulton, 1986), which has a gain bandwidth large enough to support ultrashort pulse generation. However, existing dye laser modelocking techniques were inadequate because of the much longer upper state lifetime and the smaller gain cross section of this laser, which results in negligible pulse-to-pulse dynamic gain saturation. The breakthrough came with the discovery of KLM (Spence et al., 1991) (Fig. 20b–4). Rapid progress toward shorter pulses was then achieved by reducing higher-order dispersion using thinner Ti:sapphire rystals as little as about 2 mm thick and fused quartz prisms for dispersion compensation (Fig. 20b–6 to 9). The first sub-10-fs pulses were obtained with silver cavity mirrors and minimum dispersion by operating the Ti:sapphire laser at a center wavelength of 850 nm (Zhou et al., 1994) (Fig. 20b–11). A more sophisticated approach to compensation for higher-order dispersion was developed with chirped mirrors (Szipöcs et al., 1994, 1997), and KLM pulses as short as 7.5 fs with an average output power of 25 mW have been generated (Xu et al., 1996) (Fig. 20b–13). The average output power of sub-10-fs pulses has been recently increased to about 700 mW at a pulse repetition rate of 77 MHz (Xu et al., 1997). The pulse energy is about 9 nJ with a peak power of more than 1 MW. Self-starting KLM in this pulsewidth regime has not been reported, and shaking mirrors are typically used to start modelocking.

Broadband SESAMs, however, can provide reliable self-starting modelocking even in the sub-10-fs regime. Currently, the shortest pulses generated directly from lasers have a pulse duration of 6.5 fs at 200-mW average output power (Jung et al., 1997c) (Fig. 20b–14). KLM was started with a broadband SESAM device that is a half-wavelength-thick low-finesse A-FPSA design with a silver mirror and a single 15-nm GaAs quantum well saturable absorber embedded in AlAs and AlGaAs spacer layers (Jung

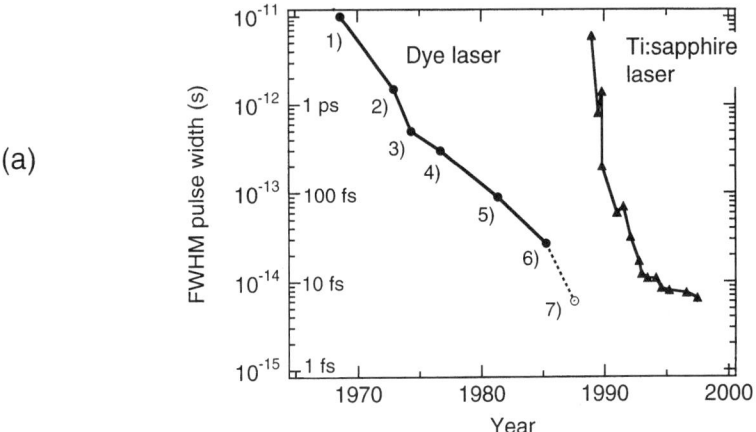

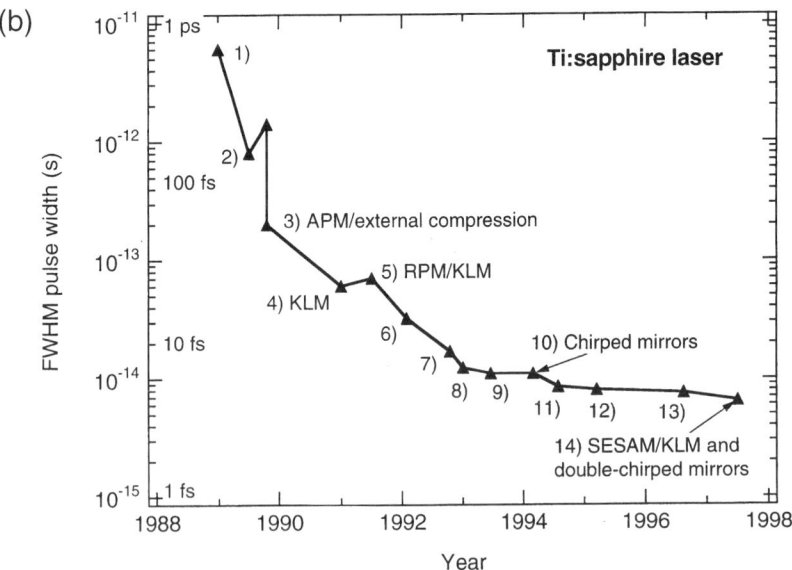

FIG. 20. History in ultrashort pulse generation. (a) Dye lasers: (1) First modelocked dye laser (Schmidt and Schafer, 1968) with 10-ps pulses (Soffer and Linn, 1968). (2) First cw modelocked dye, 1.5 ps (Ippen et al., 1972). (3) First sub-ps pulses, 0.5 ps (Shank and Ippen, 1974). (4) CPM concept in a linear cavity, 0.3 ps (Ruddock and Bradley, 1976). (5) CPM ring laser, 90 fs (Fork et al., 1981). (6) Shortest pulse directly out of a dye laser, 27 fs (Valdmanis et al., 1985). (7) Amplification and external fiber-grating pulse compressor, 6 fs (Fork et al., 1987). (b) Ti:sapphire lasers (1) Active modelocked with acousto-optic modelocker (AOM), 6 ps (Kafka et al., 1989). (2) AOM and APM, 800 fs (French et al., 1989). (3) APM, 1.4 ps with external pulse compression 200 fs (Goodberlet et al., 1989). (4) First KLM, 60 fs (Spence et al.,

et al., 1997b; Fluck *et al.*, 1996a). The modulation depth of 0.9% resulted in a modelocking buildup time of around 1 ms, which was recently improved to only about 60 µs with a larger SESAM modulation depth of approximately 3.4% (Sutter *et al.*, 1998a). Initially, pulsewidths generated in Ti:sapphire lasers with SESAMs were limited by the reflection bandwidth of the lower AlGaAs/AlAs Bragg mirror (see Fig. 15). Using an AR-coated SESAM with a single 15-nm GaAs quantum well absorber, we obtained a larger modulation depth of $\Delta R = 4.9\%$ (see Fig. 15, first row), which supported soliton modelocking over the full cavity stability regime with a modelocking buildup time of about 3 µs and a pulse duration of 34 fs (Brovelli *et al.*, 1995a; Jung *et al.*, 1997b). The AlGaAs/AlAs Bragg mirror bandwidth limitation of the AR-coated SESAM can be reduced with a high-finesse A-FPSA in which a broader dielectric mirror is used as the top reflector (Fig. 15, second row). However, there is a trade-off in that a broader bandwidth is obtained at the expense of a smaller modulation depth. The modulation depth of 0.5% supports only about 40-fs pulses with soliton modelocking, but 19-fs pulses are obtained with the assistance of KLM (Jung *et al.*, 1995, 1997b).

We recently improved the modelocking buildup time for 6.5-fs pulse generation with a larger modulation depth of about 3.4% of the broadband SESAM device (see Fig. 15, fourth row) (Sutter *et al.*, 1998a). The modelocking buildup time was only 60 µs. However, KLM was still used to support this ultrashort pulsewidth. With an even larger 6% modulation depth of an AR-coated broadband SESAM (see Fig. 15, third row), we obtained soliton modelocking over the full cavity stability regime with pulses as short as 13 fs. The modelocking buildup time was about 200 µs. A more detailed discussion is presented in Jung *et al.* (1997b), which also demonstrates that the simple equations discussed earlier give a good estimate for the modelocking buildup time and the self-starting threshold.

Another key point in this success is the improved dispersion compensation of the laser cavity using double-chirped mirrors (Kärtner *et al.*, 1997)

1991). (5) RPM-assisted KLM, 70 fs (Keller *et al.*, 1991a). (6) KLM, LaFN28 prisms, 2-cm-thick Ti:sapphire crystal, 32 fs (Huang *et al.*, 1992b) and KLM, F2 prisms, 8-mm-thick Ti:sapphire, 33 fs (Krausz *et al.*, 1992b). (7) KLM, LaKL21 prisms, 9-mm-thick Ti-sapphire, 17 fs (Huang *et al.*, 1992a). (8) KLM, fused quartz prisms, 4-mm-thick Ti:sapphire, 12.3 fs (Curley *et al.*, 1993). (9) KLM, fused quartz prisms, 4.5-mm-thick Ti:sapphire, 11 fs (Asaki *et al.*, 1993). (10) KLM, chirped mirrors (no prisms), 1.8 mm-thick,Ti:sapphire, 11 fs (Stingl *et al.*, 1994). (11) First sub-10-fs pulses: KLM, fused quartz prisms, silver mirror, 2-mm-thick Ti:sapphire, 8.5 fs (Zhou *et al.*, 1994). (12) KLM, chirped mirrors, 2-mm-thick Ti:sapphire, 8 fs (Stingl *et al.*, 1995). (13) KLM, chirped mirrors in ring resonator, 7.5 fs (Xu *et al.*, 1996). (14) SESAM-assisted KLM, self-starting, fused quartz prisms and double chirped mirrors, 2.3-mm-thick Ti:sapphire, 6.5 fs (Jung *et al.*, 1997c).

in combination with a prism pair. Prism pairs are well established for intracavity dispersion compensation (Fork et al., 1984) and offer two advantages: the pulsewidth can be varied by simply moving one of the prisms, and the laser can be tuned in wavelength by simply moving a knife edge at a position where the beam is spectrally broadened. Both properties are often desired—in spectroscopic applications, for example. However, the prism pair suffers from higher-order dispersion, which is the main limitation in pulse shortening. Therefore, the double-chirped mirrors were designed to show the inverse higher-order group delay dispersion (GDD) of the prism pair plus laser crystal to eliminate the higher-order dispersion and obtain a slightly negative but constant GDD required for an ideal soliton pulse (Fig. 21).

Chirped mirrors (Szipöcs et al., 1994) have been established in different setups for the generation of ultrashort laser pulses. According to Szipöcs et al. (1994), chirping means that the Bragg wavelength is gradually increased along the mirror, producing a negative GDD. However, no analytical explanation of the unwanted oscillations typically observed in the group delay and GDD of such a simple-chirped mirror was given. These oscillations were most minimized purely by computer optimization. Recently, we developed a theory of chirped mirrors based on an exact coupled-mode

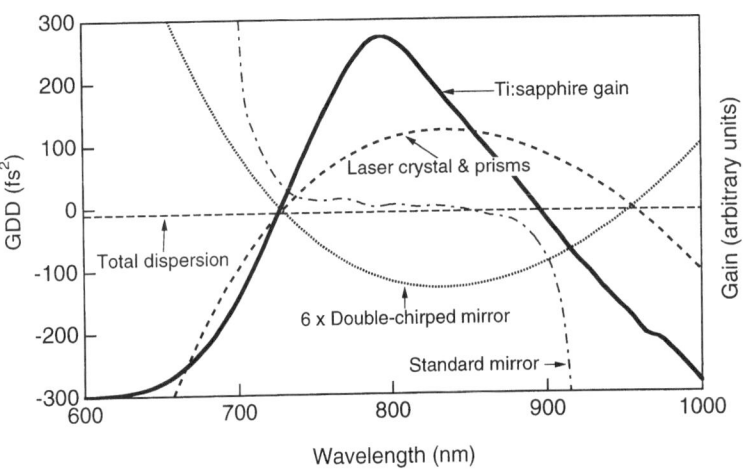

FIG. 21. Higher-order dispersion compensation requirements in sub-10-fs Ti-sapphire lasers: Ti:sapphire gain spectrum and group delay dispersion (GDD) of fused quartz prism and 2.3-mm-thick Ti:sapphire crystal, GDD of six reflections from standard high-reflecting Bragg mirrors (Spectra Physics), and GDD of six reflections from double-chirped mirrors required for stable soliton pulses.

analysis (Matuschek *et al.*, 1997). Following our theory, one has to use the double-chirp technique in combination with a broadband AR coating in order to avoid the oscillations in the GDD (Kärtner *et al.*, 1997; Matuschek *et al.*, 1998) (Fig. 22). Using our simple but very accurate analytical expressions for phase, group delay, and GDD, we designed double-chirped mirrors (DCMs) with a smooth and custom tailored GDD suitable for generating 6.5-fs pulses directly from a Ti:sapphire laser (Jung *et al.*, 1997c). With similar DCMs, we demonstrated pulses of about 30-fs duration tunable over a bandwidth of 300 nm without changing the cavity mirrors (Sutter *et al.*, 1998b). According to Fig. 22, a double-chirped mirror is a multilayer interference coating that can be considered as a composition of four sections, each with a different task. The layer materials are SiO_2 and TiO_2. The first section is the AR coating, typically composed of 10 to 14 layers. It is necessary because the theory is derived assuming ideal matching to air. The other sections represent the actual DCM structure, as derived from theory. The double-chirped section is responsible for the elimination of the oscillations in the GDD. Double chirping means that, in addition to the local Bragg wavelength, the local coupling of the incident wave to the reflected wave is independently chirped as well. The local coupling is adjusted by slowly increasing the high-index layer thickness in every pair so that the total optical thickness remains approximately $\lambda_B/2$. This corresponds to an adiabatic matching of the impedance. In the simple-chirp section, only the Bragg wavelength is chirped, whereas in the quarter-wave section the Bragg wavelength is kept constant. This method produces a high reflectivity for the long wavelengths. The AR coating, together with the rest of the mirror, is used as a starting design for a numerical optimization program. Since we start from a theoretical design that is close to the desired design goal, local optimization using a standard gradient algorithm is sufficient.

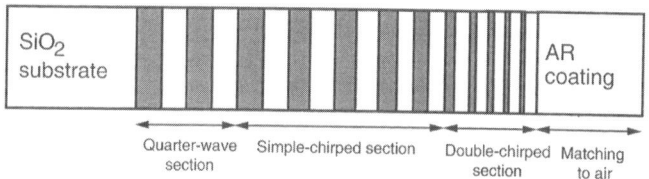

FIG. 22. Schematic design of a double-chirped mirror, which can be considered to be composed of four different sections, each with a different task.

V. Passively Q-Switched Solid State Lasers Using SESAMs

1. LASER DESIGN

Motivated by the need for compact, pulsed solid state lasers, many research efforts have been directed toward development of monolithic laser resonators. Single-frequency Q-switched performance has been demonstrated with a monolithic ring laser (MISER or NPRO) (Kane and Byer, 1985). Another approach—the microchip laser—uses a thin disk of laser material that forms a linear flat-flat resonator (Zayhowski and Mooradian, 1989). The stability of the resonator is obtained by thermal lensing and gain guiding (Longhi, 1994; Sanchez and Chardon, 1996). Single-frequency performance is achieved by making the resonator short enough so that the free spectral range of the longitudinal modes exceeds the gain bandwidth. This can be easily achieved with approximately 200-μm-thick Nd:YVO$_4$ laser material, but may be problematic for broadband laser materials such as Er:Yb:glass, which is particularly interesting for "eye-safe" ranging applications.

2. APPROACHING LIMITS: MICROCHIP LASERS

In Q-switched operation, microchip lasers can provide high peak powers while still maintaining a diffraction-limited output beam and in many cases single-axial-mode operation as well. The extremely short cavity, typically less than 1 mm, allows for Q-switched pulsewidths well below 1 ns. Previously, pulsewidths of 115 ps were demonstrated (Zayhowski and Dill, 1995a) with an actively Q-switched microchip laser. Passive Q-switching of a microchip laser with an Nd:YAG laser bonded to a thin piece of bulk Cr^{4+}:YAG saturable absorber resulted in pulses as short as 218 ps (Zayhowski and Dill, 1994; Zayhowski et al., 1995). By co-doping an Nd:YAG microchip crystal with Cr^{4+}:YAG, single-frequency 290-ps Q-switched pulses were achieved (Wang et al., 1995). An alternative method of passively Q-switching a microchip laser uses an A-FPSA. Using this method, we demonstrated single-frequency Q-switched pulses as short as 56 ps (Braun et al., 1997) and more recently 37 ps (Spühler et al., 1998) with a diode-pumped Nd:YVO$_4$ microchip laser (Fig. 23). These are the shortest Q-switched pulses ever obtained from a solid state laser, and have entered a pulsewidth regime previously limited exclusively to modelocked lasers.

Previously, an InGaAsP film grown on an InP substrate was used as an intracavity saturable absorber in a flashlamp pumped Nd:YAG laser, generating 20-ns pulses with 1.65-mJ pulse energy (Tsou et al., 1993). In

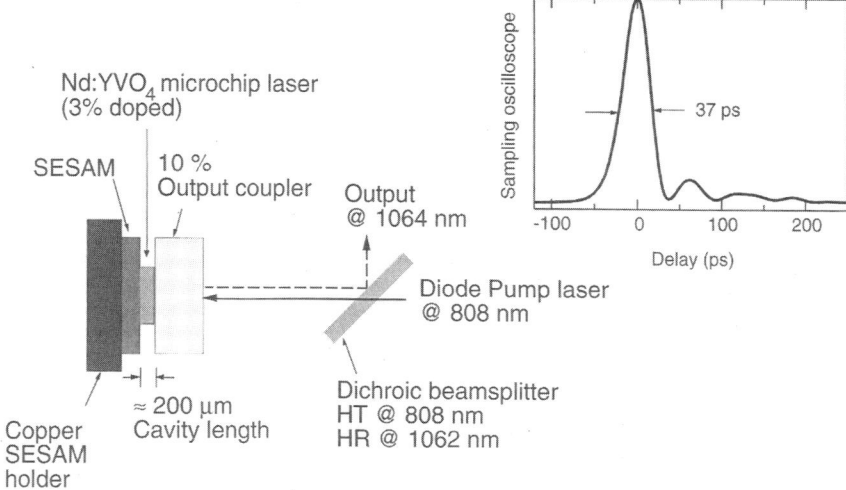

FIG. 23. Passively Q-switched Nd:YVO$_4$ microchip laser producing pulses as short as 37 ps with 53-nJ pulse energy, 160-kHz pulse repetition rate, and an average power of 8.5 mW. The SESAM design is based on an A-FPSA with 35 InGaAs/GaAs MQW saturable absorber and a 50% top reflector resulting in a modulation depth of 13%.

addition, a monolithic Nd:YAG ring laser (i.e., MISER) has been passively Q-switched with an evanescent-wave-coupled InGaAs/GaAs MQW saturable absorber mirror, generating single-frequency pulses of about 100-ns duration with a pulse energy of 0.7 µJ (Braun and Keller, 1995). Recently, two-photon and free carrier absorption in a thin, single-crystal GaAs wafer has been used for passive Q-switching of an Nd:YAG laser at 1.06 µm. Pulses as short as 7 ns with 20-µJ pulse energy have been generated. Microchip lasers have the advantage that the cavity length is very short, potentially allowing for much shorter Q-switched pulses. However, the mode area in the gain and the absorber are the same in a microchip laser. Therefore we have to adjust the modulation depth, saturation fluence, and intensity of the saturable absorber to the specific laser parameters.

The use of a SESAM for passive Q-switching of microchip lasers has several advantages. First, there is no significant increase in the cavity length, owing to the short optical penetration depth (typically less than 1 µm) (Brovelli and Keller, 1995b). This helps to minimize cavity length and therefore pulsewidth, which is directly proportional to the cavity round trip time (see Eq. 27). Second, the bandgap of the absorber layer can be adjusted for different lasing wavelengths. Finally, there is enough design freedom to allow independent adjustment of the saturation intensity $I_{sat,A}$ and the maximum modulation depth ΔR for a fixed mode size. These are the absorber parameters that determine pulsewidth (Eq. 27), pulse energy (Eq. 28), and repetition rate (Eq. 29), assuming a fully saturated absorber. Thus, the pulsewidth as well as the repetition rate can be designed in a deterministic way over several orders of magnitude (Braun et al., 1997).

At 1 μm, we used InGaAs saturable absorbers grown on an AlAs/GaAs Bragg mirror. However, for saturable absorption the indium concentration has to be increased to about 40% at 1.3 μm and to 50% at about 1.5 μm, which results in a significant lattice mismatch with the GaAs substrate. This lattice mismatch reduces the surface quality and results in higher insertion losses. Furthermore, the thickness of the absorber is limited because of strain effects. Low-temperature growth would partially relieve the lattice mismatch, but the absorber lifetime would drop to a few picoseconds. In contrast to fast saturable absorber modelocking, where a short recovery time of the absorber is needed, passive Q-switching requires a small saturation intensity—i.e., a long absorber lifetime. To overcome this problem, we chose lattice-matched InGaAsP quarternary structures grown on InP substrates by MOCVD at normal growth temperatures. The bandgap can be adjusted by the composition of the quaternary. But the difference in the refractive indices of the two materials, InP and InGaAsP, is smaller than for AlAs and GaAs. Therefore, to achieve a reflectivity of about 99% at 1.3 μm, 40 pairs of 104-nm InP and 96-nm $In_{0.73}Ga_{0.27}As_{0.57}P_{0.43}$ ($\lambda_{gap} \approx 1.27$ μm) must be grown, in comparison with 25 pairs of AlAs and GaAs layers on a GaAs substrate. The saturable absorber layer is 0.65-μm-thick bulk $In_{0.65}Ga_{0.35}As_{0.73}P_{0.27}$ ($\lambda_{gap} \approx 1.4$ μm) grown on top of the InGaAsP/InP Bragg mirror. The absorber has a dielectric SiO_2/HfO_2 top reflector, which is a high reflector for the pump and has a reflectivity of 50% at the lasing wavelength. With such an A-FPSA design, we obtained 230-ps single-frequency Q-switched pulses from an Nd:YVO$_4$ microchip laser at 1.34 μm (Fluck et al., 1997). Using this approach also at 1.5 μm, we were able to perform passive Q-switching of an Er:Yb:glass microchip laser, producing pulses as short as 1.2 ns and pulse energies up to 4 μJ (Fluck et al., 1998). However, because of the large gain bandwidth of Er:Yb:glass, single-frequency operation has not yet been obtained without an additional intracavity etalon.

VI. Conclusions and Outlook

At this point, we have a good understanding of the macroscopic parameters of saturable absorbers required to obtain stable cw modelocking or

Q-switching of solid state lasers. In addition, we have summarized simple design guidelines for diode-pumped solid state lasers. With these design guidelines, we have been able to achieve passive modelocking and Q-switching of many solid state lasers.

Our continuing research efforts are targeted toward a better microscopic understanding of the semiconductor dynamics of SESAMs and connection of such dynamics to the macroscopic properties. In any case, we can continue to benefit from fast picosecond saturable absorbers with lower nonsaturable losses and larger modulation depths. The development of broadband SESAMs is an important contribution to the achievement of reliable self-starting sub-10-fs lasers with improved cavity alignment tolerances. To date, we have demonstrated 6.5-fs self-starting pulses using SESAM-assisted KLM, and 13-fs self-starting pulses using soliton modelocking over the full cavity stability regime. Shorter pulses with soliton modelocking can be obtained only by making improvements in the broadband SESAM design. Optimization of the absorber requires a better understanding of the saturation behavior and the temporal response on a sub-10-fs timescale. In addition, adequate absorbers require more sophisticated bandgap engineering. There are possibilities for influencing the bandstructure of the absorber in the SESAM by adding indium or changing to a new material system such as InP.

Acknowledgment

The author is greatly indebted to the following colleagues and graduate students who have made major contributions to this chapter: Franz Kärtner, Kurt J. Weingarten, Uwe Siegner, Rüdiger Paschotta, Luigi Brovelli, Guodong Zhang, Francois Morier-Genoud, Michael Moser, Daniel Kopf, Bernd Braum, Isabella Jung, Regula Fluck, Clements Hönninger, Nicolai Matuschek, Dirk Sutter, Jürg aus der Au, Gabriel Spühler, and Markus Haiml. The financial support of the Swiss National Science Foundation and the Swiss Priority Program in Optics is also acknowledged.

List of Abbreviations and Acronyms

A-FPSA	antiresonant Fabry-Perot saturable absorber
AOM	acousto-optic modelocker
APM	additive pulse modelocking
AR	anti-reflection
CCD	charge coupled device
CPM	colliding pulse modelocked
CW	continuous wave

DCM	double-chirped mirrors
D-SAM	dispersive saturable absorber mirror
FWHM	full width at half maximum
GDD	group delay dispersion
HWHM	half width at half maximum
IAC	interferometric autocorrelation
KLM	Kerr lens modelocking
LT	low-temperature
MBE	molecular beam epitaxy
MISER	monolithic ring laser
MOCVD	metalo-organic chemical vapor deposition (note Ch. 1)
MWQ	multiple quantum well
NPRO	monolithic ring laser
OMM	optimized modematching
RPM	resonant passive modelocking
SAM	self-amplitude modulation
SBR	saturable Bragg reflector
SESAM	semiconductor saturable absorber mirror
SPM	self-phase modulation
TEA	transversely excited atmospheric pressure

References

Afromowitz, M. A. (1974). "Refractive Index of Ga(1-x)Al(x)As," *Solid-State Commun.* **15**, 59–63.

Agrawal, G. P. and Olsson, N. A. (1989). "Self-Phase Modulation and Spectral Broadening of Optical Pulses in Semiconductor Laser Amplifiers," *IEEE J. Quantum Electron.* **25**, 2297–2306.

Asaki, M. T., Huang, C.-P., Garvey, D., Zhou, J., Kapteyn, H. C., Murnane, M. N. (1993). "Generation of 11-fs Pulses from a Self-Mode-Locked Ti:Sapphire Laser," *Opt. Lett.* **18**, 977–979.

Aus der Au, J., Kopf, D., Morier-Genoud, F., Moser, M., and Keller, U. (1997). "60-fs Pulses from a Diode-Pumped Nd:Glass Laser," *Opt. Lett.* **22**, 307–309.

Aus der Au, J., Loesel, F. H., Morier-Genoud, F., Moser, M., and Keller, U. (1998). "Femtosecond Diode-Pumped Nd:Glass Laser with More Than 1-W Average Output Power," *Opt. Lett.* **23**, 271–273.

Barnett, B. C., Rahman, L., Islam, M. N., Chen, Y. C., Bhattacharya, P., Riha, W., Reddy, K. V., Howe, A. T., Stair, K. A., Iwamura, H., Friberg, S. R., and Mukai, T. (1995). "High-Power Erbium-Doped Fiber Laser Mode-Locked by a Semiconductor Saturable Absorber," *Opt. Lett.* **20**, 471–473.

Blow, K. J. and Nelson, B. P. (1988). "Improved Modelocking of an F-Center Laser with a Nonlinear Nonsoliton External Cavity," *Opt. Lett.* **13**, 1026–1028.

Brabec, T., Spielmann, C., and Krausz, F. (1991). "Mode Locking in Solitary Lasers," *Opt. Lett.* **16**, 1961–1963.

Brabec, T., Spielmann, C., and Krausz, F. (1992). "Limits of Pulse Shortening in Solitary Lasers," *Opt. Lett.* **17**, 748–750.

Braun, B., Hönninger, C., Zhang, G., Keller, U., Heine, F., Kellner, T., and Huber, G. (1996). "Efficient Intracavity Frequency Doubling of a Passively Modelocked Diode-Pumped Nd:LSB Laser," *Opt. Lett.* **21**, 1567–1569.

Braun, B., Kärtner, F. X., Moser, M., Zhang, G., and Keller, U. (1997). "56 ps Passively Q-Switched Diode-Pumped Microchip Laser," *Opt. Lett.* **22**, 381–383.

Braun, B. and Keller, U. (1995). "Single Frequency Q-Switched Ring Laser with an Antiresonant Fabry-Perot Saturable Absorber," *Opt. Lett.* **20**, 1020–1022.

Brovelli, L. R., Jung, I. D., Kopf, D., Kamp, M. Moser, M., Kärtner, F. X., and Keller, U. (1995a). "Self-Starting Soliton Modelocked Ti:Sapphire Laser Using a Thin Semiconductor Saturable Absorber," *Electron. Lett.* **31**, 287–289.

Brovelli, L. R. and Keller, U. (1995b). "Simple Analytical Expressions for the Reflectivity and the Penetration Depth of a Bragg Mirror Between Arbitrary Media," *Opt. Commun.* **116**, 343–350.

Brovelli, L. R., Keller, U., and Chiu, T. H. (1995c). "Design and Operation of Antiresonant Fabry-Perot Saturable Semiconductor Absorbers for Mode-Locked Solid-State Lasers," *J. Opt. Soc. Am. B* **12**, 311–322.

Brovelli, L. R., Lanker, M., Keller, U., Goossen, K. W., Walker, J. A., and Cunningham, J. E. (1995d). "An Antiresonant Fabry-Perot Quantum Well Modulator to Actively Mode-Lock and Synchronize Solid-State Lasers," *Electron. Lett.* **31**, 381–382.

Cerullo, G., De Silvestri, S., and Magni, V. (1994a). "Self-Starting Kerr Lens Mode-Locking of a Ti-Sapphire Laser," *Opt. Lett.* **19**, 1040–1042.

Cerullo, G., De Silvestri, S., Magni, V., and Pallaro, L. (1994b). "Resonators for Kerr-Lens Mode-Locked Femtosecond Ti-Sapphire Lasers," *Opt. Lett.* **19**, 807–809.

Cerullo, G., Dienes, A., and Magni, V. (1996). "Space-Time Coupling and Collapse Threshold for Femtosecond Pulses in Dispersive Nonlinear Media," *Opt. Lett.* **21**, 65–67.

Christov, I. P., Kapteyn, H. C., Murnane, M. M., Huang, C. P., and Zhou, J. (1995). "Space-Time Focusing of Femtosecond Pulses in a Ti:Sapphire Laser," *Opt. Lett.* **20**, 309–311.

Collings, B. C., Bergman, K., and Knox, W. H. (1997). "True Fundamental Solitons in a Passively Mode-Locked Short Cavity Cr^{4+}:YAG Laser," *Opt. Lett.* **22**, 1098–1100.

Collings, B. C., Stark, J. B., Tsuda, S., Knox, W. H., Cunningham, J. E., Jan, W. Y., Pathak, R., and Bergman, K. (1996). "Saturable Bragg Reflector Self-Starting Passive Modelocking of a Cr^{4+}:YAG Laser Pumped with a Diode-Pumped $Nd:YVO_4$ Laser," *Opt. Lett.* **21**, 1171–1173.

Cundiff, S. T., Knox, W. H., Ippen, E. P., and Haus, H. A. (1996). "Frequency Dependent Mode Size in Broadband Kerr-Lens Mode Locking," *Opt. Lett.* **21**, 662–664.

Curley, P. F., Spielmann, C., Brabec, T., Krausz, F., Wintner, E., and Schmidt, A. J. (1993). "Operation of a Femtosecond Ti:Sapphire Solitary Laser in the Vicinity of Zero Group-Delay Dispersion," *Opt. Lett.* **18**, 54–56.

Delfyett, P. J., Florez, L. T., Stoffel, N., Gmitter, T., Andreadakis, N. C., Silberberg, Y., Heritage, J. P., and Alphonse, G. A. (1992). "High-Power Ultrafast Laser Diodes," *IEEE J. Quantum Electron.* **28**, 2203–2219.

DeLoach, L. D., Payne, S. A., Chase, L. L., Smith, L. K., Kway, W. L., and Krupke, W. F. (1993). "Evaluation of Absorption and Emission Properties of Yb Doped Crystals for Laser Applications," *IEEE J. Quantum Electron.* **29**, 1179–1191.

DeMaria, A. J., Stetser, D. A., and Heynau, H. (1966). "Self Mode-Locking of Lasers with Saturable Absorbers," *Appl. Phys. Lett.* **8**, 174–176.

de Souza, E. A., Soccolich, C. E., Pleibel, W., Stolen, R. H., Simpson, J. R., and DiGiovanni, D. J. (1993). "Saturable Absorber Modelocked Polarization Maintaining Erbium-Doped Fiber Laser," *Electron. Lett.* **29**, 447–449.

Diels, J.-C. (1990). "Femtosecond Dye Lasers." In *Dye Lasers Principles: With Applications* (F. Duarte and L. Hillman, eds.). Academic Press, Boston, pp. 41–132.

Dymott, M. J. P. and Ferguson, A. I. (1995). "Self-Mode-Locked Diode-Pumped Cr:LiSAF Laser Producing 34-fs Pulses at 42-mW Average Power," *Opt. Lett.* **20**, 1157–1159.

Dymott, M. J. P. and Ferguson, A. I. (1997). "Pulse Duration Limitations in a Diode-Pumped Femtosecond Kerr-Lens Modelocked Cr:LiSAF Laser," *Appl. Phys.* B **65**, 227–234.

Fan, T. Y. and Sanchez, A. (1990). "Pump Source Requirements for End-Pumped Lasers," *IEEE J. Quantum Electron.* **26**, 311–316.

Fermann, M. E., Galvanauskas, A., Sucha, G., and Harter, D. (1997). "Fiber-Lasers for Ultrafast Optics," *Appl. Phys.* B **65**, 259–275.

Fluck, R., Braun, B., Gini, E., Melchior, H. and Keller, U. (1997). "Passively Q-Switched 1.34 μm Nd:YVO$_4$ Microchip Lasers using Semiconductor Saturable-Absorber Mirrors," *Opt. Lett.* **22**, 991–993.

Fluck, R., Häring, R., Paschotta, R., Gini, E., Melchior, H., and Keller, U. (1998). "Eyesafe Pulsed Microchip Laser Using Semiconductor Saturable Absorber Mirrors," *Appl. Phys. Lett.* **72**, 3273–3275.

Fluck, R., Jung, I. D., Zhang, G., Kärtner, F. X., and Keller, U. (1996a). "Broadband Saturable Absorber for 10 fs Pulse Generation," *Opt. Lett.* **21**, 743–745.

Fluck, R., Zhang, G., Keller, U., Weingarten, K. J., and Moser, M. (1996b). "Diode-Pumped Passively Mode-Locked 1.3 μm Nd:YVO$_4$ and Nd:YLF Lasers Using Semiconductor Saturable Absorbers," *Opt. Lett.* **21**, 1378–1380.

Fork, R. L., Cruz, C. H. B., Becker, P. C., and Shank, C. V. (1987). "Compression of Optical Pulses to Six Femtoseconds by Using Cubic Phase Compensation," *Opt. Lett.* **12**, 483–485.

Fork, R. L., Greene, B. I., and Shank, C. V. (1981). "Generation of Optical Pulses Shorter Than 0.1 ps by Colliding Pulse Modelocking," *Appl. Phys. Lett.* **38**, 617–619.

Fork, R. L., Martinez, O. E., and Gordon, J. P. (1984). "Negative Dispersion Using Pairs of Prisms," *Opt. Lett.* **9**, 150–152.

Fox, A. M., Miller, D. A. B., Livescu, G., Cunningham, J. E., Henry, J. E., and Jan, W. Y. (1990). "Exciton Saturation in Electrically Biased Quantum Wells," *J. Appl. Phys. Lett.* **57**, 2315.

French, P. M. W., Mellish, R., Taylor, J. R., Delfyett, P. J., and Florez, L. T. (1993). "All-Solid-State Diode-Pumped Modelocked Cr:LiSAF Laser," *Electron. Lett.* **29**, 1262–1263.

French, P. M. W., Williams, J. A. R., and Taylor, J. R. (1989). "Femtosecond Pulse Generation from a Titanium Doped Sapphire Laser Using Nonlinear External Cavity Feedback," *Opt. Lett.* **14**, 686–688.

Gibson, A. F., Kimmitt, M. F., and Norris, B. (1974). "Generation of Bandwidth-Limited Pulses from a TEA CO$_2$ Laser Using P-Type Germanium," *Appl. Phys. Lett.* **24**, 306–307.

Goodberlet, J., Wang, J., Fujimoto, J. G., and Schulz, P. A. (1989). "Femtosecond Passively Mode-Locked Ti:Sapphire Laser with a Nonlinear External Cavity," *Opt. Lett.* **14**, 1125–1127.

Gupta, S., Whitaker, J. F., and Mourou, G. A. (1992). "Ultrafast Carrier Dynamics in III-V Semiconductors Grown by Molecular-Beam Epitaxy at Very Low Substrate Temperatures," *IEEE J. Quantum Electron.* **28**, 2464–2472.

Prasad, A., Haiml, M., Jung, I. D., Kunde, J., Morier-Genoud, F., Weber, E. R., Siegner, U., and Keller, U. (1998). "Ultrafast Response Times and Enhanced Optical Nonlinearity in Beryllium Doped Low-Temperature Grown," *International Conference on Lasers and Electrooptics (CLEO'98)*, San Francisco, California, May 3–8, paper CFJ2.

Haiml, M., Prasad, A., Morier-Genoud, F., Siegner, U., Keller, U., and Weber, E. R. (1998b). "Ultrafast Response Times and Enhanced Optical Nonlinearity in Annealed and Beryllium Doped Low-Temperature Grown GaAs," 10th Conference on Semiconducting and Insulating Materials (SIMC-X 1998), June 1–5, Berkeley, CA, paper Tu2.2.

Haus, H. A. (1975a). "A Theory of Forced Mode Locking," *IEEE J. Quantum Electron.* **11**, 323–330.

Haus, H. A. (1975b). "Theory of Modelocking with a Fast Saturable Absorber," *J. Appl. Phys.* **46**, 3049–3058.

Haus, H. A. (1975c). "Theory of Mode Locking with a Slow Saturable Absorber," *IEEE J. Quantum Electron.* **11**, 736–746.

Haus, H. A. (1976). "Parameter Ranges for CW Passive Modelocking," *IEEE J. Quantum Electron.* **12**, 169–176.

Haus, H. A., Fujimoto, J. G., and Ippen, E. P. (1992). "Analytic Theory of Additive Pulse and Kerr Lens Mode Locking," *IEEE J. Quantum Electron.* **28**, 2086–2096.

Haus, H. A., Fujimoto, J. G., and Ippen, E. P. (1991b). "Structures for Additive Pulse Modelocking," *J. Opt. Soc. Am. B* **8**, 2068–2076.

Haus, H. A. and Ippen, E. P. (1991c). "Self-Starting of Passively Mode-Locked Lasers," *Opt. Lett.* **16**, 1331–1333.

Haus, H. A., Keller, U., and Knox, W. H. (1991). "A Theory of Coupled Cavity Modelocking with Resonant Nonlinearity," *J. Opt. Soc. Am. B* **8**, 1252–1258.

Haus, H. A. and Silberberg, Y. (1986). "Laser Mode Locking with Addition of Nonlinear Index," *IEEE J. Quantum Electron.* **22**, 325–331.

Hautojärvi, P., Mäkinen, J., Palko, S., Saarinen, K., Corbel, C., and Liszkay, L. (1993). "Point Defects in III-V Materials Grown by Molecular Beam Epitaxy at Low Temperature," *Mat. Sci. Eng.* **B22**, 16–22.

Hönninger, C., Morier-Genoud, F., Moser, M., Keller, U., Brovelli, L. R., and Harder, C. (1998a). "Efficient and Tunable Diode-Pumped Femtosecond Yb:Glass Lasers," *Opt. Lett.* **23**, 126–128.

Hönninger, C., Paschotta, R., Morier-Genoud, F., Moser, M., and Keller, U. (1998b). "Q-Switching Stability Limits of CW Passive Modelocking," *J. Opt. Soc. Am. B* **15**, in press.

Hönninger, C., Zhang, G., Keller, U., and Giesen, A. (1995). "Femtosecond Yb:YAG Laser Using Semiconductor Saturable Absorbers," *Opt. Lett.* **20**, 2402–2404.

Huang, C.-P., Asaki, M. T., Backus, S., Murnane, M. M., Kapteyn, H. C., and Nathel, H. (1992a). "17-fs Pulses from a Self-Mode-Locked Ti:Sapphire Laser," *Opt. Lett.* **17**, 1289–1291.

Huang, C.-P., Kapteyn, H. C., McIntosh, J. W., and Murnane, M. M. (1992b). "Generation of Transform-Limited 32-fs Pulses from a Self-Mode-Locked Ti:Sapphire Laser," *Opt. Lett.* **17**, 139–141.

Ippen, E. P., Eichenberger, D. J., and Dixon, R. W. (1980). "Picosecond Pulse Generation by Passive Modelocking of Diode Lasers," *Appl. Phys. Lett.* **37**, 267–269.

Ippen, E. P., Haus, H. A., and Liu, L. Y. (1989). "Additive Pulse Modelocking," *J. Opt. Soc. Am. B* **6**, 1736–1745.

Ippen, E. P., Shank, C. V., and Dienes, A. (1972). "Passive Modelocking of the cw Dye Laser," *Appl. Phys. Lett.* **21**, 348–350.

Islam, M. N., Sunderman, E. R., Soccolich, C. E., Bar-Joseph, I., Sauer, N., Chang, T. Y., and Miller, B. I. (1989). "Color Center Lasers Passively Mode Locked by Quantum Wells," *IEEE J. Quantum Electron.* **25**, 2454–2463.

Jung, I. D., Brovelli, L. R., Kamp, M., Keller, U., and Moser, M. (1995a). "Scaling of the Antiresonant Fabry-Perot Saturable Absorber Design Toward a Thin Saturable Absorber," *Opt. Lett.* **20**, 1559–1561.

Jung, I. D., Kärtner, F. X., Brovelli, L. R., Kamp, M., and Keller, U. (1995b). "Experimental Verification of Soliton Modelocking Using Only a Slow Saturable Absorber," *Opt. Lett.* **20**, 1892–1894.

Jung, I. D., Kärtner, F. X., Henkmann, J., Zhang, G., and Keller, U. (1997a). "High-Dynamic-Range Characterization of Ultrashort Pulses," *Appl. Phys. B* **65**, 307–310.

Jung, I. D., Kärtner, F. X., Matuschek, N., Sutter, D. H., Morier-Genoud, F., Shi, Z., Scheuer, V., Tilsch, M., Tschudi, T., and Keller, U. (1997b). "Semiconductor Saturable Absorber Mirrors Supporting Sub-10 fs Pulses," *Appl. Phys. B: Special Issue on Ultrashort Pulse Generation* **65**, 137–150.

Jung, I. D., Kärtner, F. X., Matuschek, N., Sutter, D. H., Morier-Genoud, F., Zhang, G., Keller, U., Scheuer, V., Tilsch, M., and Tschudi, T. (1997c). "Self-Starting 6.5 fs Pulses from a Ti:Sapphire Laser," *Opt. Lett.* **22**, 1009–1011.

Kafka, J. D., Alfrey, A. J., and Baer, T., eds. (1989). *Ultrafast Phenomena VI*. Springer-Verlag, Berlin.

Kaminska, M., Liliental-Weber, Z., Weber, E. R., George, T., Kortright, J. B., Smith, F. W., Tsaur, B.-Y., and Calawa, A. R. (1989). "Structural Properties of As-Rich GaAs Grown by Molecular Epitaxy at Low Temperatures," *Appl. Phys. Lett.* **54**, 1881–1883.

Kaminska, M. and Weber, E. R. (1993). "EL2 Defect in GaAs." In *Imperfections in III/V Materials*, Vol. 38 (E. R. Weber, ed.). Academic Press, Boston, pp. 59–89.

Kane, T. J. and Byer, R. L. (1985). "Monolithic, Unidirectional Single-Mode Nd:YAG Ring Laser," *Opt. Lett.* **10**, 65–67.

Kärtner, F. X., Aus der Au, J., and Keller, U. (1998). "Slow and Fast Saturable Absorbers for Modelocking of Solid-State Lasers — What's the Difference? *IEEE J. Sel. Top. Quantum Electron.* **4**, 159–168.

Kärtner, F. X., Brovelli, L. R., Kopf, D., Kamp, M., Calasso, I., and Keller, U. (1995a). "Control of Solid-State Laser Dynamics by Semiconductor Devices," *Opt. Eng.* **34**, 2024–2036.

Kärtner, F. X., Jung, I. D., and Keller, U. (1996). "Soliton Modelocking with Saturable Absorbers," *IEEE J. Sel. Top. Quantum Electron. Special Issue on Ultrafast Electronics, Photonics and Optoelectronics* **2**, 540–556.

Kärtner, F. X. and Keller, U. (1995b). "Stabilization of Soliton-Like Pulses with a Slow Saturable Absorber," *Opt. Lett.* **20**, 16–18.

Kärtner, F. X., Kopf, D., and Keller, U. (1995c). "Solitary Pulse Stabilization and Shortening in Actively Mode-Locked Lasers," *J. Opt. Soc. Am. B* **12**, 486–496.

Kärtner, F. X., Matuschek, N., Schibli, T., Keller, U., Haus, H. A., Heine, C., Morf, R., Scheuer, V., Tilsch, M., and Tschudi, T. (1997). "Design and Fabrication of Double-Chirped Mirrors," *Opt. Lett.* **22**, 831–833.

Kasper, A. and Witte, K. J. (1996). "10-fs Pulse Generation from a Uni-Directional Kerr-Lens Mode-Locked Ti:Sapphire Ring Laser," *Opt. Lett.* **21**, 360.

Kaup, D. J. (1990). "Perturbation Theory for Solitons in Optical Fibers," *Phys. Rev. A* **42**, 5689–5694.

Kean, P. N., Zhu, X., Crust, D. W., Grant, R. S. Landford, N., and Sibbett, W. (1989). "Enhanced Modelocking of Color Center Lasers," *Opt. Lett.* **14**, 39–41.

Keller, U. (1994). "Ultrafast All-Solid-State Laser Technology," *Appl. Phys. B* **58**, 347–363.

Keller, U. and Chiu, T. H. (1992a). "Resonant Passive Modelocked Nd:YLF Laser," *IEEE J. Quantum Electron.* **28**, 1710–1721.

Keller, U., Chiu, T. H., and Ferguson, J. F. (1993). "Self-Starting Femtosecond Mode-Locked Nd:Glass Laser That Uses Intracavity Saturable Absorbers," *Opt. Lett.* **18**, 1077–1079.

Keller, U., Knox, W. H., and Roskos, H. (1990). "Coupled-Cavity Resonant Passive Modelocked (RPM) Ti:Sapphire Laser," *Opt. Lett.* **15**, 1377–1379.

Keller, U., Knox, W. H., and 'tHooft, G. W. (1992b). "Ultrafast Solid-State Modelocked Lasers Using Resonant Nonlinearities," *IEEE J. Quantum Electron.* **28**, 2123–2133.

Keller, U., Miller, D. A. B., Boyd, G. D., Chiu, T. H., Ferguson, J. F., and Asom, M. T. (1992c). "Solid-State Low-Loss Intracavity Saturable Absorber for Nd:YLF Lasers: An Antireson-

ant Semiconductor Fabry-Perot Saturable Absorber," *Opt. Lett.* **17**, 505–507.
Keller, U., Nelson, L. E., and Chiu, T. H. (1992d). "Diode-Pumped, High-Repetition Rate, Resonant Passive Modelocked Nd:YLF Laser." In *Advanced Solid-State Lasers*, Vol. 13 (L. Chase and A. Pinto, eds.). Optical Society of America, Washington, D.C., pp. 94–97.
Keller, U., 'tHooft, G. W., Knox, W. H., and Cunningham, J. E. (1991a). "Femtosecond Pulses from a Continuously Self-Starting Passively Mode-Locked Ti:Sapphire Laser," *Opt. Lett.* **16**, 1022–1024.
Keller, U., Weingarten, K. J., Kärtner, F. X., Kopf, D., Braun, B., Jung, I. D., Fluck, R., Hönninger, C., Matuschek, N., and Aus der Au, J. (1996). "Semiconductor Saturable Absorber Mirrors (SESAMs) for Femtosecond to Nanosecond Pulse Generation in Solid-State Lasers," *IEEE J. Sel. Top. Quantum Electron. Special Issue on Ultrafast Electronics, Photonics and Optoelectronics* **2**, 435–453.
Keller, U., Woodward, T. K., Sivco, D. L., and Cho, A. Y. (1991b). "Coupled-Cavity Resonant Passive Modelocked Nd:Yttrium Lithium Fluoride Laser," *Opt. Lett.* **16**, 390–392.
Kellner, T., Heine, F., Huber, G., Hönninger, C., Braun, B., Morier-Genoud, F., and Keller, U. (1998). "Soliton Modelocked Nd:YAlO$_3$ Laser at 930 nm," *J. Opt. Soc. Am. B* **15**, 1663–1666.
Kim, B. G., Garmire, E., Hummel, S. G., and Dapkus, P. D. (1989). "Nonlinear Bragg Reflector Based on Saturable Absorption," *Appl. Phys. Lett.* **54**, 1095–1097.
Liu, X., Prasad, A., Nishio, J., Weber, E. R., Liliental-Weber, Z., and Walukiewicz, W. (1995). "Native Point Defects in Low-Temperature Grown GaAs," *Appl. Phys. Lett.* **67**, 279.
Kopf, D., Aus der Au, J., Keller, U., Bona, G. L., and Roentgen, P. (1995a). "A 400-mW Continuous-Wave Diode-Pumped Cr:LiSAF Laser Based on a Power-Scalable Concept," *Opt. Lett.* **20**, 1782–1784.
Kopf, D., Kärtner, F., Weingarten, K. J., and Keller, U. (1994a). "Pulse Shortening in a Nd:Glass Laser by Gain Reshaping and Soliton Formation," *Opt. Lett.* **19**, 2146–2148.
Kopf, D., Kärtner, F. X., Weingarten, K. J., and Keller, U. (1995b). "Diode-Pumped Modelocked Nd:Glass Lasers Using an A-FPSA," *Opt. Lett.* **20**, 1169–1171.
Kopf, D., Keller, U., Emanuel, M. A., Beach, R. J., and Skidmore, J. A. (1997a). "1.1-W cw Cr:LiSAF Laser Pumped by a 1-cm Diode-Array," *Opt. Lett.* **22**, 99–101.
Kopf, D., Prasad, A., Zhang, G., Moser, M., and Keller, U. (1997b). "Broadly Tunable Femtosecond Cr:LiSAF Laser," *Opt. Lett.* **22**, 621–623.
Kopf, D., Strässle, T., Zhang, G., Kärtner, F. X., Keller, U., Moser, M., Jubin, D., Weingarten, K. J., Beach, R. J., Emanuel, M. A., and Skidmore, J. A. (1996a). Diode-Pumped Femtosecond Solid State Lasers Based on Semiconductor Saturable Absorbers." In *Generation, Amplification, and Measurement of Ultrashort Laser Pulses III*. Proc. SPIE 2701, San Jose, CA, pp. 11–22.
Kopf, D., Weingarten, K. J., Brovelli, L. R., Kamp, M., and Keller, U. (1994b). "Diode-Pumped 100-fs Passively Mode-Locked Cr:LiSAF Laser Using an A-FPSA," *Opt. Lett.* **19**, 2143–2145.
Kopf, D., Weingarten, K. J., Brovelli, L. R., Kamp, M., and Keller, U. (1995c). "Sub-50-fs Diode-Pumped Mode-Locked Cr:LiSAF with an A-FPSA," *Conference on Lasers and Electro-Optics (CLEO '95)*, paper CWM2.
Kopf, D., Weingarten, K. J., Zhang, G., Moser, M., Emanuel, M. A., Beach, R. J., Skidmore, J. A., and Keller, U. (1997c). "High-Average-Power Diode-Pumped Femtosecond Cr:LiSAF Lasers," *Appl. Phys. B* **65**, 235–243.
Kopf, D., Zhang, G., Fluck, R., Moser, M., and Keller, U. (1996b). "All-in-One Dispersion-Compensating Saturable Absorber Mirror for Compact Femtosecond Laser Sources," *Opt. Lett.* **21**, 486–488.
Krausz, F., Brabec, T., and Spielmann, C. (1991). "Self-Starting Passive Modelocking," *Opt. Lett.* **16**, 235–237.

Krausz, F., Fermann, M. E., Brabec, T., Curley, P. F., Hofer, M., Ober, M. H., Spielmann, C., Wintner, E., and Schmidt, A. J. (1992a). "Femtosecond Solid-State Lasers," *IEEE J. Quantum Electron.* **28**, 2097–2122.

Krausz, F., Spielmann, C., Brabec, T., Wintner, E., and Schmidt, A. J. (1992b). "Generation of 33 fs Optical Pulses from a Solid-State Laser," *Opt. Lett.* **17**, 204–206.

Lederer, M. J., Luther-Davies, B., Tan, H. H., and Jagadish, C. (1997). "GaAs Based Anti-Resonant Fabry-Perot Saturable Absorber Fabricated by Metal Organic Vapor Phase Epitaxy and Ion Implantation," *Appl. Phys. Lett.* **70**, 3428–3430.

Liu, X., Prasad, A., Chen, W. M., Kurpiewski, A., Stoschek, A., Lilental-Weber, Z., and Weber, E. R. (1994). "Mechanism Responsible for the Semi-Insulating Properties of Low-Temperature-Grown GaAs," *Appl. Phys. Lett.* **65**, 3002–3004.

Loh, W. H., Atkinson, D., Morkel, P. R., Hopkinson, M., Rivers, A., Seeds, A. J., and Payne, D. N. (1993). "Passively Modelocked Er Fiber Laser Using a Semiconductor Nonlinear Mirror," *IEEE Photon. Technol. Lett.* **5**, 35–37.

Longhi, S. (1994). "Theory of Transverse Modes in End-Pumped Microchip Lasers," *J. Opt. Soc. Am. B* **11**, 3042–3044.

Look, D. C. (1993). "Molecular Beam Epitaxial GaAs Grown at Low Temperatures," *Thin Solid Films* **231**, 61–73.

Look, D. C., Walters, D. C., Mier, M., Stutz, C. E., and Brierley, S. K. (1992). "Native Donors and Acceptors in Molecular-Beam Epitaxial GaAs Grown at 200 °C," *Appl. Phys. Lett.* **60**, 2900–2902.

Luysberg, M., Sohn, H., Prasad, A., Specht, P., Fujioka, H., Klockenbrink, R., and Weber, E. R. (1997). "Electrical and Structural Properties of LT-GaAs: Influence of As/Ga Flux Ratio and Growth Temperature," *Mat. Res. Soc. Symp. 1997*, paper, pp. 485.

Macleod, H. A. (1985). *Thin-Film Optical Filters*. Adam Hilger, Bristol.

Manasreh, M. O., Look, D. C., Evans, K. R., and Stutz, C. E. (1990). "Infrared Absorption of Deep Defects in Molecular-Beam-Epitaxial GaAs Layers Grown at 200 °C: Observation of an EL2-like Defect," *Phys. Rev. B* **41**, 10272–10275.

Martinez, O. E., Fork, R. L., and Gordon, J. P. (1984). "Theory of Passively Modelocked Lasers Including Self-Phase Modulation and Group-Velocity Dispersion," *Opt. Lett.* **9**, 156–158.

Matuschek, N., Kärtner, F. X., Keller, U. (1997). "Exact Coupled-Mode Theories for Multilayer Interference Coatings with Arbitrary Strong Index Modulations," *IEEE J. Quantum Electron.* **33**, 295-302.

Matuschek, N., Kärtner, F. X., Keller, U. (1998). "Theory of Double-Chirped Mirrors," *IEEE. J. Sel. Top. Quantum Electron.* **4**, 159–168.

Mellish, R., French, P. M. W., Taylor, J. R., Delfyett, P. J., and Florez, L. T. (1993). "Self-Starting Femtosecond Ti:Sapphire Laser with Intracavity Multiquantum Well Absorber," *Electron. Lett.* **29**, 894–896.

Mellish, P. M., French, P. M. W., Taylor, J. R., Delfyett, P. J., and Florez, L. T. (1994). "All-Solid-State Femtosecond Diode-Pumped Cr:LiSAF Laser," *Electron. Lett.* **30**, 223–224.

Melloch, M. R., Woodall, J. M., Harmon, E. S., Otsuka, N., Pollak, F. H., Nolte, D. D., Freenstra, R. M., and Lutz, M. A. (1995). "Low-Temperature Grown III-V Materials," *Annu. Rev. Mat. Sci.* **25**, 547–600.

Miller, D. A. B., Chemla, D. S., Damen, T. C., Gossard, A. C., Wiegmann, W., Wood, T. H., and Burrus, C. A. (1984). *Phys. Rev. Lett.* **53**, 2173.

Miller, D. A. B., Chemla, D. S., Damen, T. C., Gossard, A. C., Wiegmann, W., Wood, T. H., and Burrus, C. A. (1985). "Electric Field Dependence of Optical Absorption Near the Band Gap of Quantum Well Structures," *Phys. Rev. B* **32**, 1043–1060.

Miller, D. A. B., Weiner, J. S., and Chemla, D. S. (1986). "Electric Field Dependence of Linear

Optical Properties in Quantum Well Structures: Waveguide Electroabsorption and Sum Rules," *IEEE J. Quantum Electron.* **22**, 1816–1830.
Mocker, H. W. and Collins, R. J. (1965). "Mode Competition and Self-Locking Effects in a Q-Switched Ruby Laser," *Appl. Phys. Lett.* **7**, 270–273.
Mollenauer, L. F. and Stolen, R. H. (1984). "The Soliton Laser," *Opt. Lett.* **9**, 13–15.
Moulton, P. F. (1986). "Spectroscopic and Laser Characteristics of $Ti:Al_2O_3$," *J. Opt. Soc. Am. B* **3**, 125–132.
Negus, D. K., Spinelli, L., Goldblatt, N., and Feugnet, G. (1991). "Sub-100 Femtosecond Pulse Generation by Kerr Lens Modelocking in Ti:Sapphire." In *Advanced Solid-State Lasers*, Vol. 10 (G. Dubé, L. Chase, eds.). Optical Society of America, Washington, D.C., pp. 120–124.
Nelson, L. E., Jones, D. J., Tamura, K., Haus, H. A., and Ippen, E. P. (1997). "Ultrashort-Pulse Fiber Ring Lasers," *Appl. Phys. B* **65**, 277–294.
New, G. H. C. (1974). *Opt. Commun.* **6**, 188.
Ober, M. H., Hofer, M., Keller, U., and Chiu, T. H. (1993). "Self-Starting, Diode-Pumped Femtosecond Nd:Fiber Laser," *Opt. Lett.* **18**, 1532–1534.
Payne, S. A., Chase, L. L., Newkirk, H. W., Smith, L. K., and Krupke, W. F. (1988). "Cr:LiCAF: A Promising New Solid-State Laser Material," *IEEE J. Quantum Electron.* **24**, 2243–2252.
Payne, S. A., Chase, L. L., Smith, L. K., Kway, W. L., and Newkirk, H. (1989). "Laser Performance of $LiSrAlF_6:Cr^{3+}$," *J. Appl. Phys.* **66**, 1051–1056.
Petricevic, V., Gayen, S. K., and Alfano, R. R. (1988). "Laser Action in Chromium-Doped Forsterite," *Appl. Phys. Lett.* **52**, 1040–1042.
Piché, M. and Salin, F. (1993). "Self-Mode Locking of Solid-State Lasers Without Apertures," *Opt. Lett.* **18**, 1041–1043.
Prasad, A., Haiml, M., Jung, I. D., Kunde, J., Morier-Genoud, F., Weber, E. R., Siegner, U., and Keller, U. (1998). "Ultrafast Response Times and Enhanced Optical Nonlinearity in Beryllium Doped Low-Temperature Grown GaAs," *Conference on Lasers and Electro-Optics (CLEO '98)*, paper accepted for oral presentation.
Ruddock, I. S. and Bradley, D. J. (1976). "Bandwidth-Limited Subpicosecond Pulse Generation in Modelocked cw Dye Lasers," *Appl. Phys. Lett.* **29**, 296–297.
Siegman, A. E. (1986). *Lasers*. (1986). University Science Books, Mill Valley, CA.
Salin, F., Grangier, P., Roger, G., and Brun, A. (1986). "Observation of High-Order Solitons Directly Produced by a Femtosecond Ring Laser," *Phys. Rev. Lett.* **56**, 1132–1135.
Salin, F., Grangier, P., Roger, G., and Brun, A. (1988). "Experimental Observation of Nonsymmetrical N=2 Solitons in a Femtosecond Laser," *Phys. Rev. Lett.* **60**, 569–571.
Salin, F. and Squier, J. (1992). "Gain Guiding in Solid-State Lasers," *Opt. Lett.* **17**, 1352–1354.
Salin, F., Squier, J., and Piché, M. (1991). "Modelocking of Ti:Sapphire Lasers and Self-Focusing: A Gaussian Approximation," *Opt. Lett.* **16**, 1674–1676.
Sanchez, F. and Chardon, A. (1996). "Transverse Modes in Microchip Lasers," *J. Opt. Soc. Am. B* **13**, 2869–2871.
Schmidt, W. and Schäfer, F. P. (1968). "Self-Mode-Locking of Dye-Lasers with Saturable Absorbers," *Phys. Lett.* **26A**, 558–559.
Shah, J. (1996). *Ultrafast Spectroscopy of Semiconductors and Semiconductor Nanostructures*. Springer Verlag, Berlin.
Shank, C. V. (1988). "Generation of Ultrashort Optical Pulses." In *Ultrashort Laser Pulses and Applications* (W. Kaiser, ed.). Springer Verlag, Heidelberg, Ch. 2.
Shank, C. V. and Ippen, E. P. (1974). "Subpicosecond Kilowatt Pulses from a Modelocked cw Dye Laser," *Appl. Phys. Lett.* **24**, 373–375.
Sharp, R. C., Spock, D. E., Pan, N., and Elliot, J. (1996). "190-fs Passively Modelocked Thulium Fiber Laser with a Low Threshold," *Opt. Lett.* **21**, 881–883.

Shestakov, A. V., Borodin, N. I., Zhitnyuk, V. A., Ohrimtchyuk, A. G., and Gapontsev, V. P. (1988). *Soviet Quantum Electron.* **50**, 113.
Siegman, A. E. *Lasers.* (1986). University Science Books, Mill Valley, CA.
Siegner, U., Fluck, R., Zhang, G., and Keller, U. (1996). "Ultrafast High-Intensity Nonlinear Absorption Dynamics in Low-Temperature Grown Gallium Arsenide," *Appl. Phys. Lett.* **69**, 2566–2568.
Silberberg, Y., Smith, P. W., Eilenberger, D. J., Miller, D. A. B., Gossard, A. C., and Wiegmann, W. (1984). "Passive Modelocking of a Semiconductor Diode Laser," *Opt. Lett.* **9**, 507–509.
Sizer, T., II, Woodward, T. K., Keller, U. Sauer, K., Chiu, T.-H., Sivco, D. L., and Cho, A. Y. (1994). "Measurement of Carrier Escape Rates, Exciton Saturation Intensity, and Saturation Density in Electrically Based Multiple-Quantum-Well Modulators," *IEEE J. Quantum Electron.* **30**, 399–407.
Skidmore, J. A., Emanuel, M. A., Beach, R. J., Benett, W. J., Freitas, B. L., Carlson, N. W., and Solarz, R. W. (1995). "High-Power Continuous Wave 690 nm AlGaInP Laser-Diode Arrays," *Appl. Phys. Lett.* **66**, 1163–1165.
Smith, P. W. (1970). "Mode-Locking of Lasers," *Proc. IEEE* **58**, 1342–1357.
Smith, F. W., Calawa, A. R., Chen, C.-L., Manfra, M. J., and Mahoney, L. J. (1988). *IEEE Electron. Device Lett.* **9**, 77.
Smith, L. K., Payne, S. A., Kway, W. L., Chase, L. L., and Chai, B. H. T. (1992). "Investigation of the Laser Properties of Cr^{3+}:$LiSrGaF_6$," *IEEE J. Quantum Electron.* **28**, 2612–2618.
Smith, P. W., Silberberg, Y., and Miller, D. A. B. (1985). "Modelocking of Semiconductor Diode Lasers Using Saturable Excitonic Nonlinearities," *J. Opt. Soc. Am. B* **2**, 1228–1236.
Soffer, B. H. and Linn, J. W. (1968). "Continuously Tunable Picosecond-Pulse Organic-Dye Laser," *J. Appl. Phys.* **39**, 5859–5860.
Spälter, S., Böhm, M., Burk, M., Mikulla, B., Fluck, R., Jung, I. D., Zhang, G., Keller, U., Sizmann, A., and Leuchs, G. (1997). "Self-Starting Soliton Modelocked Femtosecond Cr(4+):YAG Laser Using an Antiresonant Fabry-Perot Saturable Absorber," *Appl. Phys. B* **65**, 335–338.
Spence, D. E., Kean, P. N., and Sibbett, W. (1991). "60-fsec Pulse Generation from a Self-Mode-Locked Ti:Sapphire Laser," *Opt. Lett.* **16**, 42–44.
Spielmann, C., Curley, P. F., Brabec, T., and Krausz, F. (1994). "Ultrabroadband Femtosecond Lasers," *IEEE J. Quantum Electron.* **30**, 1100–1114.
Spühler, G. J., Paschotta, R., Fluck, R., Braun, B., Moses, M., Zhang, G., Gini, E., and Keller, U. (1998). "Experimentally Confirmed Design Guidelines for Passively Q-Switched Microchip Lasers Using Semiconductor Saturable Absorbers," *J. Opt. Soc. Am. B* **15**, in press.
Stingl, A., Lenzner, M., Spielmann, C., Krausz, F., and Szipöcs, R. (1995). "Sub-10-fs Mirror-Dispersion-Controlled Ti:Sapphire Laser," *Opt. Lett.* **20**, 602–604.
Stingl, A., Spielmann, C., and Krausz, F. (1994). "Generation of 11-fs Pulses from a Ti:Sapphire Laser without the Use of Prisms." *Opt. Lett.* **19**, 204–206.
Sutter, D. H., Jung, I. D., Kärtner, F. X., Matuschek, N., Morier-Genoud, F., Scheuer, V., Tilsch, M., Tschudi, T., and Keller, U. (1998a). "Self-Starting 6.5 fs Pulses from a Ti:Sapphire Laser Using a Semiconductor Saturable Absorber and Double-Chirped Mirrors," *IEEE J. Sel. Top. Quantum Electron.* **4**, 169–178.
Sutter, D. H., Jung, I. D., Morier-Genoud, F., Kärtner, F. X., and Keller, U. (1998b). "300 nm Tunability of 30-fs Ti:Sapphire Laser Pulses with a Single Set of Double-Chirped Mirrors," *Conference on Lasers and Electro-Optics (CLEO '98)*, paper accepted for oral presentation.
Szipöcs, R., Ferencz, K., Spielmann, C., and Krausz, F. (1994). "Chirped Multilayer Coatings for Broadband Dispersion Control in Femtosecond Lasers," *Opt. Lett.* **19**, 201–203.
Szipöcs, R. and Kohazi-Kis, A. (1997). "Theory and Design of Chirped Dielectric Laser

Mirrors," *Appl. Phys. B* **65**, 115–136.

Talwar, D. N., Manasresh, M. O., Stutz, C. E., Kaspi, R., and Evans, K. R. (1993). "Infrared Studies of Be-Doped GaAs Grown by Molecular Beam Epitaxy at Low Temperatures," *J. Electron. Mat.* **22**, 1445–1448.

Tamura, K., Jacobson, J., Ippen, E. P., Haus, H. A., and Fujimoto, J. G. (1993). "Unidirectional Ring Resonators for Self-Starting Passively Modelocked Lasers," *Opt. Lett.* **18**, 220–222.

Tsou, Y., Garmire, E., Chen, W., Birnbaum, M., and Asthana, R. (1993). "Passive Q-Switching of Nd:YAG Lasers by Use of Bulk Semiconductors," *Opt. Lett.* **18**, 1514–1516.

Tsuda, S., Knox, W. H., and Cundiff, S. T. (1996a). "High Efficiency Diode Pumping of a Saturable Bragg Reflector-Mode-Locked Cr:LiSAF Femtosecond Laser," *Appl. Phys. Lett.* **69**, 1538–1540.

Tsuda, S., Knox, W. H., Cundiff, S. T., Jan, W. Y., and Cunningham, J. E. (1996b). "Mode-Locking Ultrafast Solid-State Lasers with Saturable Bragg Reflectors," *IEEE J. Sel. Top. Quantum Electron.* **2**, 454–464.

Tsuda, S., Knox, W. H., de Souza, E. A., Jan, W. Y., and Cunningham, J. E. (1995a). "Femtosecond Self-Starting Passive Mode Locking Using an AlAs/AlGaAs Intracavity Saturable Bragg Reflector," *Conference on Lasers and Electro-Optics (CLEO '95')*, paper CWM6, pp. 254–255.

Tsuda, S., Knox, W. H., de Souza, E. A., Jan, W. Y., and Cunningham, J. E. (1995b). "Low-Loss Intracavity AlAs/AlGaAs Saturable Bragg Reflector for Femtosecond Mode Locking in Solid-State Lasers," *Opt. Lett.* **20**, 1406–1408.

Valdmanis, J. A. and Fork, R. L. (1986). "Design Considerations for a Femtosecond Pulse Laser Balancing Self Phase Modulation, Group Velocity Dispersion, Saturable Absorption, and Saturable Gain," *IEEE J. Quantum Electron.* **22**, 112–118.

Valdmanis, J. A., Fork, R. L., and Gordon, J. P. (1985). "Generation of Optical Pulses as Short as 27 fs Directly from a Laser Balancing Self-Phase Modulation, Group-Velocity Dispersion, Saturable Absorption, and Saturable Gain," *Opt. Lett.* **10**, 131–133.

van der Ziel, J. P., Tsang, W. T., Logan, R. A., Mikulyak, R. M., and Augustyniak, W. M. (1981). "Subpicosecond Pulses from Passively Modelocked GaAs Buried Optical Guide Semiconductor Lasers," *Appl. Phys. Lett.* **39**, 525–527.

Wang, H. S., Wa, P. L. K., Lefaucheur, J. L., Chai, B. H. T., and Miller, A. (1994). "CW and Self-Mode-Locked Performance of a Red Pumped Cr:LiSCAF Laser," *Opt. Commun.* **110**, 679–688.

Wang, P., Zhou, S.-H., Lee, K. K., and Chen, Y. C. (1995). "Picosecond Laser Pulse Generation in a Monolithic Self-Q-Switched Solid-State Laser," *Opt. Commun.* **114**, 439–441.

Weingarten, K. J., Braun, B., and Keller, U. (1994). "In-situ Small-Signal Gain of Solid-State Lasers Determined from Relaxation Oscillation Frequency Measurements," *Opt. Lett.* **19**, 1140–1142.

Weingarten, K. J., Keller, U. Chiu, T. H., and Ferguson, J. F. (1993). "Passively Mode-Locked Diode-Pumped Solid-State Lasers Using an Antiresonant Fabry-Perot Saturable Absorber," *Opt. Lett.* **18**, 640–642.

Whitaker, J. F. (1993). "Optoelectronic Applications of LTMBE III-V Materials," *Mat. Sci. Eng.* **B22**, 61–67.

Witt, G. D. (1993). "LTMBE GaAs: Present Status and Perspectives," *Mat. Sci. Eng.* **B22**, 9.

Xu, L., Spielmann, C., Krausz, F., and Szipöcs, R. (1996). "Ultrabroadband Ring Oscillator for Sub-10-fs Pulse Generation," *Opt. Lett.* **21**, 1259–1261.

Xu, L., Tempea, G., Poppe, A., Lenzner, M., Spielmann, C., Krausz, F., Stingl, A., and Ferencz, K. (1997). "High-Power Sub-10-fs Ti:Sapphire Oscillators," *Appl. Phys. B* **65**, 151–159.

Yin, L. W., Huang, Y., Lee, J. H., Kolbas, R. M., Trew, R. J., and Mishra, U. K. (1990). "Improved Breakdown Voltage in GaAs MESFET's Utilizing Surface Layers of GaAs

Grown at Low Temperature by MBE," *IEEE Electron. Device Lett.* **11**, 561–563.

Zayhowski, J. J. and Dill, C. III. (1994). "Diode-Pumped Passively Q-Switched Picosecond Microchip Lasers," *Opt. Lett.* **19**, 1427–1429.

Zayhowski, J. J. and Dill, C. III. (1995a). "Coupled Cavity Electro-Optically Q-Switched Nd:YVO$_4$ Microchip Lasers," *Opt. Lett.* **20**, 716–718.

Zayhowski, J. J. and Mooradian, A. (1989). "Single-Frequency Microchip Nd Lasers," *Opt. Lett.* **14**, 24–26.

Zayhowski, J. J., Ochoa, J., and Dill, C. III. (1995b). "UV Generation with Passively Q-Switched Picosecond Microchip Lasers," *Conference on Lasers and Electro-Optics (CLEO '95)*, paper CTuM2, p. 139.

Zhang, Z., Torizuka, K., Itatani, T., Kobayashi, K., Sugaya, T., and Nakagawa, T. (1997). "Self-Starting Mode-Locked Femtosecond Forsterite Laser with a Semiconductor Saturable-Absorber Mirror," *Opt. Lett.* **22**, 1006–1008.

Zhou, J., Taft, G., Huang, C.-P., Murnane, M. M., Kapteyn, H. C., and Christov, I. P. (1994). "Pulse Evolution in a Broad-Bandwidth Ti:Sapphire Laser," *Opt. Lett.* **19**, 1149–1151.

Zirngibl, M., Stulz, L. W., Stone, J., Hugi, J., DiGiovanni, D., and Hansen, P. B. (1991). "1.2 ps Pulses from Passively Modelocked Laser Diode Pumped Er-Doped Fiber Ring Lasers," *Electron. Lett.* **27**, 1734–1735.

CHAPTER 5

Transient Grating Studies of Carrier Diffusion and Mobility in Semiconductors

Alan Miller

SCHOOL OF PHYSICS AND ASTRONOMY
UNIVERSITY OF ST. ANDREWS
FIFE, SCOTLAND

I. OPTICAL NONLINEARITIES AND CARRIER TRANSPORT	287
II. TRANSIENT GRATINGS	290
III. BULK SEMICONDUCTORS	292
IV. EXCITON SATURATION IN MQWs	293
V. IN-WELL TRANSPORT IN MQWs	297
VI. CROSS-WELL TRANSPORT IN QWs	303
VII. HETERO-n-i-p-i STRUCTURES	306
VIII. CONCLUSIONS	309
LIST OF ACRONYMS	310
REFERENCES	310

I. Optical Nonlinearities and Carrier Transport

Optical nonlinearities provide a powerful means of investigating the transport properties of electrons and holes in semiconductors. Ultrashort-pulse lasers allow monitoring of carrier motion by means of both absorptive and refractive optical nonlinearities on picosecond and femtosecond time-scales. Carrier motion can be driven by either the application of an electric field or the creation of a concentration gradient. In this chapter, we describe transient grating measurements that have been used to determine carrier transport properties of bulk and quantum well semiconductors. Both in-well and cross-well transport in quantum wells are described. We limit our discussion to transient gratings in semiconductors created by local nonlinear optical interactions such as band filling and exciton saturation. Picosecond-timescale, pump-probe studies of carrier transport in quantum well p-i-n structures have also employed field screening, resulting from charge separation in combination with the quantum confined Stark effect (QCSE), as the

List of Acronyms can be located preceding the references to this chapter.

nonlinear interaction. We exclude discussion of this approach in order to concentrate on the results of the transient grating technique.

Here we are primarily interested in bandgap resonant optical nonlinearities whereby laser excitation causes changes in absorption coefficient and refractive index as a function of carrier density (Miller et al., 1981; Haug, 1988; Miller and Sibbett, 1988; Shah, 1996). The optical generation of excess carriers results in changes in both the real and imaginary parts of the susceptibility, as described in previous chapters. Absorption and refraction are linked by causality through Kramers-Kronig relations. These familiar relations in linear optics may be modified to account for optical nonlinearities when laser excitation of a quasi-equilibrium of electrons and holes results in renormalized material parameters (Hutchings et al., 1992). Thus, refractive index changes Δn and absorption changes $\Delta \alpha$ may be connected by the integral expression

$$n_{eh}(\hbar\omega) = \frac{ch}{\pi} P \int_0^\infty \frac{\sigma_{eh}(\hbar\omega')d(\hbar\omega')}{(\hbar\omega')^2 - (\hbar\omega)^2} \tag{1}$$

where P denotes the principal part of the integral. In this case, we have represented the resonant optical nonlinearities by the parameters n_{eh} and σ_{eh}, which are defined as the refractive index and the absorption change per carrier pair per unit volume, respectively. For small excess carrier densities N,

$$\Delta n = n_{eh} N$$
$$\Delta \alpha = \sigma_{eh} N \tag{2}$$

Note that *changes* in refractive index at a given frequency ω are induced by *changes* in absorption at *all* other frequencies ω'. Thus, carriers generated by optical excitation near the bandgap energy of a semiconductor can produce both absorptive and refractive optical nonlinearities as a result of band filling. Similarly, free carrier absorption owing to generated electrons and holes leads to refractive index changes.

Optical nonlinearities associated with excitons have played an important role in carrier transport studies of multiple quantum well (MQW) semiconductors. Exciton absorption is more clearly resolved at room temperature in quantum wells than in bulk semiconductors, because confinement leads to larger binding energies and enhanced oscillator strengths. Exciton absorption features can be readily saturated by photogenerated free carriers through phase space filling (PSF) and Coulomb contributions. The creation of spin-polarized electrons provides an additional means of monitoring the motion of electrons in quantum wells by means of excitonic optical nonlinearities.

5 STUDIES OF CARRIER DIFFUSION AND MOBILITY IN SEMICONDUCTORS

The optical generation of free electrons and holes or excitons is essentially instantaneous, but the decay of the excess carrier population depends on the recombination rate, the diffusion coefficient if there is a concentration gradient, or the sweep-out rate in the presence of an electric field. For instance, the excess carrier density, as a function of position and time, $N(x, t)$, after excitation with an ultrashort pulse, can be determined from the continuity equation (given here for one-dimensional motion in the absence of an electric field):

$$\frac{\delta N(x, t)}{\delta t} = -\frac{N(x, t)}{\tau_R} + D_a \nabla^2 N(x, t) \qquad (3)$$

where D_a is the ambipolar diffusion coefficient for electrons and holes and τ_R is the electron-hole recombination time. The time for recovery to the equilibrium condition can range from microseconds to femtoseconds depending on the conditions.

Optical excitation generates equal densities of electrons and holes. The different diffusion coefficients for electrons and holes initially lead to spatial separation of the two types of charge carriers, but this separation creates internal electric fields that force the attracting electrons and holes back together. The net effect is a joint "ambipolar" diffusion for both electrons and holes. Under conditions of equal (and moderate) carrier densities, the ambipolar diffusion coefficient can be expressed in terms of the individual electron and hole diffusion coefficients D_e and D_h, where

$$D_a = \frac{2 D_e D_h}{D_e + D_h} \qquad (4)$$

These diffusion coefficients D are related to mobilities μ by the Einstein relation, which is valid under Boltzmann conditions:

$$D = \mu \frac{kT}{e} \qquad (5)$$

A useful parameter is the diffusion length — i.e., the average distance a free carrier moves before recombining — defined by

$$L_D = \sqrt{D_a \tau_R} \qquad (6)$$

The diffusion length in III–V semiconductors is typically on the order of a few micrometers.

II. Transient Gratings

Carrier diffusion and mobilities have been studied in bulk and quantum well semiconductors using the transient grating technique (often referred to as degenerate four-wave mixing) in a number of different configurations. A temporary diffraction grating can be created by the generation of a modulated excess carrier population when two pulses of light interfere (Jan, 1982; Eichler et al., 1986) within the semiconductor. Optical nonlinearities associated with the generated carriers can produce both amplitude and phase gratings resulting from absorptive and refractive nonlinearities. In two-beam configurations, the beams creating the grating can be self-diffracted. However, in order to monitor the carrier motion in detail, a three-beam configuration is more useful, whereby two pulses create the grating and a third is diffracted from this temporary grating (Fig. 1). Analysis of the grating decay as a function of grating period can provide detailed information about spatial dynamics within the semiconductor on micrometer dimensions and ultrashort timescales. In this chapter, we concentrate on studies that have employed picosecond or femtosecond pulses.

Two extreme cases are usually considered for grating diffraction. One is the "thin" grating (Raman-Nath) regime in which all of the diffracted waves have an optical path difference that is negligible compared with the wavelength. Several orders of diffraction may be observed. This applies to a commonly met situation in semiconductors, when a free carrier grating is created under strongly attenuating conditions by bandgap resonant, single-photon absorption. When this condition is not satisfied, the grating is termed "thick" and the Bragg condition (phase matching) must be fulfilled for efficient generation of a diffracted beam.

A typical transient grating experiment consists of two ultrashort "excite" pulses incident synchronously at the same location on the sample, at an

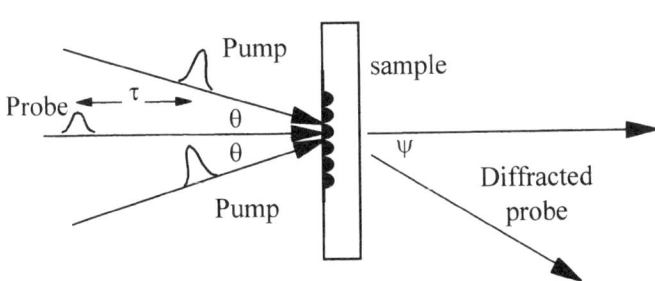

FIG. 1. Transient grating configuration using ultrashort pulses.

5 STUDIES OF CARRIER DIFFUSION AND MOBILITY IN SEMICONDUCTORS

angle θ, on either side of the normal to the surface. Interference between the pulses produces a sinusoidal intensity pattern, and, in turn, a sinusoidal modulation in carrier population density is created by interband absorption. The grating decays through a combination of carrier recombination and carrier transport processes.

For a wavelength λ, the grating spacing Λ is given by

$$\Lambda = \frac{\lambda}{2\sin\theta} \tag{7}$$

and the diffraction angle, φ is given by

$$\sin\varphi = \frac{\lambda}{\Lambda} = 2\sin\theta \tag{8}$$

The decay of the initial carrier distribution, $N(0, 0)$, is characterized by the recombination lifetime τ_R and the diffusion lifetime τ_D:

$$N(x, t) = N(0, 0)\left[\frac{1}{2} + \frac{1}{2}\cos\left(\frac{2\pi x}{\Lambda}\right)\exp\left(-\frac{t}{\tau_D}\right)\right]\exp\left(-\frac{t}{\tau_R}\right) \tag{9}$$

The diffraction efficiency η for a thin sinusoidal grating is given by a first-order Bessel function

$$\eta = \left|J_1\left(\frac{2\pi n_{eh}}{\lambda} + \frac{i\sigma_{eh}}{2}\right)Nl_\alpha\right|^2 \exp(-\alpha l) \tag{10}$$

where α is the average absorption coefficient, l is the sample thickness, and N is the density of carriers at the peak of a fringe. Also, l_α is an effective sample thickness that includes the effects of absorption:

$$l_\alpha = \frac{1 - \exp(-\alpha l)}{\alpha} \tag{11}$$

For moderate diffraction efficiencies and including only the refractive term (which usually is the dominant term) the Bessel function reduces to the simplified form

$$\eta = \left(\frac{\pi n_{eh} N l_\alpha}{\lambda}\right)^2 \exp(-\alpha l) \tag{12}$$

We note that the diffraction efficiencies are proportional to the square of the modulation depth of the refractive index. Therefore, if the grating decay is characterized by a rate constant Γ, the diffraction efficiency decays at twice this rate, and thus

$$\eta = \exp(-2\Gamma t) \qquad (13)$$

Substituting gives a grating decay rate,

$$\Gamma = \frac{1}{\tau_D} + \frac{1}{\tau_R} = \frac{4\pi^2 D_a}{\Lambda^2} + \frac{1}{\tau_R} \qquad (14)$$

Therefore, measuring the decay rate as a function of the angle between the excite beams provides a means of determining ambipolar diffusion coefficients by plotting the decay rate Γ against Λ^{-2}.

III. Bulk Semiconductors

The first semiconductor grating experiments were carried out by Woerdman and Bolger (1969, 1970a) and Woerdman (1970) in silicon using a 1.06-μm, Q-switched Nd:YAG laser. Carrier densities of 10^{17} to 10^{18} cm^{-3} and refractive index modulation of 10^{-4} were created by pulsed laser intensities of 1 MW/cm^2, resulting in diffraction efficiencies of a few percent. A series of spots could be observed behind the sample with time constants of 20 to 30 ns, consistent with a grating decay dominated by carrier diffusion (Woerdman and Bolger, 1969). Transient holograms with image reconstruction were demonstrated (Woerdman, 1970). The relatively long time constants associated with transient gratings in silicon have allowed the use of long pulse durations. Ambipolar diffusion coefficients have been measured in several studies (Jarasiunas and Vaitkus, 1977; Eichler and Massmann, 1982; Miyuski et al., 1983; Vaitkus et al., 1986; Bergner et al., 1986), resulting in values between 5 and 18.5 cm^2/s in crystalline silicon, depending on material quality, in reasonable agreement with Hall measurements. The effects of doping, implantation, and amorphization have been investigated (Miyoshi et al., 1983; Vaitkus et al., 1986).

Transient gratings in germanium were studied by Moss et al. (1981) employing 1.06-μm, picosecond pulses from a modelocked neodymium laser. The dependence of grating decay time on grating period resulted in ambipolar diffusion coefficients of 53 cm^2/s at 295 K and 142 cm^2/s at 135 K. In an extension of these studies to higher excitation levels to produce degenerate carrier conditions, a density dependence for the diffusion coefficient was found, although the commonly used degenerate version of the diffusion coefficient was not sufficient to account for the results (Smirl et al., 1982).

Several bulk III–V and II–VI semiconductors have been studied by the time-resolved, transient grating technique. Hoffman et al. (1978) investigated GaAs and InP using 6-ps pulses at a wavelength of 0.53 μm for the grating creation and time-delayed 1.06-μm probe pulses, and deduced that the method was especially sensitive to surface recombination velocity in this case.

Two-photon excitation has been used to create free carrier gratings in CdS, CdSe, ZnSe, and CdTe with ruby and Nd:glass lasers (Borshch et al., 1973; Jarasiunas and Vaitkas, 1974; Jarasiunas and Gerritsen, 1978; Kremenitski et al., 1979). Self-diffraction was used to deduce ambipolar diffusion coefficients of $3\,\text{cm}^2/\text{s}$ (300 K) and $20\,\text{cm}^2/\text{s}$ (62 K) for CdS, $65\,\text{cm}^2/\text{s}$ (13 K) for CdSe, and $2.5\,\text{cm}^2/\text{s}$ for ZnSe at room temperature.

Polycrystalline samples of ZnSe and CdTe were studied by Canto-Said et al. (1991), using two-photon excitation, in a study of higher-order optical nonlinearities. In this case, a density-dependent diffusion coefficient was observed. At carrier densities above $5 \times 10^{14}\,\text{cm}^{-3}$, the diffusion coefficient maintained a constant value of $4.5\,\text{cm}^2/\text{s}$. Below this level of excitation, the diffusion coefficient decreased rapidly to less than $1\,\text{cm}^2/\text{s}$, which is consistent with ambipolar diffusion coefficients of $0.7\,\text{cm}^2/\text{s}$ determined from Hall measurements. It was proposed that at low densities, carriers excited by two-photon absorption are trapped, but at higher laser irradiances the traps become filled.

IV. Exciton Saturation in MQWs

Transient grating studies in multiple quantum well structures have invariably utilized large excitonic optical nonlinearities (Miller et al., 1982). In this section, we describe the nature of exciton saturation in MQWs at room temperature, making use of interband selection rules and circularly polarized light to distinguish between phase space filling (PSF) and Coulomb contributions to exciton bleaching in two-dimensional structures (Schmitt-Rink et al., 1989). Spin-dependent excitonic optical nonlinearities provide additional insight into carrier transport in MQWs (Snelling et al., 1994).

Figure 2 illustrates exciton saturation spectra for three GaAs/AlGaAs MQW structures with different well widths of 4.4, 6.5, and 10 nm at room

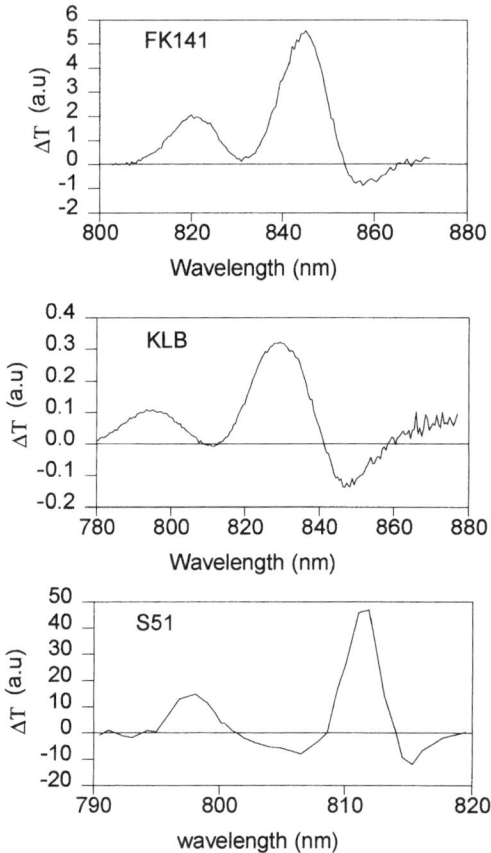

FIG. 2. Comparison of differential transmission changes for three GaAs/AlGaAs MQW structures having different quantum well widths: (a) 10 nm, (b) 6.5 nm, and (c) 4.4 nm.

temperature. These spectra were recorded in a pump-probe configuration using linearly polarized, picosecond pulses for both the pump and probe and a fixed delay of a few picoseconds between the pump and probe pulses.

Absorption saturation can be clearly observed at an average laser power of less than 1 mW for laser spot sizes on the order of 10 to 100 μm. This saturation occurs as a result of the optical generation of free carriers at 2D densities in excess of 10^{11} cm^{-2}. Resonant excitation initially creates bound electron-hole pairs—i.e., excitons. When the density of excitons approaches that required to fill space ($\sim 10^{17}$ cm^{-3} based on an exciton diameter of 30 nm for bulk GaAs), the number of additional excitons that can be created is reduced by Pauli exclusion, thus decreasing the level of absorption. Knox et al. (1985) measured a room-temperature time for ionization of excitons into free electron-hole pairs of 300 fs. Therefore, on picosecond and longer timescales, it can be assumed that the exciton saturation results from free carriers. The increase in transmission (decrease in absorption) results from a combination of PSF, Coulomb screening, and broadening (Holden et al., 1997). The negative signals (increased absorption) are caused by lifetime broadening of the exciton absorption features. Any reduction of the binding energy caused by Coulomb screening to produce a blue shift of the exciton is balanced by a red shift resulting from bandgap renormalization, so that the wavelength of the exciton peak remains unaltered. Above 3 mW, the absorption becomes progressively more difficult to saturate once the exciton feature has disappeared, because bandgap renormalization causes a red shift of the bandgap energy and thus a progressively larger density of states needs to be saturated. At high optical powers, and thus very high carrier densities, band filling causes the remaining absorption to saturate, but this is normally accompanied by a red shift owing to lattice heating when cw excitation or laser pulses are used at high repetition rates.

It is possible to create 100% spin-polarized electrons using circularly polarized light because of the interband selection rules close to $k = 0$ in quantum well structures (Fig. 3). The symbols σ_+ and σ_- represent transitions for light with right and left circular polarizations, respectively.

In pump and probe experiments, the two beams can be chosen to have either the same or opposite senses of polarization. Figure 4 shows the results of saturation measurements in GaAs/AlGaAs MQWs at room temperature with the three configurations: opposite linear polarization (OLP), same circular polarization (SCP), and opposite circular polarization (OCP). The

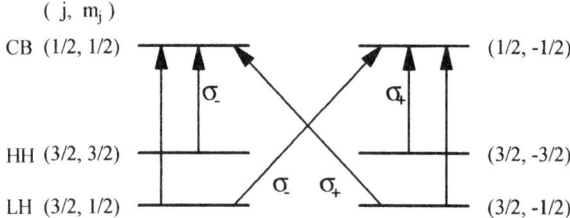

FIG. 3. The selection rules for the transitions from the heavy hole (HH) and light hole (LH) valence bands to the conduction band (CB) in MQWs. The (j, m_j) refer to the quantum numbers for angular momentum and its component along one direction. The σ_+ and σ_- transitions refer to right and left circularly polarized light and correspond to $\Delta m_j = \pm 1$, respectively, where the propagation direction is used to define m_j.

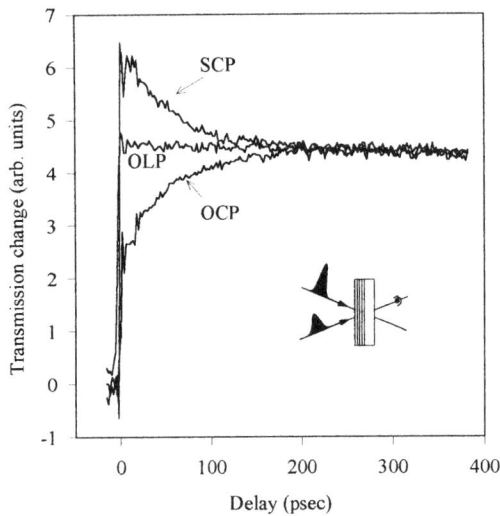

FIG. 4. The dependence of probe transmission change on optical time delay in pump-probe measurements for three polarization configurations: opposite linear polarization (OLP), same circular polarization (SCP), and opposite circular polarization (OCP). (From Norwood et al., 1995.)

experiments were carried out at the peak of the heavy hole exciton absorption, at the same power densities.

The opposite linear polarization (OLP) result is the condition usually reported, whereby the initial transmission change is essentially instantaneous and the recovery depends on the carrier recombination time (>1 ns). The SCP and OCP results show the effects of the initial spin polarization of the electrons and subsequent spin relaxation on a 100-ps timescale. Because PSF is sensitive to carrier spin, but Coulomb screening is not (within the Boltzmann approximation), the initial rapid rise in the OCP can be attributed solely to the Coulomb contribution, whereas the difference in signals between SCP and OCP is attributable to PSF. The probe beam in the OCP configuration is not sensitive to PSF until the spins randomize. The band-filling conditions for each configuration are shown in Fig. 5. For the OLP case, equal numbers of spin-up and spin-down carriers are created. However, the use of circular polarized light of the same irradiance will create the same number of carriers within a single spin band, thus enhancing the saturation signal if the probe has the same polarization (SCP) and reducing it if it has the opposite polarization (OCP).

Note that the phonon scattering responsible for the ionization of the

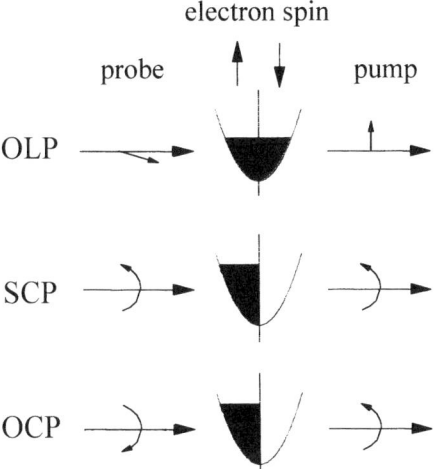

FIG. 5. The effect on phase space filling of using different polarizations of pump and probe beams. OLP, opposite linear polarization; SCP, same circular polarization; OCP, opposite circular polarization.

excitons into free carriers does not relax the spin polarization. The deduced relaxation time from Fig. 4 is $\tau_s = 55$ ps. This is the electron spin relaxation time. The hole spin is assumed to relax on femtosecond timescales owing to the mixed spin nature of the upper valence bands.

V. In-Well Transport in MQWs

Two-beam, self-diffraction measurements were initially used to deduce the in-well ambipolar diffusion coefficient in GaAs/AlGaAs MQWs (Chemla et al., 1984). More detailed information has been obtained by employing the three-beam method (Miller et al., 1989) (see Fig. 1). Figure 6 shows measured decays of diffracted probe pulses after the creation of gratings in room-temperature GaAs/AlGaAs MQWs at different grating spacings, produced by changing the angle between the pump beams. Recall (see Eq. 13) that the diffracted probe signal decays at twice the grating rate, because the diffraction efficiency depends on the square of the refractive index change (or excess carrier density). By fitting the data to a straight line, as shown in Fig. 7, the in-well ambipolar diffusion coefficient can be deduced from the slope (see Eq. 14). Values of D_a equal to 13.8 and 16.2 cm^2/s were obtained

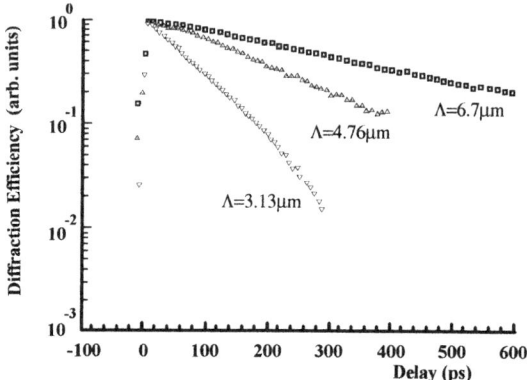

FIG. 6. Diffraction of the probe signal as a function of time for grating spacings $\Lambda = 3.13$, 4.76, and 6.7 μm.

for the two samples of different quality. The ambipolar diffusion coefficient is approximately twice the hole diffusion coefficient (see Eq. 4).

An extension of this technique is to apply an electric field to the MQW sample. Feldman *et al.* (1992) applied a lateral field to cause in-well motion of the photogenerated carriers in a transient grating experiment. The excess electrons and holes move in opposite directions to screen the applied electric field. The diffraction efficiency is then affected by (1) the field-induced

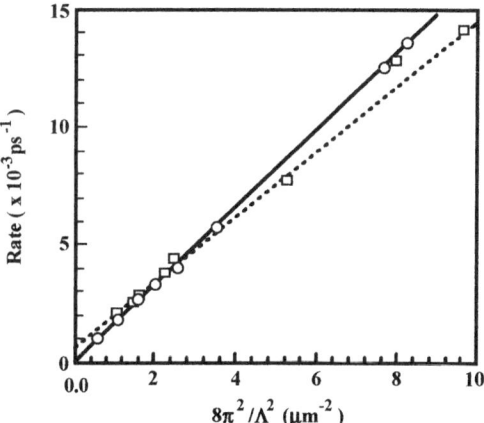

FIG. 7. Decay rates plotted to extract ambipolar diffusion coefficients. The results shown are for two different GaAs/AlGaAs MQW samples.

redistribution of carriers in k-space owing to their acceleration (carrier heating), and (2) screening of the electric field producing strong electroabsorption effects at the exciton frequency. A drift velocity of 5×10^6 cm s^{-1} was deduced from measurements of GaAs/AlAs MQWs at 10 K.

The effect of circularly polarized light on exciton saturation allows the creation of another type of transient grating in an MQW (Cameron et al., 1996). In this case, the grating is based on the spatial modulation of electron spin orientation within a uniform excess carrier distribution. Diffraction occurs by virtue of the spin-dependent free carrier contribution to exciton saturation, as described in Section IV. The decay rate of this grating is dependent on the *electron* diffusion coefficient, in contrast to the *ambipolar* diffusion, which is the normal outcome of carrier motion after optical generation.

Polarization gratings are formed when two beams with crossed linear polarizations intersect. In this case, no intensity modulation exists, and hence no amplitude grating is formed, but a periodic modulation of the electric field polarization is created across the excitation region. For equal pump intensities, the polarization changes from linear to circular to orthogonal linear to circular of the opposite sense and back to linear, as shown in Fig. 8(a). The period of this modulation is identical to that produced for amplitude gratings and is defined by the angle between the two pump beams and the wavelength of light used. The crossed-linear configuration is also used to create orientational gratings in semiconductors by means of the modulation of the linear polarization through anisotropic state filling (Smirl et al., 1982) (state filling provides preferential excitation of electrons and holes in specific k-directions) or alignment of exciton dipoles (Schultheis et al., 1986; Cundiff et al., 1992). However, Fig. 8 also shows that crossed-linear polarizations provide a modulation in circular polarization. An

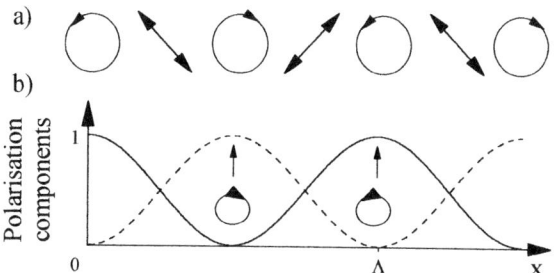

FIG. 8. (a) The polarization modulation produced when two orthogonally polarized light beams interfere and (b) amplitudes for right and left circular components of the intensity as a function of distance in the plane of the grating.

alternative description of this polarization grating is obtained by separating it into two circularly polarized components with opposite directions of rotation (Fig. 8b). The electric field modulation $E(x)$, along the grating direction x, can be written in terms of these two circular polarization components:

$$E(x) = E_0 \left[\sin\left(\frac{\pi x}{\Lambda}\right) e^{-i(\pi/4)} \frac{\hat{x} + i\hat{y}}{\sqrt{2}} + \cos\left(\frac{\pi x}{\Lambda}\right) e^{i(\pi/4)} \frac{\hat{x} - i\hat{y}}{\sqrt{2}} \right] + c.c. \quad (15)$$

where Λ is the grating spacing and $\hat{x}$ and $\hat{y}$ are unit vectors describing the polarization directions of the incident beams. Therefore, a nonlinearity that is sensitive to circular polarization will create a grating from two orthogonal linearly polarized beams.

As described in Section III, lifting of the degeneracy between the light and heavy hole bands owing to confinement in MQW semiconductors allows access to polarization-sensitive optical selection rules (see Fig. 3). Circularly polarized light, resonant with the heavy hole exciton, generates electron-hole pairs with well-defined spins. The spin orientation is maintained by the electrons for tens of picoseconds after rapid ionization of the excitons by longitudinal optical phonons (within 300 fs at room temperature) (Knox et al., 1985). Hole spins may be expected to relax on sub-picosecond timescales at room temperature because of band mixing and the mixed spin character of the valence states. Thus, excitation at the heavy hole exciton in a MQW with crossed-linear polarized pulses creates a spatial modulation of the electron spins. At sufficient laser irradiance, electrons saturate the exciton absorption by means of PSF and Coulomb screening. Since the PSF contribution is spin-dependent, a spatial modulation of electron spin orientation produces a polarization grating that can diffract a circularly polarized probe beam. Indeed, a linearly polarized probe beam can also be diffracted by this spin grating, because both right and left circular components are diffracted from the two circular polarization components of the grating.

The spin-orientational grating decay rate is determined by a combination of the spin relaxation rate and diffusion of the electrons within the quantum wells (Cameron et al., 1996). Because hole spin relaxation is rapid, wash-out of the spin grating can be interpreted in terms of the motion of electrons alone, and the decay rate can be written as

$$\Gamma = \frac{4\pi^2 D_e}{\Lambda^2} + \frac{1}{\tau_s} \quad (16)$$

where D_e is the electron diffusion coefficient and τ_s is the electron spin relaxation time. This expression is the same as for the amplitude grating (see Eq. 14) except that electron diffusion has replaced D_a, the ambipolar diffusion coefficient, and spin relaxation has replaced τ_R, the recombination time for electrons and holes.

Experiments were carried out using 1-ps pulses from a self-modelocked Ti:sapphire laser. Two pump beams and a time-delayed probe beam were configured in a standard degenerate four wave mixing (DFWM) forward traveling geometry (see Fig. 1), so that the grating modulation was produced in the plane of the quantum wells. The polarization of one of the pump beams was rotated by 90 degrees to allow the creation of the polarization grating. The average power in the pump beams was 500 μW, with approximately 50 μW in the probe beam. The spot sizes on the sample were 30 μm full width half maximum (FWHM).

Figure 9 compares the measured decays for an amplitude grating (parallel linear polarizations) and a polarization grating (orthogonal linear polarizations) with the same grating period, $\Lambda = 5$ μm, in GaAs/AlGaAs MQWs, using a linearly polarized probe beam. Both signals were characterized by a single exponential decay. The diffracted signal from the amplitude grating had a decay time of 220 ps, whereas the measured decay for the polarization grating was 13 ps. The estimated excess carrier density for the polarization grating was about 5×10^{16} cm^{-3}. Coherence spikes at zero delay in both

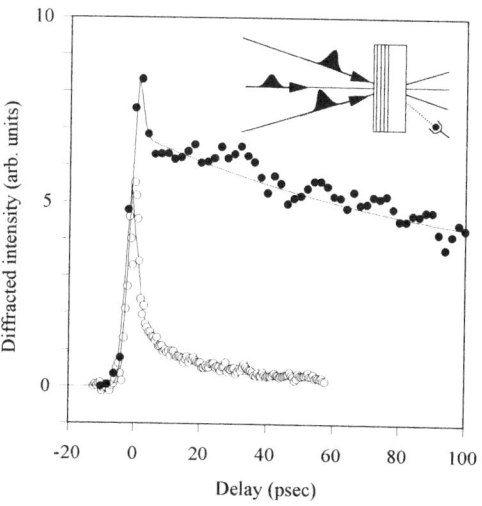

FIG. 9. The measured diffracted signal decay rate for a 5-μm amplitude grating (●) compared with the corresponding 5-μm polarization grating (○). Exponential decays with time constants of 220 and 13 ps are fitted to the two grating signals.

decay curves were attributed to diffraction of one pump beam from a grating created between the probe and the other pump beam.

For the polarization grating, the linear polarization of the diffracted probe beam was observed to be rotated by 90 degrees, as expected, because there was a π phase shift between the right and left circular polarization (spin-up and spin-down) components of the grating (see Fig. 8).

The decay rates (2Γ) of the diffracted signal obtained for several polarization and amplitude grating periods Λ are plotted in Fig. 10. The spin-relaxation result, $\tau_s = 55$ ps, is included as the infinite grating spacing limit. The amplitude and polarization grating results provide the intrawell ambipolar, D_a, and electron, D_e, diffusion coefficients, respectively, from the gradients (see Eqs. 14 and 16). A value of $D_a = 13.3$ cm^2/s is in good agreement with the results shown in Fig. 7. The deduced hole mobility of $\mu_h \sim 257$ cm^2/Vs is in excellent agreement with values for GaAs. The polarization grating results give an electron diffusion coefficient of 127 cm^2/s. This clearly shows that the holes no longer limit the diffusion rate within the wells. The electron mobility deduced from this diffusion coefficient is $\mu_e = 4924$ cm^2/Vs. This value is lower than the typical room-temperature value of 8500 cm^2/Vs for pure bulk GaAs, but, the difference can be attributed to the high background doping of the sample used here. Ionized impurity scattering significantly affects the electron mobility. Taking into

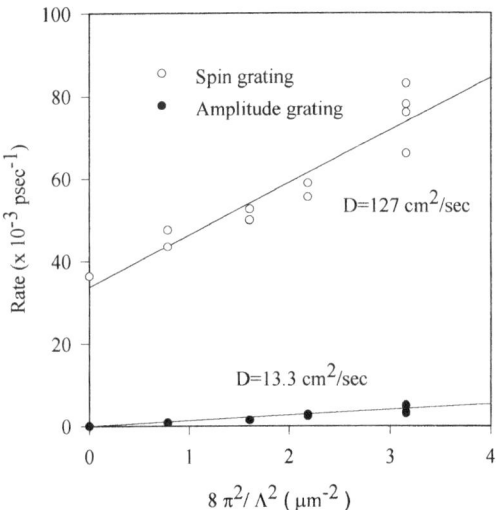

FIG. 10. Measured decay rates of the diffracted grating signal versus $8\pi^2/\Lambda^2$. Amplitude grating results (●) and polarization results (○) are plotted together for comparison.

account the background doping, a mobility of about 6000 cm²/Vs can be expected in bulk GaAs. A further decrease can be attributed to interface scattering within the quantum well layers.

The effect of barrier thickness on in-plane transport in GaAs-(Al,Ga)As superlattices using transient gratings was studied by Weinert et al. (1992). Two samples allowed a comparison of two extreme regions. Maintaining practically the same well thickness, a MQW with 5-nm-thick barriers with no coupling between the quantum wells was compared with a superlattice with 3-nm-thick barriers, whereby the minibands extended through the barriers. An ambipolar diffusion coefficient 3.5 times higher was measured in the thin barrier superlattice sample. This shows that the barrier parameters may affect the lateral transport properties because of different tunneling probabilities of carriers through the barriers. For the same quantum well thickness, smaller tunneling probabilities led to an enlarged amplitude of the free carrier wavefunction in the barriers and thus enhanced barrier-alloy disorder scattering.

Low-temperature transient grating measurements of $In_xGa_{1-x}As/GaAs$ superlattices by Williams et al. (1994) also found that the HH exciton in-plane mobility depended on barrier thickness. Three samples with barrier thicknesses between 10 and 20 nm (well widths on the order of 6 nm) exhibited higher mobilities for thinner barriers. For a 10-nm barrier thickness, the exciton mobility was measured at $6.7 \times 10^4 \, cm^2 \, V^{-1} \, s^{-1}$.

VI. Cross-Well Transport in QWs

Cross-well motion of carriers can be studied using a four-wave mixing configuration that produces a transient grating with a modulation in a direction perpendicular to the wells, as shown in Fig. 11(a). This is sometimes referred to as the short-period grating, because the period of the grating is small, $\Lambda \sim 120$ nm, with one period typically covering only a few quantum wells. The configuration shown in Fig. 11(a) is the same as that often used for phase conjugation by four-wave mixing. This grating would not normally be observed in bulk semiconductors, because diffusion over the small distance washes out the grating on femtosecond timescales. In quantum wells, on the other hand, the diffusion is restricted by the wells. Time-resolved measurements of the grating decay provide information about cross-well transport processes (Miller et al., 1989). A geometric problem occurs in this configuration when there is a small angle θ between the grating and the quantum wells (Fig. 11b). In this case, the carriers can move along the wells to wash out the grating as a result of the concentration

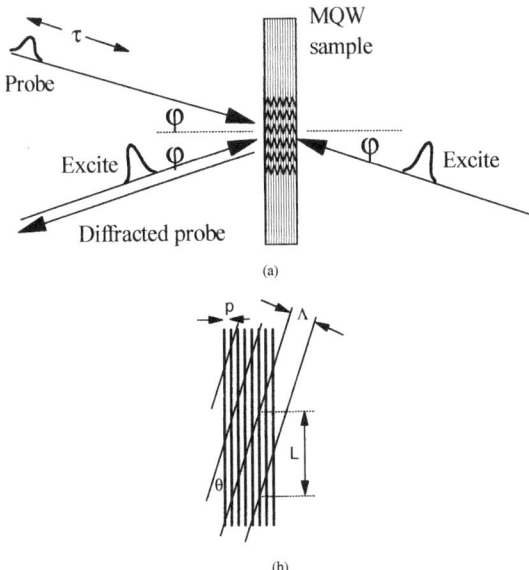

FIG. 11. (a) Counterpropagating transient grating configuration and (b) an illustration of how the quantum wells are positioned relative to the standing wave intensity maxima at some rotation of the sample.

gradient in this direction. Thus, there are two components of the grating decay owing to in-well and cross-well motion. Taking the sample rotation into account, the grating decay rate is given by

$$\Gamma = \frac{4\pi^2 D_a \sin^2\theta}{n^2 \Lambda^2} + \frac{1}{\tau_\perp} + \frac{1}{\tau_R} \qquad (17)$$

where $\tau_\perp$ is the grating decay resulting from cross-well motion.

Figure 12(a) shows the diffracted probe signals as a function of time for several angles of rotation of a MQW sample, and Fig. 12(b) plots the measured decay rates, 2Γ, versus the geometric rotation factor given in Eq. (17). Also plotted is the anticipated relaxation rate owing to intrawell diffusion using the diffusion coefficient determined from the previous measurements described in Section V. An enhancement in the diffusion rate is explained (Herbert et al., 1989) by the differential emission times of the electrons and holes from the wells (the holes have a lower barrier to cross, leaving an excess of electrons in the wells, which are more mobile than the holes).

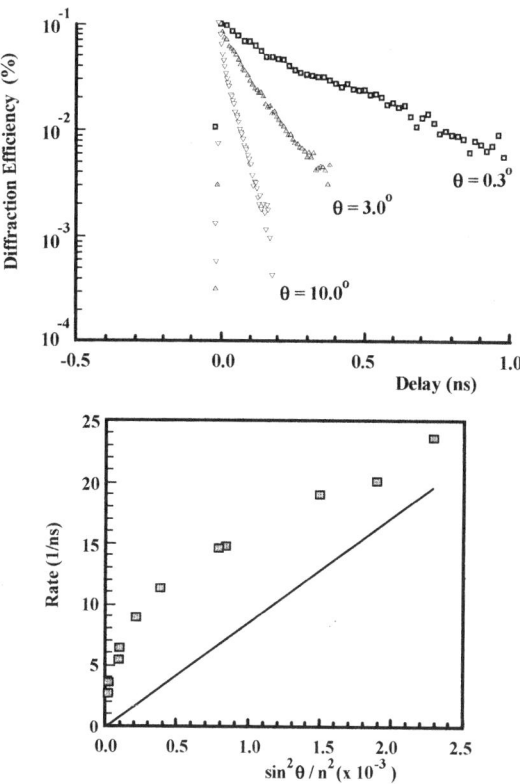

FIG. 12. (a) Diffracted probe signals at three angles of the sample and (b) measured decay rates for the diffracted probe signal for GaAs/AlGaAs MQWs in the short-period grating configuration.

Norwood *et al.* (1991) used the short-period grating technique to study perpendicular (or cross-well) transport in GaAs/AlGaAs MQWs as a function of barrier thickness ranging from 1.4 to 15 nm. These measurements (Fig. 13) were carefully carried out under conditions whereby the grating vector was perpendicular to the wells, thus avoiding any contributions from in-well transport. For thinner barriers, there was an exponential dependence of grating decay on barrier thickness, consistent with tunneling. Transport in samples with thicker barriers exhibited little dependence on barrier thickness, which was consistent with a model based on thermal emission of carriers from the wells.

In more recent work, Norwood *et al.* (1995) used the short-period grating technique to study perpendicular transport as a function of lattice tempera-

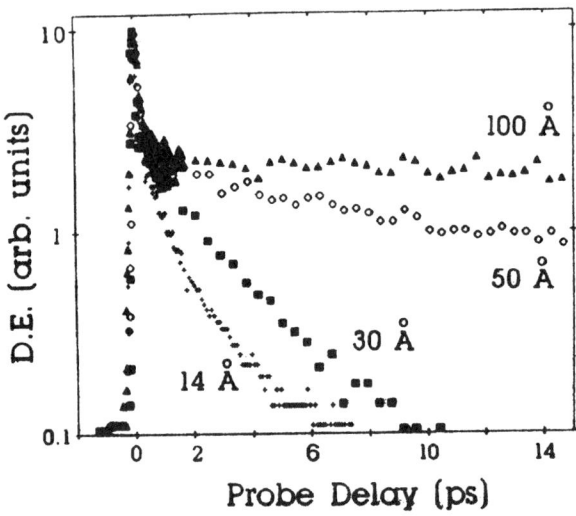

FIG. 13. Normalized diffraction efficiency from short-period gratings versus probe delay for GaAs/AlGaAs MQW structures with different barrier widths. (From Norwood, 1991.)

ture in GaAs/AlGaAs MQWs with barriers that were sufficiently thick to ensure that tunneling was minimized (Fig. 14). Below 200 K, the grating decay was found to be dominated by recombination and to be largely independent of temperature. For temperatures above 200 K, the grating decay was found to increase dramatically as the temperature increased, indicating thermal activation of the carriers over the barriers. Modeling of the results led to the conclusion that the diffusion time, rather than the thermalization time for the carriers leaving the wells, limits the speed of the perpendicular transport.

VII. Hetero-n-i-p-i Structures

Transient grating studies of hetero-n-i-p-i structures, using the same configuration as that in Fig. 1, exhibit combinations of in-well and cross-well transport and enhanced carrier mobilities. McCallum et al. (1993) studied InAs/GaAs hetero-n-i-p-i structures consisting of 12 periods of modulation-doped GaAs n-i-p-i sequence with six quantum wells in each 120-nm-wide intrinsic region (Fig. 15). The 11-nm-wide quantum wells consisted of a six-period all-binary InAs/GaAs superlattice. A similar undoped quantum well structure was used for comparison. Absorption

5 STUDIES OF CARRIER DIFFUSION AND MOBILITY IN SEMICONDUCTORS 307

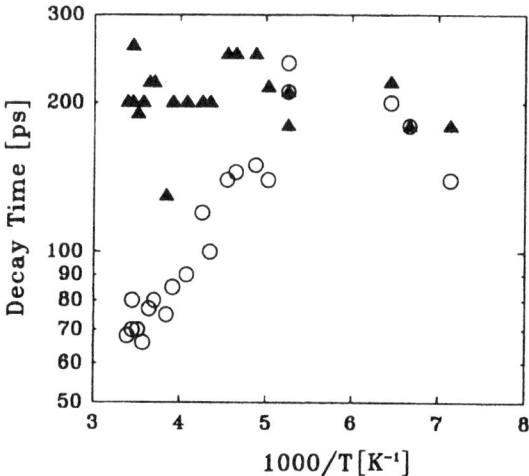

FIG. 14. The short-period grating decay time (circles) and carrier recombination lifetimes (triangles) as a function of the inverse of the lattice temperature for GaAs/AlGaAs MQWs. (From Norwood, 1995.)

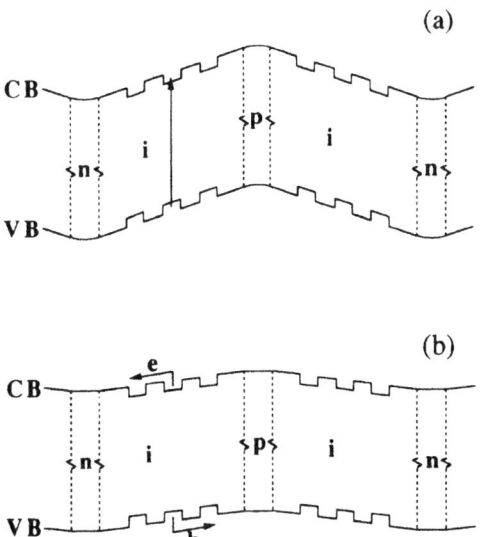

FIG. 15. Schematic of the optical response of a hetero-n-i-p-i structure: (a) photoproduction of carriers by optical excitation on heavy hole exciton resonance followed by (b) flattening of the bands as electrons and holes escape from the wells and are swept toward the doped regions. (From McCallum, 1993.)

saturation of the heavy hole exciton at 967 nm served as the optical nonlinearity that produced the transient grating.

Transient grating decays measured in the undoped sample resulted in a density-independent in-well ambipolar diffusion coefficient $D_a = 18.5 \text{ cm}^2/\text{s}$, which was in good agreement with the results described in Section V. By contrast, the hetero-n-i-p-i sample exhibited strongly carrier-density-dependent grating decay rates (Fig. 16) and effective diffusion coefficients (Fig. 17). The peak value of $180 \text{ cm}^2/\text{s}$ for the diffusion coefficient was 10 times that measured in the undoped sample.

The enhanced decay of the transient grating in the hetero-n-i-p-i structure can be understood by considering both in-well and cross-well transport of the carriers. After creation and thermalization, the built-in, cross-well electric field causes separation of the electrons and holes. Modeling showed that carrier generation and the cross-well transport in the presence of the electric field occur before significant in-well transport takes place for the typical grating spacings used in the experiments. The in-plane transport is therefore modified by the fact that there is a separation of electrons and holes. The motion is no longer ambipolar, but is enhanced by an additional drift term associated with the gradient in the n-i-p-i potential caused by an in-plane modulation of the screening of the built-in field. This was confirmed by comparison with the equivalent doped sample and also by the fact that the measured effective diffusion coefficient approached the bulk value at carrier densities sufficient to screen the field completely and to drive the hetero-n-i-p-i near flatband conditions.

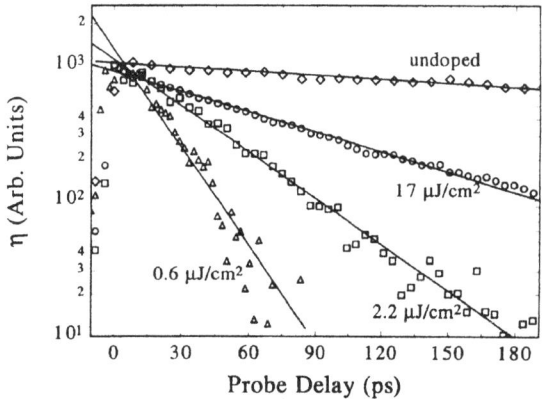

FIG. 16. Diffraction efficiency versus probe delay at a grating spacing of $5 \mu m$ for an undoped quantum well and hetero-n-i-p-i at different pump fluences. (From McCallum, 1993.)

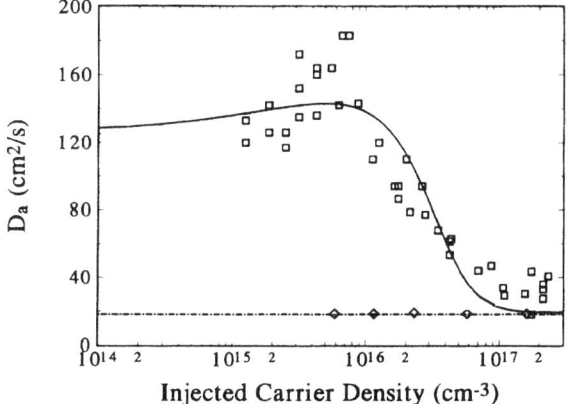

FIG. 17. Effective ambipolar diffusion coefficient as a function of average injection carrier density for a hetero-n-i-p-i sample (squares) and for an equivalent undoped multiple quantum well sample (diamonds). (From McCallum, 1993.)

The use of hetero-n-i-p-i structures therefore enhances the decay rate of the transient grating, which may be useful for increasing the speed of interaction in practical applications such as two-beam mixing. However, the separation of charge also results in a much longer carrier lifetime, so that the overall recovery time may be longer.

VIII. Conclusions

In this chapter, we have reviewed the use of optically induced transient gratings in time-resolved measurements of the motion of electrons and holes in semiconductors. Most of the observed dynamics take place on picosecond timescales because of the typically micrometer distances defined by the grating periods and also because of the specific material parameters. Although there was a great deal of interest in using this technique to study bulk semiconductors in the 1970s and early 1980s, more recent publications have concentrated almost entirely on quantum well structures.

The motion of carriers in quantum well semiconductors can be particularly complex, and the transient grating method has proved useful in determining some key parameters and timescales associated with both in-well and cross-well motion. Spin-dependent optical nonlinearities have been shown to provide additional insights into carrier transport in quantum wells, particularly in relation to determining both electron and hole mobilities. The dynamics becomes even more complicated when charge separation occurs as a result of differential thermionic emission rates from quantum wells and different mobilities for the electrons and holes.

It has not been possible to include pump-probe studies of p-i-n and n-i-p-i structures that use the quantum confined Stark effect in this chapter. These studies have provided a wealth of information on carrier transport in quantum well structures, including thermionic emission and tunneling rates as well as observations of enhanced mobilities in the presence of electric fields.

There are good prospects for developing the transient grating technique to further our understanding of carrier transport in low-dimensional semiconductors. For example, structural effects such as well and barrier heights and widths, the effects of interface roughness, and the motion of electrons and holes when subjected to applied electric and magnetic fields in transient grating configurations are likely to be explored.

List of Abbreviations

CB	conduction band
DFWM	degenerate four wave mixing
FWHM	full width half maximum
HH	heavy hole
LH	light hole
MQW	multiple quantum well
OCP	opposite circular polarization
OLP	opposite linear polarization
PSF	phase space filling
QCSE	quantum confined Stark effect
SCP	same circular polarization

References

Bergner, H., Bruckner, V., and Supianek, M. (1986). "Ultrafast Processes in Silicon Studied by Transient Gratings," *IEEE J. Quantum Electron.* **22**, 1306.

Borshch, A. A., Brodin, M. S., Ovchar, V. V., Odulov, S. G., and Soskin, M. M. (1973). *Sov. Phys. JETP Lett.* **18**, 397.

Cameron, A. R., Riblet, P., and Miller, A. (1996). "Spin Gratings and the Measurement of Electron Drift Mobility in Multiple Quantum Well Semiconductors," *Phys. Rev. Lett.* **76**, 4793.

Canto-Said, E. J., Hagan, D. J., Young, J., and Van Stryland, E. W. (1991). "Degenerate Four-Wave Mixing Measurements of High Order Nonlinearities in Semiconductors," *IEEE J. Quantum Electron.* **27**, 2274.

Chemla, D. S., Miller, D. A. B., Smith, P. W., Gossard, P. W., and Wiegmann, W. (1984). "Room Temperature Excitonic Nonlinear Absorption and Refraction in GaAs/AlGaAs Multiple Quantum Well Structures," *IEEE J. Quantum Electron.* **20**, 265.

Cundiff, S. T., Wang, H., and Steel, D. G. (1992). "Polarization-Dependent Picosecond Excitonic Nonlinearities and the Complexities of Disorder," *Phys. Rev. B* **46**, 7248.

Eichler, H. J., Gunther, P., and Pohl, D. W. (1986). *Laser Induced Dynamic Gratings.* Springer-Verlag, Berlin.

Eichler, H. J. and Massmann, F. (1982). "Diffraction Efficiency and Decay Times of Free-Carrier Gratings in Silicon," *J. Appl. Phys.* **53**, 3237.

Feldman, J., Grossmann, P., Stolz, W., Gobel, E., and Ploog, K. (1992). "Transient-Grating Experiments for the Study of Electron-Hole Separation in an Electric Field," *Semicond. Sci. Technol.* **7**, B130.

Haug, H., ed. (1988). *Optical Nonlinearities and Instabilities in Semiconductors.* Academic Press, Boston.

Herbert, D. C., Milsom, P., Miller, A., and Manning, R. J. (1989). "Orientational Dependence of Transient Gratings in Multiple Quantum Well Structures," *Semicond. Sci. Technol.* **4**, 696.

Hoffman, C. H., Jarasiunas, K., Gerritsen, H. J., and Nurmikko, A. V. (1978). "Measurement of Surface Recombination Velocity in Semiconductors by Diffraction from Picosecond Transient Free-Carrier Gratings," *Appl. Phys. Lett.* **33**, 536.

Holden, T. M., Kennedy, G. T., Cameron, A. R., Riblet, P., and Miller, A. (1997). "Exciton Saturation in Room Temperature GaAs/AlGaAs Multiple Quantum Wells," *Appl. Phys. Lett.* **71**, 936.

Hutchings, D. C., Sheik-Bahae, M., Hagan, D. J., and Van Stryland, E. W. (1992). "Kramers-Kronig Relations in Nonlinear Optics," *Opt. Quantum Electron.* **24**, 1.

Jain, R. K. (1982). "Degenerate Four-Wave Mixing in Semiconductors: Application to Phase Conjugation and to Picosecond-Resolved Studies of Transient Carrier Dynamics," *Opt. Eng.* **21**, 199.

Jarasiunas, K. and Gerritsen, H. J. (1978). "Ambipolar Diffusion Measurements in Semiconductors Using Nonlinear Transient Gratings," *Appl. Phys. Lett.* **33**, 190.

Jarasiunas, K. and Vaitkas, J. (1974). "Properties of a Laser Induced Phase Grating in CdSe," *Phys. Stat. Sol. (a)* **23**, K19.

Jarasiunas, K. and Vaitkus, J. (1977). "Investigation of Non-Equilibrium Processes by the Method of Transient Holograms," *Phys. Stat. Sol. (a)* **44**, 793.

Knox, W. H., Fork, R. L., Downer, M. C., Miller, D. A. B., Chemla, D. S., Shank, C. V., Gossard, A. C., and Wiegmann, W. (1985). "Femtosecond Dynamics of Resonantly Excited Excitons in Room Temperature GaAs Quantum Wells," *Phys. Rev. Lett.* **54**, 1306.

Kremenitski, V., Odulov, S., and Soskin, M. (1979). "Dynamic Gratings in Cadmium Telluride," *Phys. Stat. Sol. (a)* **51**, K63.

McCallum, D. S., Cartwright, A. N., Huang, X. R., Boggess, T. F., Smirl, A. L., and Hasenberg, T. C. (1993). "Enhanced Ambipolar In-Plane Transport in an InAs/GaAs Hetero-n-i-p-i," *J. Appl. Phys.* **73**, 3860.

Miller, A., Manning, R. J., Milsom, P. K., Hutchings, D. C., Crust, D. W., and Woodbridge, K. (1989). "Transient Grating Studies of Excitonic Optical Nonlinearities in GaAs/AlGaAs Multiple-Quantum-Well Structures," *J. Opt. Soc. Am. B* **6**, 567.

Miller, A., Miller, D. A. B., and Smith, S. D. (1981). "Dynamic Nonlinear Optical Processes in Semiconductors," *Adv. Phys.* **30**, 697–800.

Miller, A. and Sibbett, W. (1988). "A Perspective on Ultrafast Phenomena," *J. Mod. Opt.* **35**, 1871–1890.

Miller, D. A. B., Chemla, D. S., Eilenberger, D. J., Smith, P. W., Gossard, P. W., and Tsang,

W. T. (1982). "Large Room Temperature Optical Nonlinearity in GaAs/Ga$_{1-x}$Al$_x$As Multiple Quantum Well Structures," *Appl. Phys. Lett.* **41**, 679.

Miyoshi, T., Aoyagi, Y., Segawa, Y., Namba, S., Okamoto, H., and Hamakawa, Y. (1983). "Lifetime and Diffusion Coefficient of Carriers in X-Ray Irradiated a-Si:H," *Jpn. J. Appl. Phys.* **22**, 886.

Moss, S. C., Lindle, J. R., Mackey, H. J., and Smirl, A. L. (1981). "Measurement of the Diffusion Coefficient and Recombination Effects in Germanium by Diffraction from Optically-Induced Picosecond Transient Gratings," *Appl. Phys. Lett.* **39**, 27.

Norwood, D. P., Smirl, A. L., and Swoboda, H.-E. (1995). "Short-Period Transient Grating Measurements of Perpendicular Over-Barrier Diffusion in GaAs/AlGaAs Multiple Quantum Wells," *Appl. Phys.* **77**, 1113.

Norwood, D. P., Swoboda, H.-E., Dawson, M. D., Smirl, A. L., Anderson, D. R., and Hasenberg, T. C. (1991). "Room-Temperature Short-Period Transient Grating Measurements of Perpendicular Transport in GaAs/AlGaAs Multiple Quantum Wells," *Appl. Phys. Lett.* **59**, 219.

Schmitt, Rink, S., Chemla, D. S., and Miller, D. A. B. (1989). "Linear and Nonlinear Optical Properties of Semiconductor Quantum Wells," *Adv. Phys.* **38**, 89.

Schultheis, L. M., Kuhl, J., Honold, A., and Tu, C. W. (1986). "Picosecond Phase Coherence and Orientational Relaxation of Excitons in GaAs," *Phys. Rev. Lett.* **57**, 1797.

Shah, J. (1996). *Ultrafast Spectroscopy of Semiconductors and Semiconductor Nanostructures.* Springer Series in Solid State Sciences, Vol. 115, Springer, Berlin.

Smirl, A. L., Boggess, T. F., Wherrett, B. S., Perryman, G. P., and Miller, A. (1982). "Picosecond Optically Induced Anisotropic State Filling in Semiconductors," *Phys. Rev. Lett.* **49**, 933.

Smirl, A. L., Moss, S. C., and Lindle, J. R. (1982). "Picosecond Dynamics of High-Density Laser-Induced Transient Plasma Gratings in Germanium," *Phys. Rev. B* **25**, 2645.

Snelling, M. J., Perozzo, P., Hutchings, D. C., Galbraith, I., and Miller, A. (1994). "Investigation of Excitonic Saturation by Time-Resolved Circular Dichroism in GaAs-Al$_x$Ga$_{1-x}$As Multiple Quantum Wells," *Phys. Rev. B* **49**, 17160.

Vaitkus, J., Jarasiunas, K., Gaubas, E., Jonikas, L., Pranaitis, R., and Subacius, L. (1986). "The Diffraction of Light by Transient Gratings in Crystalline, Ion-Implanted and Amorphous Silicon," *IEEE J. Quantum Electron.* **22**, 1298.

Weinert, H., Kolenda, J., and Petrauskas, M. (1992). "Lateral Transport in GaAs-(Al,Ga)As Superlattice Investigated by Transient Grating Measurements," *Sol. State Commun.* **81**, 467.

Williams, G. V., Phillips, C. C., and Woodbridge, K. (1994). "Time-Resolved DFWM Excite-Probe and Transient Grating Studies of In$_x$Ga$_{1-x}$As/GaAs Superlattices," *Semicond. Sci. Technol.* **9**, 1096.

Woerdman, J. P. (1970a). "Diffraction of Light by Laser Generated Free Carriers in Si: Dispersion or Absorption?," *Phys. Lett. A* **32**, 305.

Woerdman, J. P. (1970b). "Formation of Transient Free Carrier Holograms in Si," *Opt. Commun.* **2**, 212.

Woerdman, J. P. and Bolger, B. (1969). "Diffraction of Light by a Laser Induced Grating in Si," *Phys. Lett. A* **30**, 164.

Index

Numbers followed by the letter f indicate figures; numbers followed by the letter t indicate tables.

A

Absorber recovery time, 222
Absorber reflectivity, 241
Acousto-optic modulator (AOM), 83, 106–107
All-optical switching, 148–153,155
Antiresonant Fabry-Perot saturable absorber (A-FPSA),
 antireflection-coated (AR-coated) SESAM, 217–218, 256–258
 basic concept of, 215–220, 243, 249–259
 high-finesse, 251–255
 low-finesse, 256–257
Applications, 145
Absorption, 85–86, 88, 104–105,110–111, 118
Additive pulse modelocking (APM), 83, 107, 225
Amplitude modulation (AM), 107, 130

B

Band structure, 86–88
Beamsplitter (BS), 101, 106
Bernard-Duraffourg condition, 88
Biexcitons,
 as giant two-photon absorption resulting from formation of, 196–198
 as a nonlinear medium, 195–201
 optical bistability with, 200–201
 transition between single excitons and, 198–200
Boltzmann conditions, 87, 289
Broadband optical communications, 84

Bulk semiconductors,
 grating experiments and, 292–293
 second-order susceptibility measurements of experimental values and, 30–34

C

Carriers
 confinement and waveguiding properties of, 91–93
 heating effects of, 97–100, 102, 104, 110, 118, 120, 132, 134, 136, 140–144, 149, 154
 recombination of, 94–95, 135
 scattering effects of, 96–97, 104, 142–143
Coherent light emission from quantum wires, 182–189
Colliding pulse modelocked (CPM), 267
Conduction bands, 86–91, 98
Control pulse, 150
Cross-gain modulation (XGM), 146–149
Cross-phase modulation (XPM), 147–148
Co-polarization, 107–108, 111, 117–118
Cross-polarization, 107–109, 112, 117–118, 127
Curves, gain, 89–91, 97–99, 114

D

Data analysis, 119–134
Degenerate four-wave mixing (DFWM), 301
Dicke's superradiance, 165–169
Diffusion effects, 95–96

313

Double-chirped mirror (DCM), 268–271
Double-resonant case, 52–54

E

Epitaxial growth techniques, 212–214
Excitons,
    in excitonic interband transitions, 47–49
    in excitonic optical nonlinearity, 169–178
    in quantum wells, 178–182
    superradiance of, 189–191
    superradiance from quantum dots and wires, 164–169
Exciton-polaritons (EP), 182–189, 201–208

F

Femtosecond modelocking, 229–230
Four-wave mixing (FWM), 141–146, 148–149
Free carrier absorption (FCA), 84, 98–100, 105, 112, 154
Free carrier interband transitions, 46
Frequency modulation (FM), 107, 113, 130

G

Group delay dispersion (GDD), 225–227, 233–234, 270–271

H

Heterodyne pump-probe, see Measurement techniques
Hetero-n-i-p-i structures, 306–309
Higher-order susceptibility, 6–12
Hybrid bond model, 25–30

K

Kerr lens modelocking (KLM), 214–215, 225–229, 238, 267–271
Kramers-Kronig (KK), 114, 117, 149, 288

L

Laser diode (LD), 31
Linear gain, 86–91
Long wavelength off-resonant limit, 50–51

M

Measurement techniques,
    heterodyne pump-probe, 106–112
    orthogonally polarized collinear pump-probe, 101–105, 112
Metalo-organic chemical vapor deposition (MOCVD), 212
Modelocking, passive
    during buildup phase, 236–239
    mechanisms, 223–227
    modulation depth and, 230–234
    recovery time and, 228–230
    saturation and, 235–236
Molecular beam epitaxy (MBE), 212, 216–217
Multiple quantum wells (MQW) 84, 96, 106, 109–111, 122, 127–129, 132–133, 137, 288, 294–295, 300–306
    exciton saturation in, 293–297
    in-well transport in, 297–303
    p-i-n modulation and, 259–260

N

Nonlinear gain, 86–87, 94–100, 136, 138–139, 154–155
Nonlinear optical responses,
    figure of merit for, 191–195
    weak localization and, 201–208

O

Opposite circular polarization (OCP), 295–297
Opposite linear polarization (OLP), 295–297
Optical Kerr effect, 171
Optical nonlinearities and carrier transport, 287–289
Optical rectification and terahertz emission,
    brief review of, 63–67
    in bulk semiconductors, 67–73
    in semiconductor quantum wells, 73–78
Optimized modematching (OMM), 261–267
Orthogonally polarized collinear pump-probe, see Measurement techniques

P

Phasematching, 30–37, 40–41

Phase space filling (PSF), 288–296, 300
Photomultiplier tube (PMT), 31
Photonic bandgap (PBG) system, 188–189
Piezoelectric transducer (PZT), 114
p-n junction, 92
Polarization sensitive delay (PSD), 152
Population inversion, 86
Probe transmission, changes in, 104–105, 108, 113, 121–122, 131–133, 138, 154
Pulse fluence, 221–222, 241
Pump-probe techniques, 101–113, 145

Q

Q-Switching,
  laser mode size in gain and absorber and, 242–242
  onset of instabilities of, 239–242
  requirements for passive, 243–245
Quantum confined Stark effect (QCSE), 287
Quantum wells,
  asymmetric, 42–49
  and cross-well transport in, 303–306
  devices, 93
  symmetric, 48–49
Quasi-phasematching (QPM), 37–40

R

Refractive index, 112, 114–116, 118–119, 124, 127, 140–141, 154–155
Resonant passive modelocking (RPM), 215

S

Saturable Bragg reflector (SBR), 258–259, see also low-finesse A-FPSA, 256–257
Saturation fluence, 222–223, 248–249
Saturation intensity, 222–223
Saturation of short pulses in active waveguides, 134–141
Same circular polarization (SCP), 295–297
Second harmonic generation (SHG), 38, 54–56, 61–62
Second-order nonlinearities and susceptibility,
  in bulk semiconductors, 4–30
  experimental demonstration processes of, 34–42
  interband theory and experimentation of, 58–63
  intersubband experimentation of, 55–58
  intersubband theory of, 49–55
  and phasematching, 30–37
  in quantum wells and superlattices, 42–63
  semiclassical theory of, 12–19
Semiconductor diode laser, 84–85, 91–93, 212
Semiconductor optical amplifiers (SOA), 84–85, 93, 104, 106, 132, 146–154
Semiconductor saturable absorber mirrors (SESAMs),
  designs of, 249–260
  of a dispersive structure, 259
  history of, 214–220
  macroscopic properties of, 221–223
  microscopic properties of, 245–249
  motivation for, 212–214
  passively modelocked solid state lasers, with use of, 260–271
  passively Q-switched solid state lasers, with use of, 272–275
Short pulses, shaping and saturation of, 134–141
Signal pulse, 150
Single-resonant case, 51–52
Soliton modelocking, 229–233
Spectral hole burning (SHB), 84, 96–97, 102–103, 105, 110, 117, 124, 132–133, 144, 149, 154
Self-phase modulation (SPM), 233
Strained layer multiple quantum well (SLMQW), 110–111, 113

T

Third-order susceptibility, 163–175
Transient gratings, 290–292
Transverse electric (TE), 84, 106, 109–113
Transverse magnetic (TM), 84, 106, 109, 111–113
Two-photon absorption (TPA), 84, 99, 104–105, 109–112, 117, 127, 131–132, 134, 149, 154

U

Ultrafast nonlinear interferometer (UNI), 152–153

V

Valence bands, 86–91, 98

W

Waveguiding, 92–93
Wavelength converters, 146–148, 155
Wavelength division multiplexed (WDM), 146

Z

Zinc-blende semiconductors, off-resonant second-order susceptibilities of, 19–25

Contents of Volumes in This Series

Volume 1 Physics of III–V Compounds

C. Hilsum, Some Key Features of III–V Compounds
Franco Bassani, Methods of Band Calculations Applicable to III–V Compounds
E. O. Kane, The k-p Method
V. L. Bonch-Bruevich, Effect of Heavy Doping on the Semiconductor Band Structure
Donald Long, Energy Band Structures of Mixed Crystals of III–V Compounds
Laura M. Roth and Petros N. Argyres, Magnetic Quantum Effects
S. M. Puri and T. H. Geballe, Thermomagnetic Effects in the Quantum Region
W. M. Becker, Band Characteristics near Principal Minima from Magnetoresistance
E. H. Putley, Freeze-Out Effects, Hot Electron Effects, and Submillimeter Photoconductivity in InSb
H. Weiss, Magnetoresistance
Betsy Ancker-Johnson, Plasma in Semiconductors and Semimetals

Volume 2 Physics of III–V Compounds

M. G. Holland, Thermal Conductivity
S. I. Novkova, Thermal Expansion
U. Piesbergen, Heat Capacity and Debye Temperatures
G. Giesecke, Lattice Constants
J. R. Drabble, Elastic Properties
A. U. Mac Rae and G. W. Gobeli, Low Energy Electron Diffraction Studies
Robert Lee Mieher, Nuclear Magnetic Resonance
Bernard Goldstein, Electron Paramagnetic Resonance
T. S. Moss, Photoconduction in III–V Compounds
E. Antoncik ad J. Tauc, Quantum Efficiency of the Internal Photoelectric Effect in InSb
G. W. Gobeli and I. G. Allen, Photoelectric Threshold and Work Function
P. S. Pershan, Nonlinear Optics in III–V Compounds
M. Gershenzon, Radiative Recombination in the III–V Compounds
Frank Stern, Stimulated Emission in Semiconductors

Volume 3 Optical of Properties III–V Compounds

Marvin Hass, Lattice Reflection
William G. Spitzer, Multiphonon Lattice Absorption
D. L. Stierwalt and R. F. Potter, Emittance Studies
H. R. Philipp and H. Ehrenveich, Ultraviolet Optical Properties
Manuel Cardona, Optical Absorption above the Fundamental Edge
Earnest J. Johnson, Absorption near the Fundamental Edge
John O. Dimmock, Introduction to the Theory of Exciton States in Semiconductors
B. Lax and J. G. Mavroides, Interband Magnetooptical Effects
H. Y. Fan, Effects of Free Carries on Optical Properties
Edward D. Palik and George B. Wright, Free-Carrier Magnetooptical Effects
Richard H. Bube, Photoelectronic Analysis
B. O. Seraphin and H. E. Bennett, Optical Constants

Volume 4 Physics of III–V Compounds

N. A. Goryunova, A. S. Borschevskii, and D. N. Tretiakov, Hardness
N. N. Sirota, Heats of Formation and Temperatures and Heats of Fusion of Compounds $A^{III}B^{V}$
Don L. Kendall, Diffusion
A. G. Chynoweth, Charge Multiplication Phenomena
Robert W. Keyes, The Effects of Hydrostatic Pressure on the Properties of III–V Semiconductors
L. W. Aukerman, Radiation Effects
N. A. Goryunova, F. P. Kesamanly, and D. N. Nasledov, Phenomena in Solid Solutions
R. T. Bate, Electrical Properties of Nonuniform Crystals

Volume 5 Infrared Detectors

Henry Levinstein, Characterization of Infrared Detectors
Paul W. Kruse, Indium Antimonide Photoconductive and Photoelectromagnetic Detectors
M. B. Prince, Narrowband Self-Filtering Detectors
Ivars Melngalis and T. C. Harman, Single-Crystal Lead-Tin Chalcogenides
Donald Long and Joseph L. Schmidt, Mercury-Cadmium Telluride and Closely Related Alloys
E. H. Putley, The Pyroelectric Detector
Norman B. Stevens, Radiation Thermopiles
R. J. Keyes and T. M. Quist, Low Level Coherent and Incoherent Detection in the Infrared
M. C. Teich, Coherent Detection in the Infrared
F. R. Arams, E. W. Sard, B. J. Peyton, and F. P. Pace, Infrared Heterodyne Detection with Gigahertz IF Response
H. S. Sommers, Jr., Macrowave-Based Photoconductive Detector
Robert Sehr and Rainer Zuleeg, Imaging and Display

Volume 6 Injection Phenomena

Murray A. Lampert and Ronald B. Schilling, Current Injection in Solids: The Regional Approximation Method
Richard Williams, Injection by Internal Photoemission
Allen M. Barnett, Current Filament Formation

R. Baron and J. W. Mayer, Double Injection in Semiconductors
W. Ruppel, The Photoconductor-Metal Contact

Volume 7 Application and Devices
Part A

John A. Copeland and Stephen Knight, Applications Utilizing Bulk Negative Resistance
F. A. Padovani, The Voltage-Current Characteristics of Metal-Semiconductor Contacts
P. L. Hower, W. W. Hooper, B. R. Cairns, R. D. Fairman, and D. A. Tremere, The GaAs Field-Effect Transistor
Marvin H. White, MOS Transistors
G. R. Antell, Gallium Arsenide Transistors
T. L. Tansley, Heterojunction Properties

Part B

T. Misawa, IMPATT Diodes
H. C. Okean, Tunnel Diodes
Robert B. Campbell and Hung-Chi Chang, Silicon Junction Carbide Devices
R. E. Enstrom, H. Kressel, and L. Krassner, High-Temperature Power Rectifiers of $GaAs_{1-x}P_x$

Volume 8 Transport and Optical Phenomena

Richard J. Stirn, Band Structure and Galvanomagnetic Effects in III–V Compounds with Indirect Band Gaps
Roland W. Ure, Jr., Thermoelectric Effects in III–V Compounds
Herbert Piller, Faraday Rotation
H. Barry Bebb and E. W. Williams, Photoluminescence I: Theory
E. W. Williams and H. Barry Bebb, Photoluminescence II: Gallium Arsenide

Volume 9 Modulation Techniques

B. O. Seraphin, Electroreflectance
R. L. Aggarwal, Modulated Interband Magnetooptics
Daniel F. Blossey and Paul Handler, Electroabsorption
Bruno Batz, Thermal and Wavelength Modulation Spectroscopy
Ivar Balslev, Piezopptical Effects
D. E. Aspnes and N. Bottka, Electric-Field Effects on the Dielectric Function of Semiconductors and Insulators

Volume 10 Transport Phenomena

R. L. Rhode, Low-Field Electron Transport
J. D. Wiley, Mobility of Holes in III–V Compounds
C. M. Wolfe and G. E. Stillman, Apparent Mobility Enhancement in Inhomogeneous Crystals
Robert L. Petersen, The Magnetophonon Effect

Volume 11 Solar Cells

Harold J. Hovel, Introduction; Carrier Collection, Spectral Response, and Photocurrent; Solar Cell Electrical Characteristics; Efficiency; Thickness; Other Solar Cell Devices; Radiation Effects; Temperature and Intensity; Solar Cell Technology

Volume 12 Infrared Detectors (II)

W. L. Eiseman, J. D. Merriam, and R. F. Potter, Operational Characteristics of Infrared Photodetectors
Peter R. Bratt, Impurity Germanium and Silicon Infrared Detectors
E. H. Putley, InSb Submillimeter Photoconductive Detectors
G. E. Stillman, C. M. Wolfe, and J. O. Dimmock, Far-Infrared Photoconductivity in High Purity GaAs
G. E. Stillman and C. M. Wolfe, Avalanche Photodiodes
P. L. Richards, The Josephson Junction as a Detector of Microwave and Far-Infrared Radiation
E. H. Putley, The Pyroelectric Detector–An Update

Volume 13 Cadmium Telluride

Kenneth Zanio, Materials Preparations; Physics; Defects; Applications

Volume 14 Lasers, Junctions, Transport

N. Holonyak, Jr. and M. H. Lee, Photopumped III–V Semiconductor Lasers
Henry Kressel and Jerome K. Butler, Heterojunction Laser Diodes
A Van der Ziel, Space-Charge-Limited Solid-State Diodes
Peter J. Price, Monte Carlo Calculation of Electron Transport in Solids

Volume 15 Contacts, Junctions, Emitters

B. L. Sharma, Ohmic Contacts to III–V Compounds Semiconductors
Allen Nussbaum, The Theory of Semiconducting Junctions
John S. Escher, NEA Semiconductor Photoemitters

Volume 16 Defects, (HgCd)Se, (HgCd)Te

Henry Kressel, The Effect of Crystal Defects on Optoelectronic Devices
C. R. Whitsett, J. G. Broerman, and C. J. Summers, Crystal Growth and Properties of $Hg_{1-x}Cd_xSe$ alloys
M. H. Weiler, Magnetooptical Properties of $Hg_{1-x}Cd_x$Te Alloys
Paul W. Kruse and John G. Ready, Nonlinear Optical Effects in $Hg_{1-x}Cd_x$Te

Volume 17 CW Processing of Silicon and Other Semiconductors

James F. Gibbons, Beam Processing of Silicon
Arto Lietoila, Richard B. Gold, James F. Gibbons, and Lee A. Christel, Temperature Distribu-

tions and Solid Phase Reaction Rates Produced by Scanning CW Beams
Arto Leitoila and James F. Gibbons, Applications of CW Beam Processing to Ion Implanted Crystalline Silicon
N. M. Johnson, Electronic Defects in CW Transient Thermal Processed Silicon
K. F. Lee, T. J. Stultz, and James F. Gibbons, Beam Recrystallized Polycrystalline Silicon: Properties, Applications, and Techniques
T. Shibata, A. Wakita, T. W. Sigmon, and James F. Gibbons, Metal-Silicon Reactions and Silicide
Yves I. Nissim and James F. Gibbons, CW Beam Processing of Gallium Arsenide

Volume 18 Mercury Cadmium Telluride

Paul W. Kruse, The Emergence of $(Hg_{1-x}Cd_x)Te$ as a Modern Infrared Sensitive Material
H. E. Hirsch, S. C. Liang, and A. G. White, Preparation of High-Purity Cadmium, Mercury, and Tellurium
W. F. H. Micklethwaite, The Crystal Growth of Cadmium Mercury Telluride
Paul E. Petersen, Auger Recombination in Mercury Cadmium Telluride
R. M. Broudy and V. J. Mazurczyck, (HgCd)Te Photoconductive Detectors
M. B. Reine, A. K. Soad, and T. J. Tredwell, Photovoltaic Infrared Detectors
M. A. Kinch, Metal-Insulator-Semiconductor Infrared Detectors

Volume 19 Deep Levels, GaAs, Alloys, Photochemistry

G. F. Neumark and K. Kosai, Deep Levels in Wide Band-Gap III–V Semiconductors
David C. Look, The Electrical and Photoelectronic Properties of Semi-Insulating GaAs
R. F. Brebrick, Ching-Hua Su, and Pok-Kai Liao, Associated Solution Model for Ga-In-Sb and Hg-Cd-Te
Yu. Ya. Gurevich and Yu. V. Pleskon, Photoelectrochemistry of Semiconductors

Volume 20 Semi-Insulating GaAs

R. N. Thomas, H. M. Hobgood, G. W. Eldridge, D. L. Barrett, T. T. Braggins, L. B. Ta, and S. K. Wang, High-Purity LEC Growth and Direct Implantation of GaAs for Monolithic Microwave Circuits
C. A. Stolte, Ion Implantation and Materials for GaAs Integrated Circuits
C. G. Kirkpatrick, R. T. Chen, D. E. Holmes, P. M. Asbeck, K. R. Elliott, R. D. Fairman, and J. R. Oliver, LEC GaAs for Integrated Circuit Applications
J. S. Blakemore and S. Rahimi, Models for Mid-Gap Centers in Gallium Arsenide

Volume 21 Hydrogenated Amorphous Silicon
Part A

Jacques I. Pankove, Introduction
Masataka Hirose, Glow Discharge; Chemical Vapor Deposition
Yoshiyuki Uchida, di Glow Discharge
T. D. Moustakas, Sputtering
Isao Yamada, Ionized-Cluster Beam Deposition
Bruce A. Scott, Homogeneous Chemical Vapor Deposition

Frank J. Kampas, Chemical Reactions in Plasma Deposition
Paul A. Longeway, Plasma Kinetics
Herbert A. Weakliem, Diagnostics of Silane Glow Discharges Using Probes and Mass Spectroscopy
Lester Gluttman, Relation between the Atomic and the Electronic Structures
A. Chenevas-Paule, Experiment Determination of Structure
S. Minomura, Pressure Effects on the Local Atomic Structure
David Adler, Defects and Density of Localized States

Part B

Jacques I. Pankove, Introduction
G. D. Cody, The Optical Absorption Edge of a-Si:H
Nabil M. Amer and Warren B. Jackson, Optical Properties of Defect States in a-Si:H
P. J. Zanzucchi, The Vibrational Spectra of a-Si:H
Yoshihiro Hamakawa, Electroreflectance and Electroabsorption
Jeffrey S. Lannin, Raman Scattering of Amorphous Si, Ge, and Their Alloys
R. A. Street, Luminescence in a-Si:H
Richard S. Crandall, Photoconductivity
J. Tauc, Time-Resolved Spectroscopy of Electronic Relaxation Processes
P. E. Vanier, IR-Induced Quenching and Enhancement of Photoconductivity and Photoluminescence
H. Schade, Irradiation-Induced Metastable Effects
L. Ley, Photoelectron Emission Studies

Part C

Jacques I. Pankove, Introduction
J. David Cohen, Density of States from Junction Measurements in Hydrogenated Amorphous Silicon
P. C. Taylor, Magnetic Resonance Measurements in a-Si:H
K. Morigaki, Optically Detected Magnetic Resonance
J. Dresner, Carrier Mobility in a-Si:H
T. Tiedje, Information about band-Tail States from Time-of-Flight Experiments
Arnold R. Moore, Diffusion Length in Undoped a-Si:H
W. Beyer and J. Overhof, Doping Effects in a-Si:H
H. Fritzche, Electronic Properties of Surfaces in a-Si:H
C. R. Wronski, The Staebler-Wronski Effect
R. J. Nemanich, Schottky Barriers on a-Si:H
B. Abeles and T. Tiedje, Amorphous Semiconductor Superlattices

Part D

Jacques I. Pankove, Introduction
D. E. Carlson, Solar Cells
G. A. Swartz, Closed-Form Solution of I–V Characteristic for a a-Si:H Solar Cells
Isamu Shimizu, Electrophotography
Sachio Ishioka, Image Pickup Tubes

P. G. LeComber and W. E. Spear, The Development of the a-Si:H Field-Effect Transistor and Its Possible Applications
D. G. Ast, a-Si:H FET-Addressed LCD Panel
S. Kaneko, Solid-State Image Sensor
Masakiyo Matsumura, Charge-Coupled Devices
M. A. Bosch, Optical Recording
A. D'Amico and G. Fortunato, Ambient Sensors
Hiroshi Kukimoto, Amorphous Light-Emitting Devices
Robert J. Phelan, Jr., Fast Detectors and Modulators
Jacques I. Pankove, Hybrid Structures
P. G. LeComber, A. E. Owen, W. E. Spear, J. Hajto, and W. K. Choi, Electronic Switching in Amorphous Silicon Junction Devices

Volume 22 Lightwave Communications Technology
Part A

Kazuo Nakajima, The Liquid-Phase Epitaxial Growth of IngaAsp
W. T. Tsang, Molecular Beam Epitaxy for III–V Compound Semiconductors
G. B. Stringfellow, Organometallic Vapor-Phase Epitaxial Growth of III–V Semiconductors
G. Beuchet, Halide and Chloride Transport Vapor-Phase Deposition of InGaAsP and GaAs
Manijeh Razeghi, Low-Pressure Metallo-Organic Chemical Vapor Deposition of $Ga_xIn_{1-x}AsP_{1-y}$ Alloys
P. M. Petroff, Defects in III–V Compound Semiconductors

Part B

J. P. van der Ziel, Mode Locking of Semiconductor Lasers
Kam Y. Lau and Ammon Yariv, High-Frequency Current Modulation of Semiconductor Injection Lasers
Charles H. Henry, Special Properties of Semiconductor Lasers
Yasuharu Suematsu, Katsumi Kishino, Shigehisa Arai, and Fumio Koyama. Dynamic Single-Mode Semiconductor Lasers with a Distributed Reflector
W. T. Tsang, The Cleaved-Coupled-Cavity (C^3) Laser

Part C

R. J. Nelson and N. K. Dutta, Review of InGaAsP InP Laser Structures and Comparison of Their Performance
N. Chinone and M. Nakamura, Mode-Stabilized Semiconductor Lasers for 0.7–0.8- and 1.1–1.6-μm Regions
Yoshiji Horikoshi, Semiconductor Lasers with Wavelengths Exceeding 2 μm
B. A. Dean and M. Dixon, The Functional Reliability of Semiconductor Lasers as Optical Transmitters
R. H. Saul, T. P. Lee, and C. A. Burus, Light-Emitting Device Design
C. L. Zipfel, Light-Emitting Diode-Reliability
Tien Pei Lee and Tingye Li, LED-Based Multimode Lightwave Systems
Kinichiro Ogawa, Semiconductor Noise-Mode Partition Noise

Part D

Federico Capasso, The Physics of Avalanche Photodiodes
T. P. Pearsall and M. A. Pollack, Compound Semiconductor Photodiodes
Takao Kaneda, Silicon and Germanium Avalanche Photodiodes
S. R. Forrest, Sensitivity of Avalanche Photodetector Receivers for High-Bit-Rate Long-Wavelength Optical Communication Systems
J. C. Campbell, Phototransistors for Lightwave Communications

Part E

Shyh Wang, Principles and Characteristics of Integrable Active and Passive Optical Devices
Shlomo Margalit and Amnon Yariv, Integrated Electronic and Photonic Devices
Takaoki Mukai, Yoshihisa Yamamoto, and Tatsuya Kimura, Optical Amplification by Semiconductor Lasers

Volume 23 Pulsed Laser Processing of Semiconductors

R. F. Wood, C. W. White, and R. T. Young, Laser Processing of Semiconductors: An Overview
C. W. White, Segregation, Solute Trapping, and Supersaturated Alloys
G. E. Jellison, Jr., Optical and Electrical Properties of Pulsed Laser-Annealed Silicon
R. F. Wood and G. E. Jellison, Jr., Melting Model of Pulsed Laser Processing
R. F. Wood and F. W. Young, Jr., Nonequilibrium Solidification Following Pulsed Laser Melting
D. H. Lowndes and G. E. Jellison, Jr., Time-Resolved Measurement During Pulsed Laser Irradiation of Silicon
D. M. Zehner, Surface Studies of Pulsed Laser Irradiated Semiconductors
D. H. Lowndes, Pulsed Beam Processing of Gallium Arsenide
R. B. James, Pulsed CO_2 Laser Annealing of Semiconductors
R. T. Young and R. F. Wood, Applications of Pulsed Laser Processing

Volume 24 Applications of Multiquantum Wells, Selective Doping, and Superlattices

C. Weisbuch, Fundamental Properties of III–V Semiconductor Two-Dimensional Quantized Structures: The Basis for Optical and Electronic Device Applications
H. Morkoc and H. Unlu, Factors Affecting the Performance of (Al,Ga)As/GaAs and (Al,Ga)As/InGaAs Modulation-Doped Field-Effect Transistors: Microwave and Digital Applications
N. T. Linh, Two-Dimensional Electron Gas FETs: Microwave Applications
M. Abe et al., Ultra-High-Speed HEMT Integrated Circuits
D. S. Chemla, D. A. B. Miller, and P. W. Smith, Nonlinear Optical Properties of Multiple Quantum Well Structures for Optical Signal Processing
F. Capasso, Graded-Gap and Superlattice Devices by Band-Gap Engineering
W. T. Tsang, Quantum Confinement Heterostructure Semiconductor Lasers
G. C. Osbourn et al., Principles and Applications of Semiconductor Strained-Layer Superlattices

Volume 25 Diluted Magnetic Semiconductors

W. Giriat and J. K. Furdyna, Crystal Structure, Composition, and Materials Preparation of Diluted Magnetic Semiconductors

W. M. Becker, Band Structure and Optical Properties of Wide-Gap $A^{II}_{1-x}Mn_xB^{IV}$ Alloys at Zero Magnetic Field

Saul Oseroff and Pieter H. Keesom, Magnetic Properties: Macroscopic Studies

Giebultowicz and T. M. Holden, Neutron Scattering Studies of the Magnetic Structure and Dynamics of Diluted Magnetic Semiconductors

J. Kossut, Band Structure and Quantum Transport Phenomena in Narrow-Gap Diluted Magnetic Semiconductors

C. Riquaux, Magnetooptical Properties of Large-Gap Diluted Magnetic Semiconductors

J. A. Gaj, Magnetooptical Properties of Large-Gap Diluted Magnetic Semiconductors

J. Mycielski, Shallow Acceptors in Diluted Magnetic Semiconductors: Splitting, Boil-off, Giant Negative Magnetoresistance

A. K. Ramadas and R. Rodriquez, Raman Scattering in Diluted Magnetic Semiconductors

P. A. Wolff, Theory of Bound Magnetic Polarons in Semimagnetic Semiconductors

Volume 26 III–V Compound Semiconductors and Semiconductor Properties of Superionic Materials

Zou Yuanxi, III–V Compounds

H. V. Winston, A. T. Hunter, H. Kimura, and R. E. Lee, InAs-Alloyed GaAs Substrates for Direct Implantation

P. K. Bhattachary and S. Dhar, Deep Levels in III–V Compound Semiconductors Grown by MBE

Yu. Yu. Gurevich and A. K. Ivanov-Shits, Semiconductor Properties of Supersonic Materials

Volume 27 High Conducting Quasi-One-Dimensional Organic Crystals

E. M. Conwell, Introduction to Highly Conducting Quasi-One-Dimensional Organic Crystals

I. A. Howard, A Reference Guide to the Conducting Quasi-One-Dimensional Organic Molecular Crystals

J. P. Pouquet, Structural Instabilities

E. M. Conwell, Transport Properties

C. S. Jacobsen, Optical Properties

J. C. Scott, Magnetic Properties

L. Zuppiroli, Irradiation Effects: Perfect Crystals and Real Crystals

Volume 28 Measurement of High-Speed Signals in Solid State Devices

J. Frey and D. Ioannou, Materials and Devices for High-Speed and Optoelectronic Applications

H. Schumacher and E. Strid, Electronic Wafer Probing Techniques

D. H. Auston, Picosecond Photoconductivity: High-Speed Measurements of Devices and Materials

J. A. Valdmanis, Electro-Optic Measurement Techniques for Picosecond Materials, Devices, and Integrated Circuits.

J. M. Wiesenfeld and R. K. Jain, Direct Optical Probing of Integrated Circuits and High-Speed Devices

G. Plows, Electron-Beam Probing

A. M. Weiner and R. B. Marcus, Photoemissive Probing

Volume 29 Very High Speed Integrated Circuits: Gallium Arsenide LSI

M. Kuzuhara and T. Nazaki, Active Layer Formation by Ion Implantation
H. Hasimoto, Focused Ion Beam Implantation Technology
T. Nozaki and A. Higashisaka, Device Fabrication Process Technology
M. Ino and T. Takada, GaAs LSI Circuit Design
M. Hirayama, M. Ohmori, and K. Yamasaki, GaAs LSI Fabrication and Performance

Volume 30 Very High Speed Integrated Circuits: Heterostructure

H. Watanabe, T. Mizutani, and A. Usui, Fundamentals of Epitaxial Growth and Atomic Layer Epitaxy
S. Hiyamizu, Characteristics of Two-Dimensional Electron Gas in III–V Compound Heterostructures Grown by MBE
T. Nakanisi, Metalorganic Vapor Phase Epitaxy for High-Quality Active Layers
T. Nimura, High Electron Mobility Transistor and LSI Applications
T. Sugeta and T. Ishibashi, Hetero-Bipolar Transistor and LSI Application
H. Matsueda, T. Tanaka, and M. Nakamura, Optoelectronic Integrated Circuits

Volume 31 Indium Phosphide: Crystal Growth and Characterization

J. P. Farges, Growth of Discoloration-free InP
M. J. McCollum and G. E. Stillman, High Purity InP Grown by Hydride Vapor Phase Epitaxy
T. Inada and T. Fukuda, Direct Synthesis and Growth of Indium Phosphide by the Liquid Phosphorous Encapsulated Czochralski Method
O. Oda, K. Katagiri, K. Shinohara, S. Katsura, Y. Takahashi, K. Kainosho, K. Kohiro, and R. Hirano, InP Crystal Growth, Substrate Preparation and Evaluation
K. Tada, M. Tatsumi, M. Morioka, T. Araki, and T. Kawase, InP Substrates: Production and Quality Control
M. Razeghi, LP-MOCVD Growth, Characterization, and Application of InP Material
T. A. Kennedy and P. J. Lin-Chung, Stoichiometric Defects in InP

Volme 32 Strained-Layer Superlattices: Physics

T. P. Pearsall, Strained-Layer Superlattices
Fred H. Pollack, Effects of Homogeneous Strain on the Electronic and Vibrational Levels in Semiconductors
J. Y. Marzin, J. M. Gerárd, P. Voisin, and J. A. Brum, Optical Studies of Strained III–V Heterolayers
R. People and S. A. Jackson, Structurally Induced States from Strain and Confinement
M. Jaros, Microscopic Phenomena in Ordered Suprlattices

Volume 33 Strained-Layer Superlattices: Materials Science and Technology

R. Hull and J. C. Bean, Principles and Concepts of Strained-Layer Epitaxy
William J. Schaff, Paul J. Tasker, Marc C. Foisy, and Lester F. Eastman, Device Applications of Strained-Layer Epitaxy

S. T. Picraux, B. L. Doyle, and J. Y. Tsao, Structure and Characterization of Strained-Layer Superlattices
E. Kasper and F. Schäffer, Group IV Compounds
Dale L. Martin, Molecular Beam Epitaxy of IV–VI Compounds Heterojunction
Robert L. Gunshor, Leslie A. Kolodziejski, Arto V. Nurmikko, and Nobuo Otsuka, Molecular Beam Epitaxy of II–VI Semiconductor Microstructures

Volume 34 Hydrogen in Semiconductors

J. I. Pankove and N. M. Johnson, Introduction to Hydrogen in Semiconductors
C. H. Seager, Hydrogenation Methods
J. I. Pankove, Hydrogenation of Defects in Crystalline Silicon
J. W. Corbett, P. Deák, U. V. Desnica, and S. J. Pearton, Hydrogen Passivation of Damage Centers in Semiconductors
S. J. Pearton, Neutralization of Deep Levels in Silicon
J. I. Pankove, Neutralization of Shallow Acceptors in Silicon
N. M. Johnson, Neutralization of Donor Dopants and Formation of Hydrogen-Induced Defects in n-Type Silicon
M. Stavola and S. J. Pearton, Vibrational Spectroscopy of Hydrogen-Related Defects in Silicon
A. D. Marwick, Hydrogen in Semiconductors: Ion Beam Techniques
C. Herring and N. M. Johnson, Hydrogen Migration and Solubility in Silicon
E. E. Haller, Hydrogen-Related Phenomena in Crystalline Germanium
J. Kakalios, Hydrogen Diffusion in Amorphous Silicon
J. Chevalier, B. Clerjaud, and B. Pajot, Neutralization of Defects and Dopants in III–V Semiconductors
G. G. DeLeo and W. B. Fowler, Computational Studies of Hydrogen-Containing Complexes in Semiconductors
R. F. Kiefl and T. L. Estle, Muonium in Semiconductors
C. G. Van de Walle, Theory of Isolated Interstitial Hydrogen and Muonium in Crystalline Semiconductors

Volume 35 Nanostructured Systems

Mark Reed, Introduction
H. van Houten, C. W. J. Beenakker, and B. J. van Wees, Quantum Point Contacts
G. Timp, When Does a Wire Become an Electron Waveguide?
M. Büttiker, The Quantum Hall Effects in Open Conductors
W. Hansen, J. P. Kotthaus, and U. Merkt, Electrons in Laterally Periodic Nanostructures

Volume 36 The Spectroscopy of Semiconductors

D. Heiman, Spectroscopy of Semiconductors at Low Temperatures and High Magnetic Fields
Arto V. Nurmikko, Transient Spectroscopy by Ultrashort Laser Pulse Techniques
A. K. Ramdas and S. Rodriguez, Piezospectroscopy of Semiconductors
Orest J. Glembocki and Benjamin V. Shanabrook, Photoreflectance Spectroscopy of Microstructures
David G. Seiler, Christopher L. Littler, and Margaret H. Wiler, One- and Two-Photon Magneto-Optical Spectroscopy of InSb and $Hg_{1-x}Cd_xTe$

Volume 37 The Mechanical Properties of Semiconductors

A.-B. Chen, Arden Sher and W. T. Yost, Elastic Constants and Related Properties of Semiconductor Compounds and Their Alloys
David R. Clarke, Fracture of Silicon and Other Semiconductors
Hans Siethoff, The Plasticity of Elemental and Compound Semiconductors
Sivaraman Guruswamy, Katherine T. Faber and John P. Hirth, Mechanical Behavior of Compound Semiconductors
Subhanh Mahajan, Deformation Behavior of Compound Semiconductors
John P. Hirth, Injection of Dislocations into Strained Multilayer Structures
Don Kendall, Charles B. Fleddermann, and Kevin J. Malloy, Critical Technologies for the Micromachining of Silicon
Ikuo Matsuba and Kinji Mokuya, Processing and Semiconductor Thermoelastic Behavior

Volume 38 Imperfections in III/V Materials

Udo Scherz and Matthias Scheffler, Density-Functional Theory of sp-Bonded Defects in III/V Semiconductors
Maria Kaminska and Eicke R. Weber, El2 Defect in GaAs
David C. Look, Defects Relevant for Compensation in Semi-Insulating GaAs
R. C. Newman, Local Vibrational Mode Spectroscopy of Defects in III/V Compounds
Andrzej M. Hennel, Transition Metals in III/V Compounds
Kevin J. Malloy and Ken Khachaturyan, DX and Related Defects in Semiconductors
V. Swaminathan and Andrew S. Jordan, Dislocations in III/V Compounds
Krzysztof W. Nauka, Deep Level Defects in the Epitaxial III/V Materials

Volume 39 Minority Carriers in III–V Semiconductors: Physics and Applications

Niloy K. Dutta, Radiative Transitions in GaAs and Other III–V Compounds
Richard K. Ahrenkiel, Minority-Carrier Lifetime in III–V Semiconductors
Tomofumi Furuta, High Field Minority Electron Transport in p-GaAs
Mark S. Lundstrom, Minority-Carrier Transport in III–V Semiconductors
Richard A. Abram, Effects of Heavy Doping and High Excitation on the Band Structure of GaAs
David Yevick and Witold Bardyszewski, An Introduction to Non-Equilibrium Many-Body Analyses of Optical Processes in III–V Semiconductors

Volume 40 Epitaxial Microstructures

E. F. Schubert, Delta-Doping of Semiconductors: Electronic, Optical, and Structural Properties of Materials and Devices
A. Gossard, M. Sundaram, and P. Hopkins, Wide Graded Potential Wells
P. Petroff, Direct Growth of Nanometer-Size Quantum Wire Superlattices
E. Kapon, Lateral Patterning of Quantum Well Heterostructures by Growth of Nonplanar Substrates
H. Temkin, D. Gershoni, and M. Panish, Optical Properties of Ga1-$_x$In$_x$As/InP Quantum Wells

Volume 41 High Speed Heterostructure Devices

F. Capasso, F. Beltram, S. Sen, A. Pahlevi, and A. Y. Cho, Quantum Electron Devices: Physics and Applications
P. Solomon, D. J. Frank, S. L. Wright, and F. Canora, GaAs-Gate Semiconductor–Insulator–Semiconductor FET
M. H. Hashemi and U. K. Mishra, Unipolar InP-Based Transistors
R. Kiehl, Complementary Heterostructure FET Integrated Circuits
T. Ishibashi, GaAs-Based and InP-Based Heterostructure Bipolar Transistors
H. C. Liu and T. C. L. G. Sollner, High-Frequency-Tunneling Devices
H. Ohnishi, T. More, M. Takatsu, K. Imamura, and N. Yokoyama, Resonant-Tunneling Hot-Electron Transistors and Circuits

Volume 42 Oxygen in Silicon

F. Shimura, Introduction to Oxygen in Silicon
W. Lin, The Incorporation of Oxygen into Silicon Crystals
T. J. Schaffner and D. K. Schroder, Characterization Techniques for Oxygen in Silicon
W. M. Bullis, Oxygen Concentration Measurement
S. M. Hu, Intrinsic Point Defects in Silicon
B. Pajot, Some Atomic Configurations of Oxygen
J. Michel and L. C. Kimerling, Electical Properties of Oxygen in Silicon
R. C. Newman and R. Jones, Diffusion of Oxygen in Silicon
T. Y. Tan and W. J. Taylor, Mechanisms of Oxygen Precipitation: Some Quantitative Aspects
M. Schrems, Simulation of Oxygen Precipitation
K. Simino and I. Yonenaga, Oxygen Effect on Mechanical Properties
W. Bergholz, Grown-in and Process-Induced Effects
F. Shimura, Intrinsic/Internal Gettering
H. Tsuya, Oxygen Effect on Electronic Device Performance

Volume 43 Semiconductors for Room Temperature Nuclear Detector Applications

R. B. James and T. E. Schlesinger, Introduction and Overview
L. S. Darken and C. E. Cox, High-Purity Germanium Detectors
A. Burger, D. Nason, L. Van den Berg, and M. Schieber, Growth of Mercuric Iodide
X. J. Bao, T. E. Schlesinger, and R. B. James, Electrical Properties of Mercuric Iodide
X. J. Bao, R. B. James, and T. E. Schlesinger, Optical Properties of Red Mercuric Iodide
M. Hage-Ali and P. Siffert, Growth Methods of CdTe Nuclear Detector Materials
M. Hage-Ali and P Siffert, Characterization of CdTe Nuclear Detector Materials
M. Hage-Ali and P. Siffert, CdTe Nuclear Detectors and Applications
R. B. James, T. E. Schlesinger, J. Lund, and M. Schieber, $Cd_{1-x}Zn_xTe$ Spectrometers for Gamma and X-Ray Applications
D. S. McGregor, J. E. Kammeraad, Gallium Arsenide Radiation Detectors and Spectrometers
J. C. Lund, F. Olschner, and A. Burger, Lead Iodide
M. R. Squillante, and K. S. Shah, Other Materials: Status and Prospects
V. M. Gerrish, Characterization and Quantification of Detector Performance
J. S. Iwanczyk and B. E. Patt, Electronics for X-ray and Gamma Ray Spectrometers
M. Schieber, R. B. James, and T. E. Schlesinger, Summary and Remaining Issues for Room Temperature Radiation Spectrometers

Volume 44 II–IV Blue/Green Light Emitters: Device Physics and Epitaxial Growth

J. Han and R. L. Gunshor, MBE Growth and Electrical Properties of Wide Bandgap ZnSe-based II–VI Semiconductors

Shizuo Fujita and Shigeo Fujita, Growth and Characterization of ZnSe-based II–VI Semiconductors by MOVPE

Easen Ho and Leslie A. Kolodziejski, Gaseous Source UHV Epitaxy Technologies for Wide Bandgap II–VI Semiconductors

Chris G. Van de Walle, Doping of Wide-Band-Gap II–VI Compounds—Theory

Roberto Cingolani, Optical Properties of Excitons in ZnSe-Based Quantum Well Heterostructures

A. Ishibashi and A. V. Nurmikko, II–VI Diode Lasers: A Current View of Device Performance and Issues

Supratik Guha and John Petruzello, Defects and Degradation in Wide-Gap II–VI-based Structures and Light Emitting Devices

Volume 45 Effect of Disorder and Defects in Ion-Implanted Semiconductors: Electrical and Physiochemical Characterization

Heiner Ryssel, Ion Implantation into Semiconductors: Historical Perspectives

You-Nian Wang and Teng-Cai Ma, Electronic Stopping Power for Energetic Ions in Solids

Sachiko T. Nakagawa, Solid Effect on the Electronic Stopping of Crystalline Target and Application to Range Estimation

G. Müller, S. Kalbitzer and G. N. Greaves, Ion Beams in Amorphous Semiconductor Research

Jumana Boussey-Said, Sheet and Spreading Resistance Analysis of Ion Implanted and Annealed Semiconductors

M. L. Polignano and G. Queirolo, Studies of the Stripping Hall Effect in Ion-Implanted Silicon

J. Stoemenos, Transmission Electron Microscopy Analyses

Roberta Nipoti and Marco Servidori, Rutherford Backscattering Studies of Ion Implanted Semiconductors

P. Zaumseil, X-ray Diffraction Techniques

Volume 46 Effect of Disorder and Defects in Ion-Implanted Semiconductors: Optical and Photothermal Characterization

M. Fried, T. Lohner and J. Gyulai, Ellipsometric Analysis

Antonios Seas and Constantinos Christofides, Transmission and Reflection Spectroscopy on Ion Implanted Semiconductors

Andreas Othonos and Constantinos Christofides, Photoluminescence and Raman Scattering of Ion Implanted Semiconductors. Influence of Annealing

Constantinos Christofides, Photomodulated Thermoreflectance Investigation of Implanted Wafers. Annealing Kinetics of Defects

U. Zammit, Photothermal Deflection Spectroscopy Characterization of Ion-Implanted and Annealed Silicon Films

Andreas Mandelis, Arief Budiman and Miguel Vargas, Photothermal Deep-Level Transient Spectroscopy of Impurities and Defects in Semiconductors

R. Kalish and S. Charbonneau, Ion Implantation into Quantum-Well Structures

Alexandre M. Myasnikov and Nikolay N. Gerasimenko, Ion Implantation and Thermal Annealing of III-V Compound Semiconducting Systems: Some Problems of III-V Narrow Gap Semiconductors

Volume 47 Uncooled Infrared Imaging Arrays and Systems

R. G. Buser and M. P. Tompsett, Historical Overview
P. W. Kruse, Principles of Uncooled Infrared Focal Plane Arrays
R. A. Wood, Monolithic Silicon Microbolometer Arrays
C. M. Hanson, Hybrid Pyroelectric-Ferroelectric Bolometer Arrays
D. L. Polla and J. R. Choi, Monolithic Pyroelectric Bolometer Arrays
N. Teranishi, Thermoelectric Uncooled Infrared Focal Plane Arrays
M. F. Tompsett, Pyroelectric Vidicon
T. W. Kenny, Tunneling Infrared Sensors
J. R. Vig, R. L. Filler and Y. Kim, Application of Quartz Microresonators to Uncooled Infrared Imaging Arrays
P. W. Kruse, Application of Uncooled Monolithic Thermoelectric Linear Arrays to Imaging Radiometers

Volume 48 High Brightness Light Emitting Diodes

G. B. Stringfellow, Materials Issues in High-Brightness Light-Emitting Diodes
M. G. Craford, Overview of Device issues in High-Brightness Light-Emitting Diodes
F. M. Steranka, AlGaAs Red Light Emitting Diodes
C. H. Chen, S. A. Stockman, M. J. Peanasky, and C. P. Kuo, OMVPE Growth of AlGaInP for High Efficiency Visible Light-Emitting Diodes
F. A. Kish and R. M. Fletcher, AlGaInP Light-Emitting Diodes
M. W. Hodapp, Applications for High Brightness Light-Emitting Diodes
I. Akasaki and H. Amano, Organometallic Vapor Epitaxy of GaN for High Brightness Blue Light Emitting Diodes
S. Nakamura, Group III-V Nitride Based Ultraviolet-Blue-Green-Yellow Light-Emitting Diodes and Laser Diodes

Volume 49 Light Emission in Silicon: from Physics to Devices

David J. Lockwood, Light Emission in Silicon
Gerhard Abstreiter, Band Gaps and Light Emission in Si/SiGe Atomic Layer Structures
Thomas G. Brown and Dennis G. Hall, Radiative Isoelectronic Impurities in Silicon and Silicon-Germanium Alloys and Superlattices
J. Michel, L. V. C. Assali, M. T. Morse, and L. C. Kimerling, Erbium in Silicon
Yoshihiko Kanemitsu, Silicon and Germanium Nanoparticles
Philippe M. Fauchet, Porous Silicon: Photoluminescence and Electroluminescent Devices
C. Delerue, G. Allan, and M. Lannoo, Theory of Radiative and Nonradiative Processes in Silicon Nanocrystallites
Louis Brus, Silicon Polymers and Nanocrystals

Volume 50 Gallium Nitride (GaN)

J. I. Pankove and T. D. Moustakas, Introduction
S. P. DenBaars and S. Keller, Metalorganic Chemical Vapor Deposition (MOCVD) of Group III Nitrides
W. A. Bryden and T. J. Kistenmacher, Growth of Group III-A Nitrides by Reactive Sputtering
N. Newman, Thermochemistry of III-N Semiconductors
S. J. Pearton and R. J. Shul, Etching of III Nitrides

S. M. Bedair, Indium-based Nitride Compounds
A. Trampert, O. Brandt, and K. H. Ploog, Crystal Structure of Group III Nitrides
H. Morkoc, F. Hamdani, and A. Salvador, Electronic and Optical Properties of III–V Nitride based Quantum Wells and Superlattices
K. Doverspike and J. I. Pankove, Doping in the III-Nitrides
T. Suski and P. Perlin, High Pressure Studies of Defects and Impurities in Gallium Nitride
B. Monemar, Optical Properties of GaN
W. R. L. Lambrecht, Band Structure of the Group III Nitrides
N. E. Christensen and P. Perlin, Phonons and Phase Transitions in GaN
S. Nakamura, Applications of LEDs and LDs
I. Akasaki and H. Amano, Lasers
J. A. Cooper, Jr., Nonvolatile Random Access Memories in Wide Bandgap Semiconductors

Volume 51A Identification of Defects in Semiconductors

George D. Watkins, EPR and ENDOR Studies of Defects in Semiconductors
J.-M. Spaeth, Magneto-Optical and Electrical Detection of Paramagnetic Resonance in Semiconductors
T. A. Kennedy and E. R. Glaser, Magnetic Resonance of Epitaxial Layers Detected by Photoluminescence
K. H. Chow, B. Hitti, and R. F. Kiefl, μSR on Muonium in Semiconductors and Its Relation to Hydrogen
Kimmo Saarinen, Pekka Hautojärvi, and Catherine Corbel, Positron Annihilation Spectroscopy of Defects in Semiconductors
R. Jones and P. R. Briddon, The Ab Initio Cluster Method and the Dynamics of Defects in Semiconductors

Volume 51B Identification of Defects in Semiconductors

Gordon Davies, Optical Measurements of Point Defects
P. M. Mooney, Defect Identification Using Capacitance Spectroscopy
Michael Stavola, Vibrational Spectroscopy of Light Element Impurities in Semiconductors
P. Schwander, W. D. Rau, C. Kisielowski, M. Gribelyuk, and A. Ourmazd, Defect Processes in Semiconductors Studied at the Atomic Level by Transmission Electron Microscopy
Nikos D. Jager and Eicke R. Weber, Scanning Tunneling Microscopy of Defects in Semiconductors

Volume 52 SiC Materials and Devices

Kenneth Järrendahl and Robert F. Davis, Materials Properties and Characterization of SiC
V. A. Dmitriev and M. G. Spencer, SiC Fabrication Technology: Growth and Doping
V. Saxena and A. J. Steckl, Building Blocks for SiC Devices: Ohmic Contacts, Schottky Contacts, and p-n Junctions
Michael S. Shur, SiC Transistors
C. D. Brandt, R. C. Clarke, R. R. Siergiej, J. B. Casady, A. W. Morse, S. Sriram, and A. K. Agarwal, SiC for Applications in High-Power Electronics
R. J. Trew, SiC Microwave Devices

J. Edmond, H. Kong, G. Negley, M. Leonard, K. Doverspike, W. Weeks, A. Suvorov, D. Waltz, and C. Carter, Jr., SiC-Based UV Photodiodes and Light Emitting Diodes
Hadis Morkoç, Beyond Silicon Carbide! III–V Nitride-Based Heterostructures and Devices

Volume 53 Cumulative Subject and Author Index Including Tables of Contents for Volume 1–50

Volume 54 High Pressure in Semiconductor Physics I

William Paul, High Pressure in Semiconductor Physics: A Historical Overview
N. E. Christensen, Electronic Structure Calculations for Semiconductors under Pressure
R. J. Neimes and M. I. McMahon, Structural Transitions in the Group IV, III-V and II-VI Semiconductors Under Pressure
A. R. Goni and K. Syassen, Optical Properties of Semiconductors Under Pressure
Pawel Trautman, Michal Baj, and Jcek M. Baranowski, Hydrostatic Pressure and Uniaxial Stress in Investigations of the EL2 Defect in GaAs
Ming-fu Li and Peter Y. Yu, High-Pressure Study of DX Centers Using Capacitance Techniques
Tadeusz Suski, Spatial Correlations of Impurity Charges in Doped Semiconductors
Noritaka Kuroda, Pressure Effects on the Electronic Properties of Diluted Magnetic Semiconductors

Volume 55 High Pressure in Semiconductor Physics II

D. K. Maude and J. C. Portal, Parallel Transport in Low-Dimensional Semiconductor Structures
P. C. Klipstein, Tunneling Under Pressure: High-Pressure Studies of Vertical Transport in Semiconductor Heterostructures
Evangelos Anastassakis and Manuel Cardona, Phonons, Strains, and Pressure in Semiconductors
Fred H. Pollak, Effects of External Uniaxial Stress on the Optical Properties of Semiconductors and Semiconductor Microstructures
A. R. Adams, M. Silver, and J. Allam, Semiconductor Optoelectronic Devices
S. Porowski and I. Grzegory, The Application of High Nitrogen Pressure in the Physics and Technology of III-N Compounds
Mohammad Yousuf, Diamond Anvil Cells in High Pressure Studies of Semiconductors

Volume 56 Germanium Silicon

J. C. Bean, Growth Techniques and Procedures
D. E. Savage, F. Liu, V. Zielasek, and M. G. Lagally, Fundamental Crystal Growth Mechanisms
R. Hull, Misfit Strain Accommodation in SiGe Heterostructures
M. J. Shaw and M. Jaros, Fundamental Physics of Strained Layer GeSi: Quo Vadis?
F. Cerdeira, Optical Properties
S. A. Ringel and P. N. Grillot, Electronic Properties and Deep Levels in Germanium-Silicon
J. C. Campbell, Optoelectronics in Silicon and Germanium Silicon
K. Eberl, K. Brunner, and O. G. Schmidt, $Si_{1-y}C_y$ and $Si_{1-x-y}Ge_xC_y$ Alloy Layers

Volume 57 Gallium Nitride (GaN) II

Richard J. Molnar, Hydride Vapor Phase Epitaxial Growth of III-V Nitrides
T. D. Moustakas, Growth of III-V Nitrides by Molecular Beam Epitaxy
Zuzanna Liliental-Weber, Defects in Bulk GaN and Homoepitaxial Layers
Chris G. Van de Walle and Noble M. Johnson, Hydrogen in III-V Nitrides
W. Götz and N. M. Johnson, Characterization of Dopants and Deep Level Defects in Gallium Nitride
Bernard Gill, Stress Effects on Optical Properties
Christian Kisielowski, Strain in GaN Thin Films and Heterostructures
Joseph A. Miragliotta and Dennis K. Wickenden, Nonlinear Optical Properties of Gallium Nitride
B. K. Meyer, Magnetic Resonance Investigations on Group III-Nitrides
M. S. Shur and M. Asif Khan, GaN and AlGaN Ultraviolet Detectors
C. H. Qiu, J. I. Pankove, and C. Rossington, III-V Nitride-Based X-ray Detectors

Volume 58 Nonlinear Optics in Semiconductors I

Alan Kost, Resonant Optical Nonlinearities in Semiconductors
Elsa Garmire, Optical Nonlinearities in Semiconductors Enhanced by Carrier Transport
D. S. Chemla, Ultrafast Transient Nonlinear Optical Processes in Semiconductors
Mansoor Sheik-Bahae and Eric W. Van Stryland, Optical Nonlinearities in the Transparency Region of Bulk Semiconductors
James E. Millerd, Mehrdad Ziari and Afshin Partovi, Photorefractivity in Semiconductors

ISBN 0-12-752168-2